Engineering Glass Structures

Jan Belis · Johannes Kuntsche · Christian Louter ·
Jens Henrik Nielsen · Mauro Overend ·
Frank Schneider · Jens Schneider · Sebastian Schula

Engineering Glass Structures

Springer

Jan Belis
Ghent University
Ghent, Belgium

Christian Louter
Faculty of Civil Engineering
and Geosciences
Delft University of Technology
Delft, The Netherlands

Mauro Overend
Faculty of Architecture and Built
Environment
Delft University of Technology
Delft, The Netherlands

Jens Schneider
Technische Universität Wien
Vienna, Austria

Johannes Kuntsche
Darmstadt University of Applied Sciences
Darmstadt, Germany

Jens Henrik Nielsen
COWI A/S (DK)
Kongens Lyngby, Denmark

Frank Schneider
MPA Darmstadt
State Materials Testing Institute Darmstadt
Darmstadt, Germany

Sebastian Schula
SGS Engineering GmbH
Frankfurt, Germany

ISBN 978-3-032-03232-4 ISBN 978-3-032-03233-1 (eBook)
https://doi.org/10.1007/978-3-032-03233-1

This Springer imprint is published by the registered company Springer Nature Switzerland AG
The registered company address is: Gewerbestrasse 11, 6330 Cham, Switzerland

We dedicate this book to our beloved partners:
An, Ina, Valesca, Tine, Natalie, Simone, Meike and Magdalena.

Preface

Understanding the engineering aspects of architectural glass design requires knowledge of many interlinked factors, such as the material properties of glass, interlayers and other associated materials; the technologies to produce and process monolithic and multi-layered glass and glass components; the fundamentals of linear elastic fracture mechanics and viscoelasticity; the principles of strength and pre-stressing of glass; the loads and load actions on a glass structure; the detailing of structural connections; the principles of structural safety and robustness in glass design; and the maintenance, durability and forensics of glass.

Providing such knowledge is exactly the aim of this book. It is intended for a wide audience with an existing sound understanding of structural, civil or architectural engineering. In education, it can serve as a textbook for senior-level university students. In research, it can offer Ph.D. candidates an head start in their studies. In practice, it offers solid background information to professionals in the field.

This first English edition provides an updated and expanded content compared to the prior German edition, with a special focus on the forthcoming Eurocode on the design of glass structures. The book also contains shaded sections, indicating more detailed information that may not be required by the typical reader.

Ghent, Belgium	Jan Belis
Darmstadt, Germany	Johannes Kuntsche
Delft, The Netherlands	Christian Louter
Copenhagen, Denmark	Jens Henrik Nielsen
Delft, The Netherlands	Mauro Overend
Darmstadt, Germany	Frank Schneider
Vienna, Austria	Jens Schneider
Darmstadt, Germany	Sebastian Schula
December 2025	

Acknowledgements We are grateful for the input, assistance and feedback of: Telesilla Bristogianni, Rebecca Hartwell, Isamu Lautenschlager, Martin Meyland Jensen, Faidra Oikonomopoulou, Michel Prassas, Max Schneider, Paul Wissler and Marco Zaccaria.

Competing Interests The authors have no competing interests to declare that are relevant to the content of this manuscript.

Contents

Part I Introduction

Part II Glass Products and Processing

About the Authors

Jan Belis is a Professor of Glass Structures and Construction at the Faculty of Engineering and Architecture of Ghent University (BE) and holds a Ph.D. degree from Ghent University on Structural Glass. He is an Editor-in-Chief of the *Glass Structures and Engineering* journal and was Chair of COST Action TU0905 *Structural Glass—Novel Design Methods and Next-Generation Products*. He is co-organiser of the international *Challenging Glass Conference* series and has actively been contributing to the development of the Eurocode EN 19100 on Design of Glass Structures.

Johannes Kuntsche is a Professor of Structural Analysis and Mechanics at the Faculty of Civil and Environmental Engineering at Darmstadt University of Applied Sciences (h_da) since 2021. Beforehand, he was working several years as project leader in different engineering offices for structural design. Johannes Kuntsche studied civil engineering and subsequently received his Ph.D. on the topic of Laminated Glass in 2015 at TU Darmstadt.

Christian Louter is a Professor of Structural Design and Building Engineering at the Faculty of Civil Engineering and Geosciences at TU Delft (NL). Christian holds a Ph.D. degree from TU Delft on the topic of Structural Glass and has worked at EPFL, TU Delft and TU Dresden before re-joining TU Delft in 2022. Next to his research and teaching activities, Christian is an Organiser of the international *Challenging Glass Conference series*, an Editor-in-Chief of the *Glass Structures and Engineering* journal (Springer) and Committee Member of Eurocode CEN/TC250/WG1 Design of Glass Structures.

Jens Henrik Nielsen is currently employed as a leading specialist at COWI A/S in Denmark. Before this, he was an Associate Professor at the Civil and Mechanical Engineering Department at the Ttechnical University of Denmark (DTU). His research interests include numerical modelling and glass, with a particular interest in tempered glass and products made from this. He has been teaching Finite Element courses at both B.Sc. and M.Sc. levels for many years and used to be the main responsible for a course on glass and glass structures at DTU. Jens was appointed Head of study for the M.Sc. in Civil Engineering at DTU between 2016 and 2021 and is a member of the scientific committee for several international conferences and an Editor-in-Chief for the *Glass Structures and Engineering* journal (Springer).

Mauro Overend is a Professor of Structural Design and Mechanics at the Faculty of Architecture and the Built Environment at TU Delft (NL). He is a chartered engineer with several years of consulting engineering, teaching and research experience in the fields of structural engineering and facade engineering. His research and teaching interests are the interface of structural engineering, materials engineering and building physics which underpin the performance of glass in buildings and sustainable structures in general. Mauro has led interdisciplinary teams on glass-related research at the University of Cambridge and TU Delft. He has won several international awards, and his work has found applications in international design standards and in some of the most challenging buildings globally.

Frank Schneider is Deputy Head at the State Materials Testing Institute Darmstadt (MPA) in the competence field 'Glass'. He graduated from Darmstadt University of Technology in 1999 and obtained his Ph.D. from the same university in 2005. Subsequently, he held a position of Director Research and Development at OKALUX GmbH, and he was active as Chairman of the German Structural Glass Association (Fachverband Konstruktiver Glasbau e.V.).

Jens Schneider is a Professor of Structural Analysis at TU Darmstadt, Germany, and Rector of TU Wien in Vienna, Austria. His work includes construction projects focusing on glass, facade and sustainable building practices. His research interests encompass glass and polymers in structures, fracture mechanics, viscoelasticity, bridge dynamics and energy efficiency in the built environment. Schneider is committed to bridging academia with industry, exemplified by his work in several standard committees, as an engineering consultant and by the 'ETA-factory' on TU Darmstadt's campus, which promotes energy-efficient innovations.

Sebastian Schula has many years of experience in the planning and execution of facades with a special focus on glass, structural glass application and structural glazing. In addition, he has a very high level of expertise in the field of expert assessment of facade constructions within the scope of approval and permission procedures as well as the assessment of cases of damage. Furthermore, he has extensive experience in the field of quality assessment. He completed his Ph.D. on the topic of 'Characterisation of the scratch sensitivity of glasses in civil engineering' with distinction. He is partner at the facade engineering office SGS GmbH as well as lecturer for 'Glass Construction' at the Technische Universitat Darmstadt, expert author and speaker.

Abbreviations

1C	One-Component
2C	Two-Component
ANG	Annealed Glass
ASI	Amorphous Silicon
CD	Chemical Deposition
CdTe	Cadmium Telluride
CIGS	Copper Indium Gallium di-Selenid
CIP	Cast in Place
CIS	Copper Indium Diselenid
CNC	Computer Numerical Control
CRI	Colour Rendering Index
CST	Compressive Shear Test
CVD	Chemical Vapour Deposition
CWFT	Classified Without Further Testing
DGU	Double Glass Unit
DIC	Digital Image Correlation
DMA	Dynamic Mechanical Analysis
DMTA	Dynamic-Mechanical Thermal Analysis
DoL	Depth of Layer
EDX	Energy Dispersive X-ray
EPDM	Ethylene Propylene Diene Monomer Rubber
ESEM	Environmental Scanning Electron Microscope
ETFE	Ethylene Tetrafluoroethylene
EVA	Ethylene Vinyl Acetate
EVAC	Ethylene Vinyl Acetate
FTG	Fully Tempered Glass
HF	Hydrofluoric
HSG	Heat strengthened glass
IGU	Insulated Glass Unit
ITT	Initial Type Test
LiSS	Laser Induced Stress Separation Method

MLBA	Multiple Laser Beam Absorption Method
PA6	Polyamide
PACVD	Plasma Aided Chemical Vapour Deposition
PC	Poly Carbonate
PEEK	Polyether Ether Ketone
PEI	Polyetherimide
PIB	Polyisobutylene
PMMA	Poly(methyl methacrylate), Acrylic
POM	Polyoxymethylene
PP	Polypropylene
PSU	Polysulfone
PTFE	Polytetrafluoroethylene
PUR	Polyurethane
PVB	Polyvinyl Butyral
PVC	Polyvinyl Chloride
PVD	Physical Vapour Deposition
Ra	Colour Rendering Index
RH	Relative Humidity
SEM	Scanning Electron Microscope
SHGC	Solar Heat Gain Coefficient
TAR	Transmission, Absorbtion and Reflection
TEM	Environmental Scanning Electron Microscope
TET	Solar Energy Transmittance
TGU	Triple Glass Unit
TPS	Thermoplastic Spacer
TPU	Thermoplastic Polyurethane
TTG	Thermally Toughened Glass
TTS	Time Temperature Superposition
UV	Ultraviolet
VAC	Vinyl Acetate
VGU	Vacuum Glass Unit
VIG	Vacuum Insulating Glass
WLF	Williams-Landel-Ferry
ZWLCT™	Zero Width Laser Cutting Technology

List of Figures

List of Tables

Part I
Introduction

Chapter 1
Material Properties of Glass and Associated Polymers

This chapter deals with the fundamental material properties of glass and polymers, that often used in glass engineering. After a general introduction in Sect. 1.1, *the physical nature of glass, including its molecular structure and thermodynamics, is discussed in Sect.* 1.2. *From there onwards, the focus is on glass used in the building industry; Sects.* 1.3 *and* 1.4 *discuss its chemical composition and its main properties, respectively. Finally, Sects.* 1.5 *and* 1.6 *describe the polymers that are frequently used in conjunction with glass with a focus on their salient properties and testing methods.*

1.1 What Is Glass?

Glass is one of the oldest man-made materials and the fascination with its brittleness and transparency continues to this day. From the first prehistoric arrowheads crafted from volcanic glass to modern high-performance glass in buildings made from a sequence of controlled industrialised processes, the material has undergone significant development which is still ongoing. Despite the increasing use of glass as a construction material, current research has yet to answer several questions.

The main component of glass typically used in architectural and automotive applications is silicon oxide (i.e. silica): it makes up over 60% of the Earth's crust and is available in almost unlimited quantities. This relatively cheap and abundant material is transformed into an engineered high performance material by an elaborate and ingenious manufacturing processes. In everyday language, the word glass has different meanings, depending on the context. For instance, "glass" may refer to a drinking vessel, or to a constituent part of a window, or to spectacles (glasses).

Tammann (1861–1938) [2], one of the pioneers in glass research, defined glass as follows:

J. Belis et al., *Engineering Glass Structures*,
https://doi.org/10.1007/978-3-032-03233-1_1

> The glass state is the frozen state of a supercooled liquid that has solidified without crystallising.

Another definition is given in [45]:

> Glass is an inorganic product of fusion that has cooled to a rigid condition without crystallisation.

Such definitions may not adequately capture more modern materials such as aerogels leading to more recent additions to these traditional definitions, see e.g. [49].

However, for the intents and purposes of this book, the above-mentioned definitions characterise the physical and chemical properties of the glass end-product in a useful way, and are therefore broadly correct.

In comparison to other construction materials, glass has:

- a Young's modulus comparable to that of other construction materials;
- a coefficient of thermal expansion comparable to that of other common construction materials;
- a relatively constant material properties at service temperatures in typical construction applications;
- an almost perfect linear-elastic response up to fracture;
- a high brittleness (i.e. a low fracture toughness of around $K_{\mathrm{Ic}} < 1.0\,\mathrm{MPa\,m}^{1/2}$);
- a relatively high compressive strength;
- a relatively low tensile strength and large scatter of strength values due to a high brittleness and its sensitivity to surface conditions (including unavoidable microcracks, surface flaws, surface treatment, etc.);
- transmittance, reflectance and absorbance of electromagnetic radiation, including the visible part (380–780 nm) that vary with the angle of incident radiation;
- chemical resistance to a large number of substances.

1.2 Physical Nature of Glass

1.2.1 Molecular Structure

Many engineering materials have a crystalline molecular structure. However, glass is an exception. At room temperature, the three-dimensional arrangement of molecules has no long-range order and is analogous to that of a liquid. This non-crystalline solid is known as an amorphous solid. Consequently, the term 'glass' may generally refer to this amorphous state, independent of its chemical composition. However, in the context of this book, we refer to glass as a particular engineering material.

While glass is appropriately described by Tammann as *the frozen state of a supercooled liquid*, this description can be misleading. For example, glass is sometimes conceived as a viscous liquid that can flow at room temperature, which is incorrect. Tammann's model describing the irregular structure of glass is still widely used,

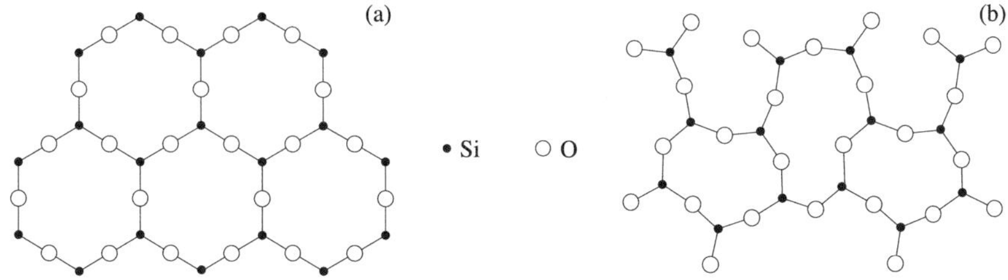

Fig. 1.1 Two-dimensional schematic representation of **a** a crystalline SiO_2 network (quarz) with existing short and long range order and **b** silica glass (non-crystalline SiO_2 network) with existing short but no long range order, according to [1]

since it captures many macroscopically observable glass properties, such as fracture and strength, which can be relatively easily explained, see Fig. 1.1.

More recently, this molecular structure was observed experimentally in an electron microscope by Huang et al. when in 2012 a two atom thick layer of silica glass was accidentally made in the laboratory during a graphene-growing experiment [36].

1.2.2 Does Glass Flow?

There are several anecdotal observations that medieval glass windows are noticeably thicker at the bottom than at the top and associated claims that this proves that glass has flowed due to gravity. Although glass does indeed flow at *high temperatures*, it was shown by Zanotto [17, 18] that at room temperature it would take approximately 10×10^{32} years (more than 7 trillion times the estimated age of the universe) for it to flow to an extent that it would be observable by the human eye. Consequently, it is much more likely that the glass in ancient windows was originally of non-uniform thickness due to the imperfect production methods used at the time, see Sect. 2.1, and it is plausible that these panes were assembled and mounted with their thicker and heavier edges at the bottom. Thus, the flow of window glass is negligible at ambient temperature.

1.2.3 Thermodynamics of Glass

The thermodynamic nature of glass, from its molten to its solid state, has yet to be fully understood. However, it is possible to explain the principal processes that occur during glass formation.

The first step is to set out the strain energy in the solid material followed by the processes that occur during heating and cooling from the thermodynamic and energy perspectives. This forms the theoretical basis for describing the processes of glass formation in this chapter.

These processes also form the basis of the thermal pre-stressing process of glass explained in Chap. 6.

1.2.3.1 Strain Energy in Solid State

The molecules or atoms of an unstressed solid are in equilibrium i.e. they are located in a fixed position relative to each other with a certain interatomic distance. In order to disturb this equilibrium energy must be added to the system to force the atoms out of their equilibrium position. This can be done, for example, through external mechanical work. In the uniaxial stress state, an applied force and corresponding stress, σ, creates a relative expansion or contraction of the solid known as the strain, ϵ, in the direction of the force, which extends the inter-atomic bonds. If the solid is linear-elastic and isotropic, any mechanical work performed on it will be stored as elastic (potential) energy in the volume of the body, V. This stored energy, W_p, in its uniaxial stress state can be represented by the strain energy in the volume of the body:

$$W_\mathrm{p} = \int_V \left(\int_\varepsilon \sigma \cdot \mathrm{d}\varepsilon \right) \mathrm{d}V = V \int_\varepsilon \sigma \cdot \mathrm{d}\varepsilon = V \int_\varepsilon E\varepsilon \cdot \mathrm{d}\varepsilon = \frac{1}{2} V E \varepsilon^2 = \frac{1}{2} V \sigma \varepsilon, \tag{1.1}$$

where the second equality sign indicates uniaxial loading (constant stress state) and the third implies linear elasticity. If the external load is removed, the atoms, driven by the stored energy, revert to their original position. If no interatomic bonds are damaged, the process is completely reversible and there is a linear relationship between the applied force and the resulting deformation. In other words, it obeys Hooke's law, so glass is also known as a linear elastic solid. The material's macroscopic resistance to uniaxial deformation is known as the modulus of elasticity or Young's modulus, E.

1.2.3.2 Melting Process

As ambient temperature increases, thermal energy is added to the solid. At temperatures above absolute zero, $T > 0\,\mathrm{K}$, the atoms in the solid start oscillating about their static position. The amplitude of the oscillations increases with temperature and as the particles require more space, the solid expands. This process takes place almost linearly over wide temperature ranges and is represented macroscopically by the coefficient of thermal expansion, α_T. The kinetic energy of the particles and the amplitude of the oscillations increase as the applied heat, ΔQ, increases. The melting point (or melting temperature), T_m, is reached when the kinetic energy is large enough to break the atomic bonds. This leads to mechanical instability; the solid body loses its cohesion and melts. The breaking of the bonds at T_m requires a large amount of added energy, ΔQ, whereby the temperature, and also the free enthalpy,

H, of the body, remains, constant ($\Delta H = 0$). The enthalpy, H, of a body is the sum of its internal energy, W_i, and the product of volume, V, and pressure, p:

$$H = W_i + V \cdot p\,. \tag{1.2}$$

The internal energy, W_i, of a body represents the entire available energy in the system. Simultaneously, during the melting process, the entropy, S, in the system increases, because the order in the system decreases. As a result of such a first-order phase transition, a clear separation of the properties of the solid and liquid states occurs. The material properties change rapidly during the transition. The energy required to separate the molecules during the melting process is equal to the bonding energy of the molecules. The phase transition itself is given by the thermodynamic equilibrium [9]:

$$\Delta H = \Delta Q - T_f \cdot \Delta S = 0\,. \tag{1.3}$$

whereby

ΔH Change in free enthalpy during the phase transition $[J]$;
ΔQ Change in enthalpy (amount of heat added) $[J]$;
T_f Melting temperature $[K]$;
ΔS Entropy change $[J/K]$.

Crystalline silicon dioxide, SiO_2 (silica), has a melting temperature, T_m, of approximately 1860 °C. Typically, silicon dioxide is also mixed together with other materials to improve its processing ability (e.g. to reduce the melting temperature) and to influence the properties of the final glass product (e.g. to change colour). In silica glasses, the melting point, T_m, is typically between 1300 and 1600 °C, depending on the composition of the glass.

1.2.3.3 Liquid State

Atoms, molecules and molecule groups can move relatively freely in an ideal liquid state. However, in contrast to a gaseous state, they are still connected to each other by intermolecular interactions (cohesion forces). The bonds break and rebuild constantly, whereby the bonding partners are constantly changing. These intermolecular interactions create friction between the neighbouring particles and the transferred force is dependent on the speed of motion. The viscosity, η, is the observable, macroscopic characteristic of this molecular agility.

In an ideal Newtonian liquid, the liquid state can be characterised by its viscosity, $\eta(T)$, which is typically strongly dependent on the temperature, T, as a measure of its resistance to flow. This simplifies the relationship between the shear stress, τ, and the rate of shear strain, $\dot{\gamma}$, to be linear, as shown in Eq. 1.4.

$$\tau(T) = \eta(T) \cdot \dot{\gamma} \tag{1.4}$$

At high strain rates, e.g. high-pressure moulding of liquid glass, non-Newtonian behaviour occurs because unrestricted movement is not possible. This leads to viscosities that are an order of magnitude lower than the Newtonian viscosity [10].

1.2.3.4 Supercooled Liquid and Glass Transition

If a liquid is cooled down to below its melting point, as a rule, crystallisation takes place because the crystalline form is a lower energy state of matter (lower entropy). The shape of the resulting crystal lattice depends on the type of chemical bond and the bonding partners involved. At a certain particle distance at the atomic scale, equilibrium is reached between the opposing attraction and repulsive forces of two adjacent particles. This equilibrium distance gives rise to the symmetrical three-dimensional lattice structure in a three-dimensional space. All particles are prone to move to this favourable position in the system. In this process, energy known as the latent heat of crystallisation is released. Crystallisation is a reversible process and the energy released during crystallisation is also a measure of the work that must be done to break the bond. The resulting crystallites nucleate at several locations in the liquid and grow over time, eventually meeting at their grain boundaries.

The molecules, however, require some time to build up to a crystalline solid body to occupy the positions corresponding to the lowest energy state of matter in the crystal lattice. If the liquid is cooled down rapidly, crystallisation can be prevented and the melt remains in the liquid state, even at temperatures below T_m. In this unstable state, the liquid is referred to as a supercooled melt or supercooled liquid.

In systems that tend to be glass-forming, the formation of a crystal lattice is also prevented by the viscosity of the molten state. The so-called glass transition corresponds to a significant reduction of the molecular mobility and an associated increase in viscosity. This prevents the particles from moving into their preferred grid position. The system therefore finds itself in a thermodynamically metastable state, which is maintained through increased viscosity. During further cooling, the bonds gradually close and the supercooled liquid solidifies into a non-crystalline, amorphous solid body in a disordered state. In theory, any material can take this state if crystallisation is avoided during cooling in its molten state. However, this can be difficult to achieve without glass forming agents such as SiO_2.

But there are also a multitude of systems that promote the formation of glass under normal circumstances. Strong glass agents such as SiO_2 (silica glass), B_2O_3 (borate glass) and P_2O_5 (phosphate glass) have the tendency to form supercooled melts, even when they are cooled down relatively slowly from a molten state.

Unlike crystallisation, the phase transition between glass states occurs continuously and without abrupt changes in the material properties; this is also referred to as the second order phase transition. The transition range is often simplified and expressed as a single temperature known as the glass transition temperature, T_g, see Fig. 1.2. The slopes of the straight lines in Fig. 1.2 can be interpreted as the thermal expansion coefficients, $\alpha_{T,l}$ and $\alpha_{T,g}$, in the liquid and glassy (solid) states, respectively. The temperature at the intersection of the straight lines is referred to as T_g.

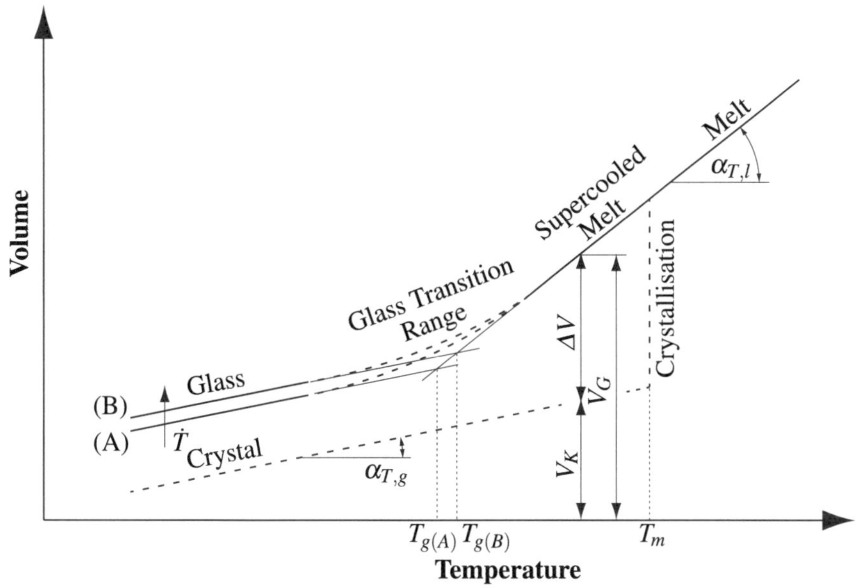

Fig. 1.2 The evolution of forming solid glass: volume as a function of temperature. Curve (A) is cooled slower than curve (B)

However, this value depends on the cooling rate, $\dot{T}$, and this in fact represents a transition range as indicated by the dashed lines.

Viscosity values between molten and solid states in the transition area are important for the processing of glass. In addition to the transition temperature, T_g, other salient temperatures are: the *processing temperature* (or: *working point*), T_w; the *softening temperature* (or: *softening point*), T_{so}; the *upper cooling temperature* (or: *annealing point*), T_a; and the *lower cooling temperature* (or: *strain point*), T_{st}, see Table 1.1.

The resulting material properties also depend on the cooling rate in the glass transition range. Slow cooling allows the molecules to enter into their preferred position. The solidified glass has a higher degree of long range order and thereby a higher density, and the transition takes place within lower temperature ranges. Fast cooling, on the other hand, leads to a lower density and long range order, since the frozen state is more open and random, as illustrated in Fig. 1.2. The transformation range corresponds to higher temperatures, as the rapidly rising viscosity leads to quick solidification. This phenomenon can also have an influence on the formation of residual stresses in the thermal pre-stressing of the glass and is referred to as structural relaxation, see e.g. Sect. 6.2.5.3.

Table 1.1 Typical viscosity values [16], designations and associated temperature for a soda-lime silica glass [15], a borosilicate glass [44] and a quartz glass [43]

Viscosity		Descriptor	Temperature		
[Pa s]	[dPa s]	[−]	Soda-lime silica [°C]	Borosilicate [°C]	Quartz [°C]
10^3	10^4	Working Temperature T_w	1 030	1 260	2 100
$10^{6.6}$	$10^{7.6}$	Softening Point T_{so}	725	825	1 730
10^{12}	10^{13}	Annealing Point T_a	550	560	1 180
$10^{12.3}$	$10^{13.3}$	Transition Temperature T_g	530	525	1 140
$10^{13.5}$	$10^{14.5}$	Strain Point T_{st}	506	518	1 075

1.3 Composition of Glass Used in the Building Industry

1.3.1 Chemical Composition

Almost all glass used in the building industry is silica glass, the most common being *soda lime silica glass* made mainly of quartz (i.e. silica) sand, lime and soda, see Table 1.2. Silica sand, in its almost pure SiO_2 form, is used as a network former in the manufacturing of glass. The silica sand contains small quantities of iron oxide Fe_2O_3, which are sufficient to give glass its characteristically green tint, see Table 1.6. Silica has a relatively high melting temperature of about 1700 °C. The high melting point and viscosity of silica is reduced by adding sodium oxide in the form of soda (sodium carbonate, Na_2CO_3) and the sodium-oxygen atoms, known as network modifier, enter the silicon-oxygen network. Sodium can also be fed as a nitrate or sulphate into the melt, as well as potash (K_2CO_3), thereby adding potassium oxide to the molten solid, which acts just like sodium oxide as a network modifier and flux agent. Calcium and magnesium enter the network structure in the form of dolomite ($CaMg[CO_3]_2$), thereby making the structures more complex thereby delaying crystallisation. Feldspar ($NaAlSi_3O_8$) and Na_2O, as well as aluminium oxide (Al_2O_3), are also added to increase the hardness and durability of the glass. In addition, cullet from its own production and/or from waste glass is added to the mixture.

In *Borosilicate glass* boron trioxide (B_2O_3) is added as a network former instead of Silicon dioxide, SiO_2, see e.g. Table 1.2. The boron-based network results in

Table 1.2 Chemical composition of soda-lime silica and borosilicate glass, as well as chemical function of the ingredients within the molten glass

Constituents [−]	[−]	Soda-lime silica glass[a] [%]	Borosilicate glass[b] [%]	Chemical function[c] [−]
Silicon dioxide	SiO_2	69–74	70-87	Nf
Calcium oxide	CaO	5–14	–	Nm
Boron trioxide	B_2O_3	–	7–15	Nf
Sodium oxide	Na_2O	10–16	0–8	Nm
Magnesium oxide	MgO	0–6	–	Nm
Potassium oxide	K_2O	–	0–8	Nm
Aluminium oxide	Al_2O_3	–	0–5	Nf & Nm
Other		0-5	0-8	

[a] According to EN 572-1 [34]
[b] According to EN 1748-1-1 [21]
[c] Nf = Network-former, Nm = Network-modifier

a significantly lower thermal expansion coefficient and better chemical resistance, thereby making borosilicate glass a suitable choice for fire resistant glazing.

Depending on the composition, the thermal expansion coefficient of the solid glass ranges from $\alpha_T = 0.5 \times 10^{-6}\,K^{-1}$ for quartz glass to $\alpha_T = 4.0 \times 10^{-6}\,K^{-1}$ for borosilicate glass and $\alpha_T = 9.0 \times 10^{-6}\,K^{-1}$ for soda lime-silica glass. Based on the magnitude of its coefficient of thermal expansion, glass is sometimes referred to as either "soft" glass ($\alpha_T > 6.0 \times 10^{-6}\,K^{-1}$) or "hard" glass ($\alpha_T < 6.0 \times 10^{-6}\,K^{-1}$).

1.3.2 Atomic Structure, Symmetry and Order of Glass

Silicon dioxide (SiO_2) forms the network of soda-lime-silica glass, and the type of bonding that occurs is significantly dependent on the interactions of the outer valence electrons with the bonding partners. In the formation of a chemical bond, different types of bonds are established, depending on the number of valence electrons and the electrical charge of the bonding partners. These bonds determine the physical and mechanical properties of the resulting connection. Oxygen has six valence electrons and tends to acquire two further electrons in its outer electron shell, which it shares in a bonding partnership with other elements, including with another atom of oxygen, thereby taking on the more stable noble gas configuration. Silicon has four valence electrons, and can therefore both give away and take on electrons to achieve a noble gas configuration. The four valence electrons of the silicon bond tend to take on a tetrahederon-shaped sp^3 hybrid orbital determining the basic shape of the molecular structure, see Fig. 1.3.

However, since only two oxygen atoms are available for each SiO_2 molecule, neighbouring molecules must share oxygen atoms in order to be able to take on the

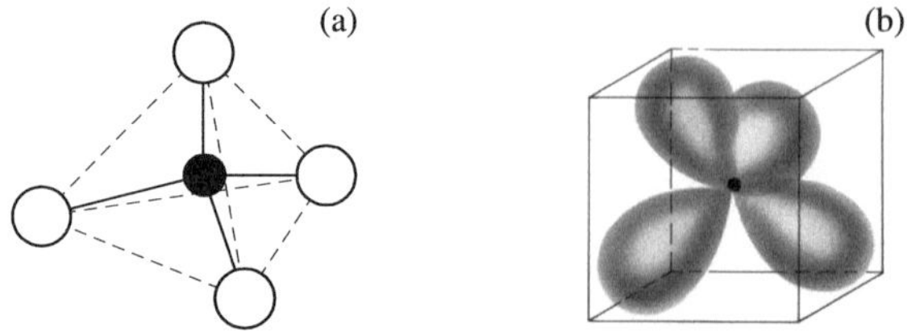

Fig. 1.3 **a** Three-dimensional representation of the SiO_4 tetrahedron; **b** The valence electrons of the central silicon atom preferring a tetraeder-shaped sp^3 hybrid orbital [6]

sp^3 hybrid orbital configuration. This is how the oxygen bridge bonds are built and the network is formed.

For complex systems, there are usually several bonding states that overlap. The resulting basic molecular structure is determined by the strongest bond in the system; the macroscopic material properties by the weakest. Glass can only form when there is a very strong interrelationship between atomic bonds, since only then is a sufficiently high viscosity present around the melting point, which, in turn, is necessary for network formation. In the SiO_2 molecule, a covalent bond with a strong ionic character occurs, which has double bonding properties. Furthermore, Van der Waals forces are also established between the molecules, strengthening the bond. Such complex bonding combinations are typical of systems that form three-dimensional networks. As a result of its double bonding nature, an exceptionally strong bond is formed between silicon and oxygen. This is reflected in its high melting point and by a shorter atomic distance between silicon and oxygen than would be found in a pure ionic bond. Once formed, the bond proves to be extremely stable. This results in a locally rigid structure with fixed atomic distances and rigid bond angles when the molten solid is cooled down. The resulting SiO_4 tetrahedron structure leads to a significant increase in the viscosity of the molten solid, and ultimately to the formation of glass. Whether or not a substance is suitable for the formation of glass is strongly dependent on the type of interatomic bond. The high strength of the atomic bond leads to a theoretical glass strength in the order of 32 000 MPa, which, in practice, is never achieved due to numerous strength-reducing factors [24], see also Sect. 4.2.2.

An ideal SiO_2 system can also be perceived as a regular tetrahedron, when a silicon atom is surrounded by four oxygen atoms, each of which is connected to two adjacent tetrahedrons. Different types of crystal lattice can be created by different periodic sequences of the tetrahedron. Besides the well-known quartz, pure SiO_2, also exists naturally in seven other crystal configurations, only differing in the arrangement of the local tetrahedron structure. Complex, irregular, three-dimensional networks emerge from the SiO_4, with each oxygen atom being shared between two tetrahedra, when glass solidifies, see Fig. 1.5.

The amorphous structure of the glass is not without any order. The local symmetry can even be understood as a driving force in glass formation. In the case of an ideal crystal, the regularity of the fine atomic structures continues up to macroscopic orders of magnitude. Many naturally grown crystals allow their outer symmetry to determine their inner atomic structure. The crystal grows periodically from out of its atomic cell in all directions. Crystals therefore have both a short range order (regularity of adjacent particle structures) and a long range order (regularity of the particle structures over long distances). The degree of order of glass forming systems, on the other hand, is basically similar to that of liquids and is less than that of crystalline solids. A crystal, through its hexagonal, densely packed sphere, achieves a maximum packing density of 74.05%. Glass and liquids, in a densely packed sphere, only reach a packing density of approximately 63.7%. These disordered systems are macroscopically isotropic have no far-reaching remote order, see Fig. 1.4. However, local regular arrangements can be found in neighbouring particles. Layers of hexagonally arranged particles rarely occur, but three-dimensional tetrahedron and octahedron structures, that locally reach a maximum filling space, are more likely.

The regular tetrahedron has a uniform O-Si-O-bond angle of $\gamma_{\text{O-Si-O}} = 109.47^\circ$ between the atoms involved, see Fig. 1.5a. The angle between three oxygen atoms in the tetrahedron edges is $\gamma_{\text{O-O-O}} = 60^\circ$. This local order is also present in the amorphous state, as the three-dimensional tetrahedral is the most energetically advantageous configuration of the system. If five tetrahedrons are assembled with a common edge, it should be noted that the resulting

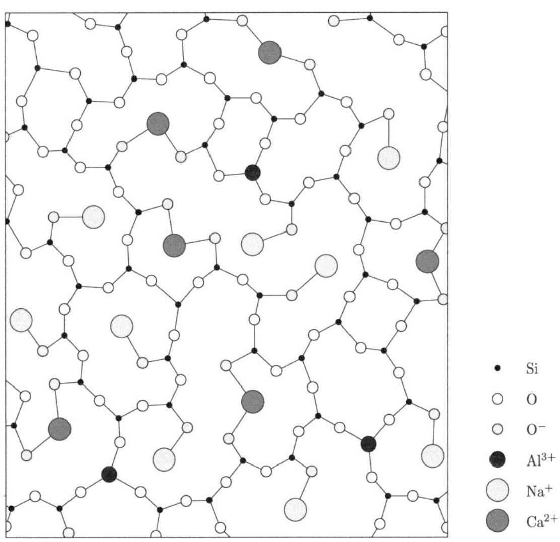

Fig. 1.4 Level representation of an irregular SiO_2 network with alkaline components. The fourth valences of the SI atoms protrude out of the drawing area

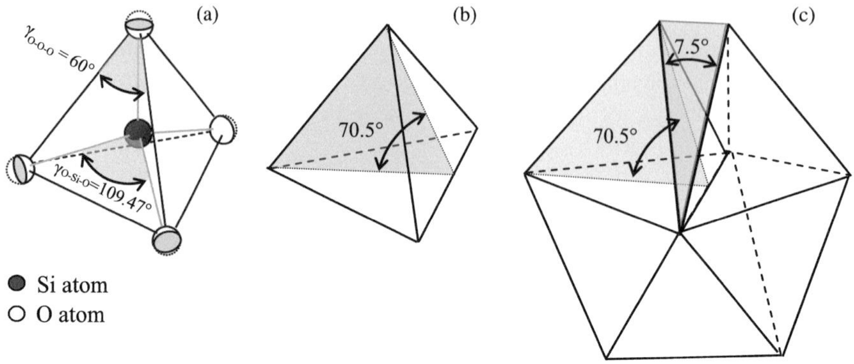

Fig. 1.5 SiO_4 **a** tetrahedron with central silicon atom (black) and four half oxygen atoms (white); **b** Schematic representation of the regular tetrahedron structure; **c** incomplete body consisting of five regular tetrahedrons

geometric structure is unable to fill the three-dimensional space completely, see Fig. 1.5c.

Each angle between the adjacent surfaces is 70.5°, see Fig. 1.5b. A gap with an angle of 7.5° appears, see Fig. 1.5c. It is not possible, without local distortions of the tetrahedron, to build up a wide-reaching periodic lattice to fill up the space completely. A glass-forming system, however, tends to maintain its local structure due to the high strength of the atomic bond. The long range order is then lost at the expense of the dominant local order. A long-range order is, despite the regular tetrahedron, no longer recognisable. More regular tetrahedrons can be attached to the gaps of the icosahedron that has been formed. In this way, coherent structures arise from regular tetrahedrons. This long chain that shares similarities with polymeric structures has almost the ideal physical and mechanical properties of the SiO_2 molecules that have resulted from the chemical bonds. In the same way, however, weak points will spread in a threadlike way through the glass. The network of the glass is thus not built up by disordered atoms, but rather is created by the stringing together of regular SiO_4-tetrahedrons. This is a notable difference from Tammann's classic model [2], whereby the properties in the glass structure follow a homogeneous distribution. In reality, the glass structure has very strong local variations in density, stiffness and strength. In addition, there are large internal stresses, which significantly affect strength. The properties of the macroscopic body are determined by these distributions. The strength of the glass body is dependent on the minimum local strength in the structure, while the measurable stiffness of the glass is an average between the high stiffness of the ideal tetrahedron structures and the low stiffness of the structural gaps. This makes it difficult to reliably infer results of the measurement of the macroscopic specimen body onto the properties, or even onto the build up, of the micro-structure. Because of the dominant local order in glass, it can therefore not result in the formation of a maximally dense packing sphere. To this end, the regular tetrahedral should be abandoned in favour of a long-range order. Instead, the resulting gaps and flaws result in strong local density fluctuations. For this reason, the atoms in the aperiodic glass structure also do not reach a maximum global filling of the space. Crystal lattice-forming systems, on the other hand, are able to abandon their local order in favour of the remote order to achieve a greater density throughout.

It is therefore possible to classify materials according to Table 1.3 in a way that goes beyond consideration of the internal order. A solid can be either crystalline or glassy (i.e. amorphous). The difference between the two is their internal ordering states, see Table 1.3. In this sense, for example, polymers could also be considered glassy solids, see Sect. 1.5.

The dominance of the closed order and the resulting structure is also the cause of the high viscosity of molten glass. At the melting point T_m, we first get a SiO_4-tetrahedron, and finally, larger rigid units of tetrahedron lattices, which increase the viscosity of the molten glass due to their spatial expansion and associated limited movement capacity. Thus, there exists at the atomic level great similarity between the crystalline structure and the glass structure. Deviations reveal themselves only in the position and orientation of the next particle or the next tetrahedron. Differences between glass and crystals are particularly evident in the distribution of the bond angles, torsion angles and atomic distances. In glass, these parameters scatter slightly more than in the crystal. The mean values of the atomic distances are only slightly larger in glass than in crystal, but the dispersion of the particle distances is larger in glass. The O-Si-O bond angles are not subject to large dispersion and have the same value in both glass and crystal, as they occur within the rigid SiO_4 tetrahedron. The Si-O-Si bond angle, on the other hand, indicates the position of two tetrahedrons and is subject to relatively strong scattering in the glass. The torsion angles are subject to very strong scattering from 0° to 180° in the glass, depending on the Si-O-Si bond angles, and thus markedly demonstrate the difference between glass and crystal.

The atomic distance (specified in angstroms, $1\ \text{Å} = 1 \times 10^{-10}$ m) between silicon and oxygen is lower with $r_{\text{Si-O}} \approx 1.61$ Å than the sum of the ionic radius of Si^{4+}-cation ($r_{\text{Si}} = 0.40$ Å) and O_2-anion ($r_{\text{O}} = 1.40$ Å) with $r_{\text{Si-O}} = 1.80$ Å. This is due to the complex and particularly strongly mixed bond between oxygen and silicon, see Table 1.4.

Table 1.3 Classification of the substance state

State	Cohesion	Retains its shape	Short-range order	Long-range order
Crystalline solid	Yes	Yes	Yes	Yes (dominant)
Glassy solid (amorphous)	Yes	Yes	Yes (dominant)	No
Liquid	Yes	No	No	No
Gas	No	No	No	No

Table 1.4 Composition of the atomic distances and bonding angles for silicate glass [23]

	Atomic distance		Bond angle
Si-O	$r_{\text{Si-O}} \approx 1.61$ Å	Si-O-Si	$\gamma_{\text{Si-O-Si}} \approx 144° \pm 10°$
O-O	$r_{\text{O-O}} \approx 2.65$ Å	O-Si-O	$\gamma_{\text{O-Si-O}} \approx 109.47°$
Si-Si	$r_{\text{Si-Si}} \approx 3.06$ Å	Torsion Angle α_1 and α_2	$0° < \alpha_i < 180°$

1.3.3 Chemical Resistance—Glass Corrosion

Glass is well known for its high chemical resistance to water and other solutions. This is true for the range of temperatures commonly encountered in building applications. Without this characteristic, its versatile application in construction would be unthinkable, see Table 1.5.

Table 1.5 Chemical resistance of soda-lime silica glass

Reagents	Resistance
Non-oxidising acids	+
oxidising acids	+
SiO_2-Releasing acids	o/−
Salts	+
Water	+
Non-oxidising bases	o/−
oxidising bases	o/−
Aliphatic hydrocarbons	+
Aromatic hydrocarbons	+
Chlorine chlorofluorocarbons	+
Alcohols	+
Ester	+
Ketones	+
Oils and Fats	+

+ resistant, o conditionally resistant, − non-resistant

Hydrofluoric based acids (HF) are one of the few chemicals known to react with glass over a short timescale [11]. In addition, glass also has a limited resistance to *water*, but with significantly lower reaction speeds. This process is also known as glass corrosion. The chemical equilibrium of water is

$$2H_2O \rightleftharpoons H_3O^+ + HO^- , \tag{1.5}$$

so that on glass surfaces, besides water molecules, protons and hydroxide anion may also be present. Studies have shown that the energy required to break down a silicon oxygen bond between two adjacent tetrahedrons is significantly reduced in the presence of water and aqueous solutions [4, 5, 8, 13].

The contact of water and aqueous solutions with the SiO network of glasses can cause selective leaching and glass corrosion [11, 14]. The reactions between glass and aqueous solutions take place through three chemical mechanisms:

- *Hydration:* Water molecules diffuse into the glass without any chemical interaction between the water and the glass network. The diffusion is limited by the ratio of the molecular size of the water or aqueous solution and the opening diameter of the glass network.
- *Ion exchange:* Between water molecules and alkali or alkaline-earth-metal silica glasses in the form of selective leaching in the area of alkaline network converters. This process is also known as the leaching of the glass. In the case of soda lime silica glass, calcium and sodium ions are embedded in the glass network. The reaction between the glass network and the protons of the solution Eq. 1.5 is negligible, but the network converters have a certain freedom of movement and enter the surrounding solution. This leads to a migration of cations within the glass network and an ion exchange between the cations of the glass network and the protons of the solution, at the interface between glass and water. For example, the chemical reaction equation is:

$$\begin{aligned} &\equiv \mathrm{Si} - \mathrm{O^-Na^+} \quad \mathrm{Na^+O^-} - \mathrm{Si} \equiv \; + 2\,\mathrm{H_3O^+} \\ &\rightarrow \; \equiv \mathrm{Si} - \mathrm{OH} \quad \mathrm{OH} - \mathrm{Si} \equiv + \mathrm{H_2O} \; + \; 2\,\mathrm{Na^+} . \end{aligned} \tag{1.6}$$

- *Hydrolysis:* Water molecules can break down the molecular bonds of the glass network and can dissolve the glass. In particular, silica glass is susceptible to nucleophilic attacks by hydroxide (OH^-, see e.g. Eq. 1.5), which can dissolve the glass network. The chemical equilibrium of this reaction is:

$$\equiv \mathrm{Si} - \mathrm{O} - \mathrm{Si(OH)_3} \; + \; \mathrm{OH^-} \;\; \rightarrow \;\; \equiv \mathrm{Si} - \mathrm{O^-} \; + \; \mathrm{Si(OH)_4} . \tag{1.7}$$

Fig. 1.6 Glass corrosion due to alkali leaching of a glass surface through prior contact of the rainwater with fresh concrete surfaces

If all four bonds of a SiO_4 tetrahedron have reacted in this manner, a soluble silica $Si(OH)_4$ (hydroxil group) emerges. The solubility of the SiO_2 network in water can be described by the classical theory by Charles and Hillig [3] with the structural equation:

$$\equiv \mathrm{Si} - \mathrm{O} - \mathrm{Si} \equiv \; + \mathrm{H_2O} \;\; \rightarrow \;\; \equiv \mathrm{Si} - \mathrm{OH} \; + \mathrm{HO} \; - \; \mathrm{Si} \equiv \tag{1.8}$$

Formally, this reaction is called a transformation of a siloxane compound[1] to a silanol compound.[2] The reaction described in Eq. 1.8 is of important for the time-dependent mechanical strength of glass, since water near crack tips causes static fatigue in the form of subcritical crack growth (stress corrosion), see Sect. 5.3.

The above-mentioned ion exchange depletes the alkalis at the surface of the soda lime-silica glass, while protons and water are stored in the glass surface. This results in a gel layer[3] approximately 100 nm thick at a formation rate proportional to the square root of time [11]. In areas with a low ratio of water volume to glass surfaces, but especially in areas where the gel layer is not washed off, OH^--ions accumulate over time. At this stage, glass corrosion is already apparent and irreversible, see Fig. 1.6.

---

[1] In siloxane compounds, the silicon atoms are connected by exactly one oxygen atom (Si-O-Si).

[2] Silanol compounds are chemical compounds consisting of a silicon backbone and hydrogen.

[3] A gel is a solid phase that is permeated by a liquid phase.

1.4 Physical Properties of Glass Used in the Building Industry

1.4.1 Fundamental Optical and Other Electromagnetic Properties: Transparency of Glass

Light transmission, in the spectral range visible to humans (380–780 nm), is an essential feature of glass, see Fig. 1.7. Any electromagnetic radiation (including visible light) that is incident on the surface of a medium, is reflected, absorbed and transmitted by the medium, see Fig. 1.8. When light is absorbed by a medium, the absorbed component is converted into other forms of energy, such as heat, electrical or chemical energy through the interaction of light with energy-absorbing components in the

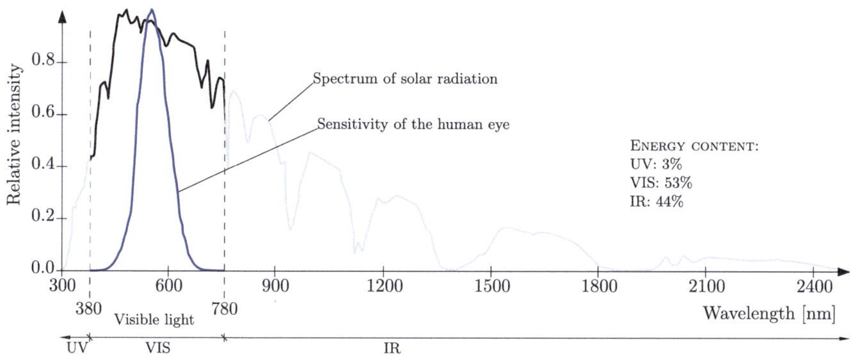

Fig. 1.7 Spectrum of the solar radiation in comparison to the sensitivity of the human eye after [38]

Fig. 1.8 Transmission *T*, reflection *R* and absorption *A* on a glass pane where incident radiation I = 100%

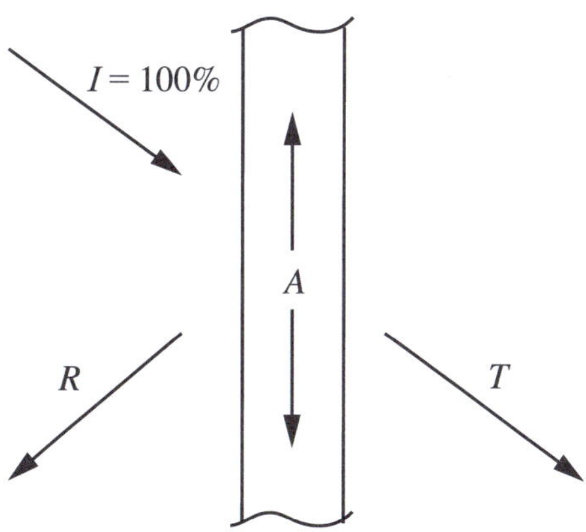

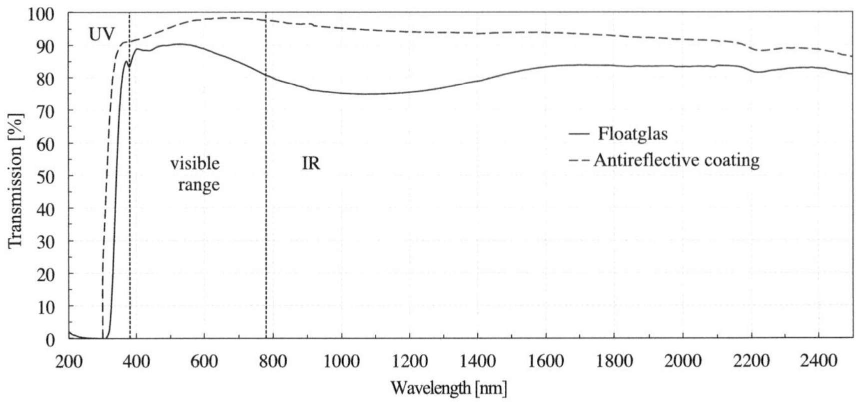

Fig. 1.9 Radiation permeability of a commercially available float glass of soda lime silica glass in comparison to a typical low iron-oxide, anti-reflective coated float glass

medium. The greatest interaction between incoming photons and the medium occurs when free electrons are present in the medium. Therefore, metals are impermeable to light. In glass, on the other hand, all electrons are tightly bonded, and therefore cannot normally interact with incoming photons. The photons are therefore transmitted, which is perceived by the human eye as transparency. In soda lime silica glass, electrons can be excited by light with high energy, consequently commercially available flat glass is impermeable in high-energy ultraviolet (UV) range. In fact it shows a pronounced absorption at wavelengths below $\approx$300 nm, which prevents transmission, see Fig. 1.9.

The position of the absorption peak depends on the binding strength of the O_2 ions in the glass and shifts in the long-wave area depending on the composition and the alkali content. In the infrared area (IR), more or less pronounced absorption bands appear in soda lime silica glass which are caused by an interaction of the light and components of the glass (Si-O-group). The position of the absorption bands and their intensity are again dependent on the glass composition. All silica glasses are impermeable to infrared rays with a wavelength above 5 000 nm due to the natural oscillations of the Si-O groups which absorb the electromagnetic waves.

Special compositions of the glass or coatings can be used to selectively influence the absorption within certain spectral ranges, see e.g. Sect. 2.3.5. An example of this are sun protection glasses with low transmission within the infrared and ultraviolet spectral ranges. The absorption capacity of the glass is therefore dependent on its composition. Glass is usually colourless in the visible spectral range. However, glass will absorb some energy, even in the visible range, if it contains small quantities of outer transition metals (Cu, Ti, V, Cr, Mn, Fe, Co or Ni). This manifests itself as colourisation of the light reflected and transmitted through the glass (also known at tint). The specific colour depends on the type of trace metal present, as shown in Table 1.6.

Table 1.6 Colouring agents and tints of glass [16]

Trace metals		Glass colour
Copper	Cu^{+}	Red
Copper	Cu^{2+}	Light blue
Chrome	Cr^{3+}	Green
Chrome	Cr^{6+}	Yellow
Manganese	Mn^{3+}	Purple
Iron	Fe^{3+}	Yellow-brown
Iron	Fe^{2+}	Blue-green
Cobalt	Co^{2+}	Intense blue
Cobalt	Co^{3+}	Green
Nickel	Ni^{2+}	Depending on the glass matrix, grey-brown, red to violet
Vanadium	V^{3+}	Green, brown
Titanium	Ti^{3+}	Purple
Neodymium	Nd^{3+}	Red-violet
Praseodymium	Pr^{3+}	Light green

1.4.2 Electromagnetic Radiation in the Built Environment

Soda lime-silica glass has a high transmittance in the optical and near infrared range (780–3000 nm) of the solar spectrum but is virtually opaque to mid (3000–50000 nm) and long wave (> 50000 nm) infrared radiation (e.g. for 6 mm clear glass visible transmittance $T_{vis} \approx 88\,\%$; short wave energy transmittance $\approx 91\,\%$; long wave energy transmittance $\approx 3\,\%$). The corollary of this is that when glass is used in the building envelope (e.g. windows), the shortwave solar energy transmitted through the glass is absorbed by objects in the internal environment (e.g. furniture, finishes, etc.) which as a result radiate energy in the long wave infra-red, but this energy cannot be re-transmitted back though the glass and is consequently 'trapped' in the indoor space leading to a rise in internal temperature, see Fig. 1.10.

As discussed in Sect. 1.4.1 and shown in Fig. 1.11, incident solar radiation, E_I, on glass is dissipated into:

- Direct solar energy transmittance, E_T;
- Solar energy reflectance, E_R;
- Solar energy absorptance, E_A,

where the absorbed energy is re-radiated, partly outwards (E_{RO}) and partly inwards (E_{RI}).

From these components it is possible to define an important glazing characteristic known as the total Solar Energy Transmittance (TET), solar factor, or g-value = $(E_T + E_{RI})/E_I$. Note the Solar Heat Gain Coefficient (SHGC), popular in North America, is a similar measure, but includes the solar energy transmittance of the

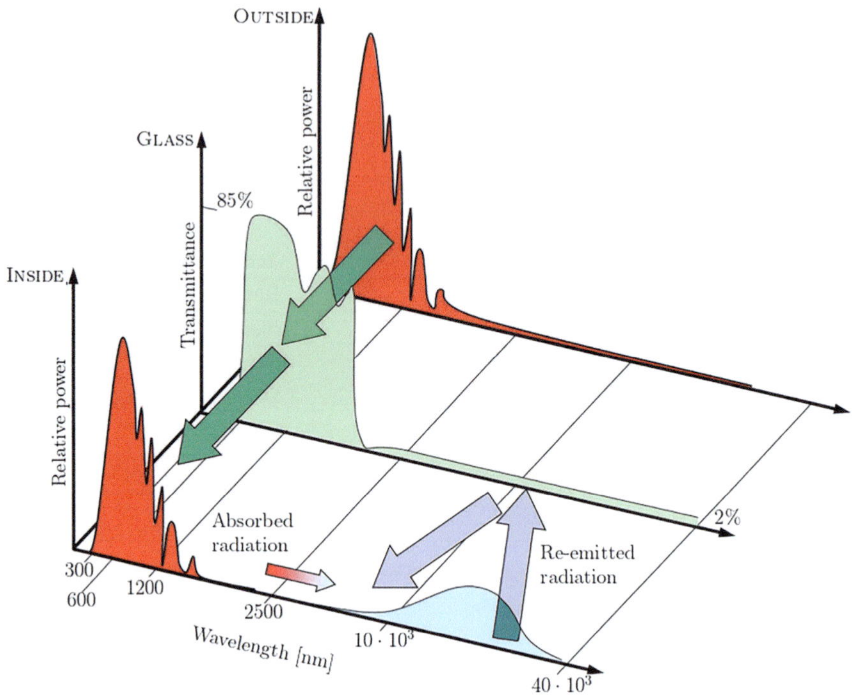

Fig. 1.10 Selective radiative transmittance of glass

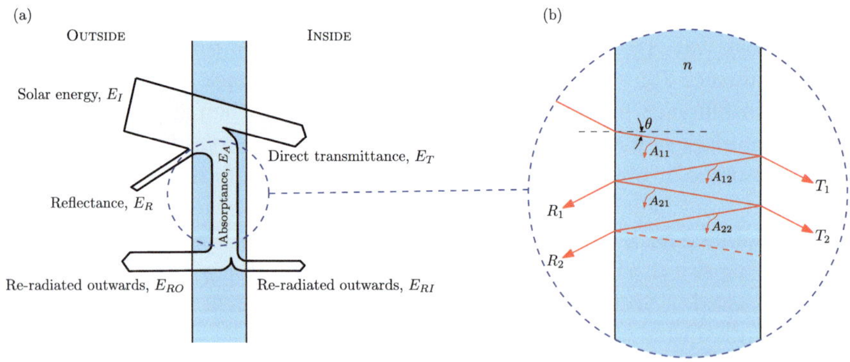

Fig. 1.11 Energy flow through monolithic glass: **a** simplified model, **b** realistic model

framing materials as well as the glass. Therefore the "centre of glazing SHGC" = g-value.

Other useful characteristics are:

- Light Transmittance, T_{vis}, which is the proportion of visible light that is transmitted through the glass;

- Light Reflectance which is the proportion of visible light at near normal incidence that is reflected by the glass;
- Colour Rendering Index (Ra) expresses the colour rendering properties of glass in transmission;
- Selectivity index $S = E_{T\text{visible}}/(E_T + E_{RI})$.

As the incident solar radiation (which is an electromagnetic wave) travels through glass it undergoes reflection and refraction at both the air-glass and glass-air interfaces and absorption whilst travelling through glass Fig. 1.11. These phenomena are described by Snell's Law (refraction angle), Fresnel's Law (proportion of reflected or transmitted energy) and Beer's Law (energy absorbed in glass). The values of transmitted, reflected and absorbed energy may be calculated theoretically. Figure 1.12 illustrates this for an air-glass interface.

The transmitted, reflected coefficients for an electromagnetic wave polarised parallel to the plane of incidence travelling through a monolithic glass sheet (i.e. accounting for flow through air-glass and glass-air interfaces is illustrated in Fig. 1.12) and given by:

$$T_{\|} = \frac{(1-r)^2(1-a_{\|})}{1-r^2(1-a_{\|})^2} \tag{1.9}$$

$$A_{\|} = \frac{a_{\|}(1-r)\left[1+r(1-a_{\|})\right]}{1-r^2(1-a_{\|})^2} \tag{1.10}$$

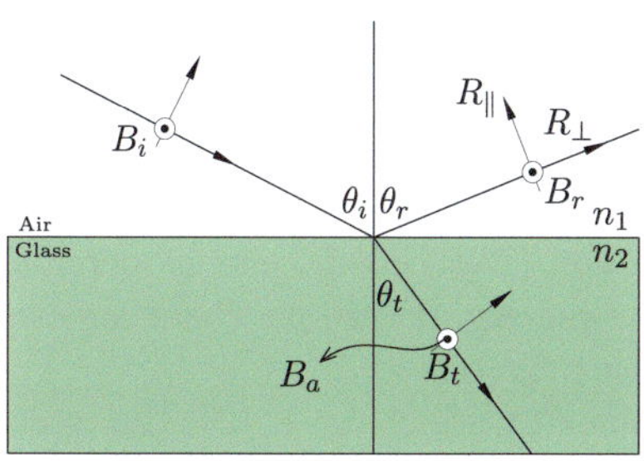

Beer's law: $B_a = B_i a_{\|}$

$a_{\|} = 1 - e^{(-kh/\cos\theta_t)}$

$a_{\|}$ = fraction of incident energy absorbed for radiation polarised parallel to plane of incidence
k = glass extinction coefficient
h = thickness of glass

Fresnel's law: $B_r = B_i R$

$$R_{\|} = \frac{\tan^2(\theta_i - \theta_t)}{\tan^2(\theta_i + \theta_t)}$$

$$R_{\perp} = \frac{\sin^2(\theta_i - \theta_t)}{\sin^2(\theta_i + \theta_t)}$$

$R_{\|}$ = fraction of incident energy reflected for radiation polarised parallel to plane of incidence
$R_{\perp}$ = fraction of incident energy reflected for radiation polarised perpendicular to plane of incidence

Snell's law: $n_1 \sin\theta_i = n_2 \sin\theta_t$

Light passing from one medium to another is refracted according to Snell's Law

Fig. 1.12 Reflection, absorption and transmission at air-glass interface

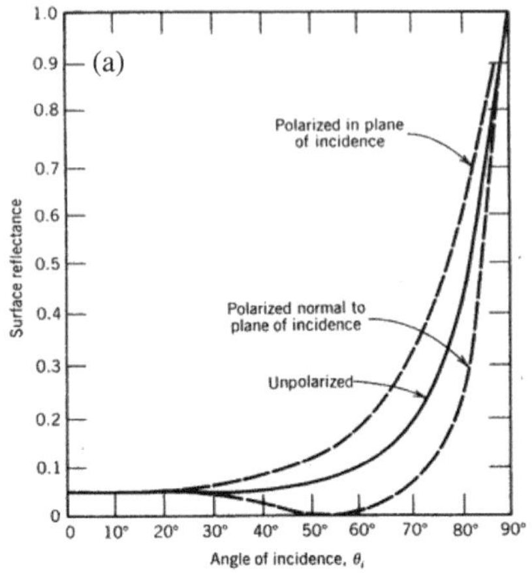

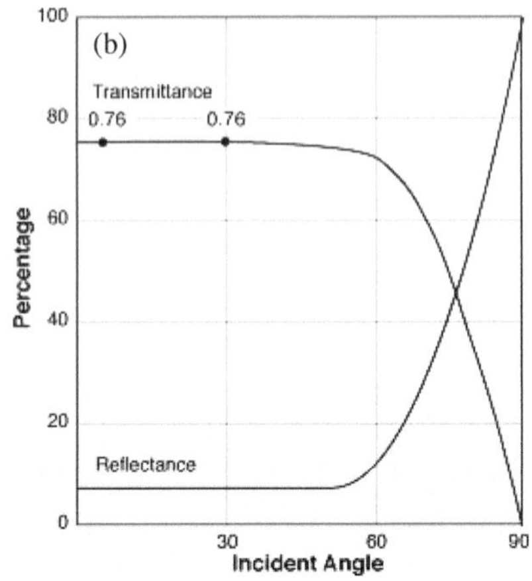

Fig. 1.13 Influence of angle of incidence on **a** reflectance of polarized light; **b** reflectance and transmittance of non-polarised light

$$R_{\parallel} = \frac{r(1-r)^2(1-a_{\parallel})}{1-r^2(1-a_{\parallel})} + r \tag{1.11}$$

It is important to note that:

- The sum of reflected and transmitted light must equal the total incident light (since these are fractions $R + T + A = 1$).
- The angle of polarisation of the incident electromagnetic wave with respect to the plane of the incident material has an effect on the respective fractions of light that are transmitted or reflected. The coefficients for non-polarised waves are taken as the mean of the parallel and perpendicular coefficients, see Fig. 1.13a.
- The angle of incidence of the electromagnetic wave has a very significant influence on the flow of electromagnetic energy through glass, see Fig. 1.13b. A single set of Transmission, Absorbtion and Reflection (TAR) coefficients is therefore insufficient for modelling the performance of glazing. Rather, the characteristics of TAR coefficients against angle of incidence are required.
- A total internal reflection of light occurs at an exit angle of $\alpha_{tot} \geq 42°$. This effect is used specifically in glass fibres for data transmission, where light is zigzagged through the fibre. The glass fibres are coated with a glass that has a lower refractive index or with a plastic.

A further complication encountered when modelling the flow of an electromagnetic wave through glass is that the TAR characteristics are not only a function of angle of incidence as described above, but also vary with wavelength, see Fig. 1.14. However the TAR characteristics for glass remain relatively constant between 400 and 2600 nm and are therefore assumed to be such for routine thermal calculations.

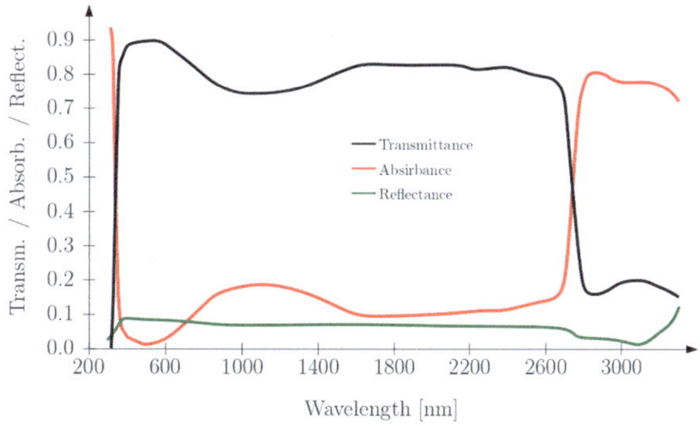

Fig. 1.14 Influence of wavelength on TAR characteristics of float glass

1.4.3 General Physical and Mechanical Properties

The characteristic material properties of a substance are governed by its internal structure. Statistical simulations are often used in order to draw conclusions about the macroscopic properties of atomic fine structures.

Many macroscopically observable properties, such as stiffness or modulus of elasticity, the thermal expansion coefficient, the density, or the refractive index but also the activation energy, and the diffusion coefficient correspond roughly to the mean value of the microscopic properties. Behaviour is different when it comes to strength. Like other brittle materials, the tensile strength of glass is determined by weakest link principle where the strength correlates with the minimum local strength of a material. For glass, in particular, the tensile strength is governed by the tensile forces and local surface conditions.

For most practical engineering applications, the deformation of glass under stress is independent of time. In fact, when using glass as a building material, an engineer will only be interested in the properties within the temperature range found in practical building conditions. Glass behaves in an almost linear-elastic manner within the practical temperature range until it fractures. Figure 1.15 shows an idealised stress-strain diagram of glass compared to steel and concrete. The salient mechanical and physical properties of soda lime silica glass, borosilicate, quartz glass and aluminosilicate glass are listed in Table 1.7.

However, some time-dependant behaviours are observable at a micro (molecular) scale, for instance:

- At temperatures above the transformation temperature, T_g, the viscous properties of the material can be exploited for thermally pre-stressing the glass (see e.g. Sect. 6.2).
- The material properties of glass, even at temperatures below T_g, are not completely time-independent. Relaxation processes, for example, reduce the intrinsic stresses in thermally pre-stressed glass, even at temperatures as low as 200 °C, as is apparent during the heat-soak test [47], (see also Sect. 6.2.6).

- A subcritical crack grows under continuous load and leads to fracture, depending on the duration of the load (see Sect. 5.3).
- A slightly delayed elastic behaviour in soda lime silica glass due to the mobility of the alkali-earth-metals in the stressed silicon dioxide network [23].

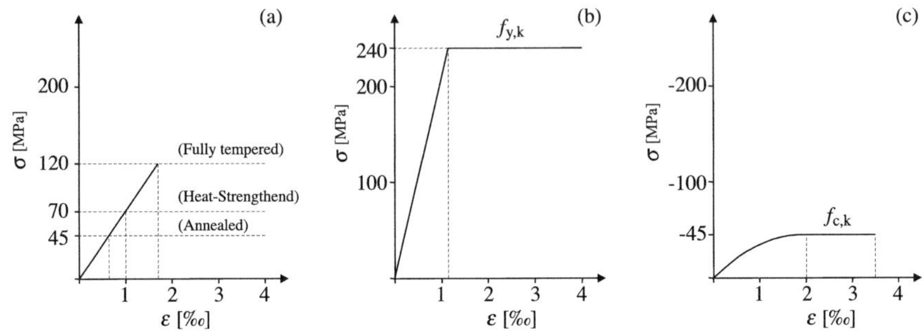

Fig. 1.15 **a** Ideal stress-strain behaviour of glass in comparison to **b** steel (S235) and **c** concrete (C45/55), in uniaxial tests

Table 1.7 Physical and mechanical properties of different glasses at room temperature, according to [21, 34, 42, 46]

Properties	Soda-lime silica	Boro silicate	Quartz	Alumino-silicate
Density, ρ [kg m^{-3}]	2 500	2 200–2 300	2 200	2 480
Young's modulus, E [MPa]	70 000	63 000	75 000	74 000
Poisson's ratio, ν [−]	0.23	0.20	0.17	0.215
Thermal expansion coefficient, $\alpha_{T,20/300}$ $\left[K^{-1}\right]$	$9.0 \cdot 10^{-6}$	$3.3 \cdot 10^{-6}$	$0.5 \cdot 10^{-6}$	$8.8 \cdot 10^{-6}$
Specific heat capacity, c_p $\left[J\,kg^{-1}K^{-1}\right]$	720–800	830	1 052	840
Heat conductivity, λ $\left[W\,m^{-1}K^{-1}\right]$	1.05	1.2	1.47	0.96
Refractive index, n [−]	1.52	1.47	1.46	1.51
Knoopsche hardness, [GPa]	6	4.5–6	5.8–6.2	5.3

1.5 Properties and Testing of Polymers

1.5.1 Introduction

The terms polymer and plastic are often used as synonyms, although the following distinction is used in this book: a polymer[4] consists of macromolecules, which are composed of reactive molecules, called monomers. Plastics can be defined as "synthetic-organic materials whose essential components contain macro-molecules" [32]. Thus, polymers are raw materials of plastics that often contain other substances (additives) in addition to the macromolecules. A macromolecule can be defined as a molecule with a minimum of 1 000 atoms [32]. In contrast to metals or ceramics, polymers are formed linearly or branched out and characterised by covalent bonds (electron pair bonding) between the atoms. This chemical bond is independent of bonding energy up to the decomposition temperature. The different chains of macromolecules are bonded to one another by weaker physical bonds (hydrogen bridge bond, Van der Waals forces) and/or by stronger ionic bonds. These bonds have an influence on the mobility of the molecule chains, and thus the mechanical properties at macroscale, which can be strongly temperature-dependent, including temperatures commonly found in practical building applications [37]. Macromolecules are a few microns long and usually contain between 10^3 to 10^5 monomers. The polymerisation grade of a polymer refers to the average number of monomers in a macromolecule [37].

The two most common elements in polymers are carbon and hydrogen. The macromolecules and the long branched polymer chains cannot be as closely packed as the molecules of other materials (e.g. metals), leading to a low molecular weight and an associated low density. Other properties that apply to most polymers are ease of formability due to low processing temperatures, low thermal and electrical conductivity, relatively good chemical resistance, temperature-dependent mechanical behaviour, gas permeability, recyclability, impact resistance, creep and ageing effects. In addition, polymers can be UV-resistant and transparent. A major advantage in the manufacturing of plastics over other materials is that of engineering a particular stiffness and/or strength [32].

Other important properties, such as solubility, swelling, thermal behaviour, mechanical behaviour and translucency, are strongly dependent on the type of polymer and often must be characterised for each individual product.

In comparison to the classical materials used in the construction industry (steel, concrete, timber and glass), the mechanical behaviour (e.g. strength and stiffness) of polymers is significantly influenced by temperature and strain rate at ambient temperature. In addition, the stress-strain behaviour for large deformations is usually non-linear, involving inelastic deformations. Furthermore, many polymers exhibit tension-compression asymmetry in their strength and stiffness responses.

---

[4] 'Polymer' means 'many units' in Greek.

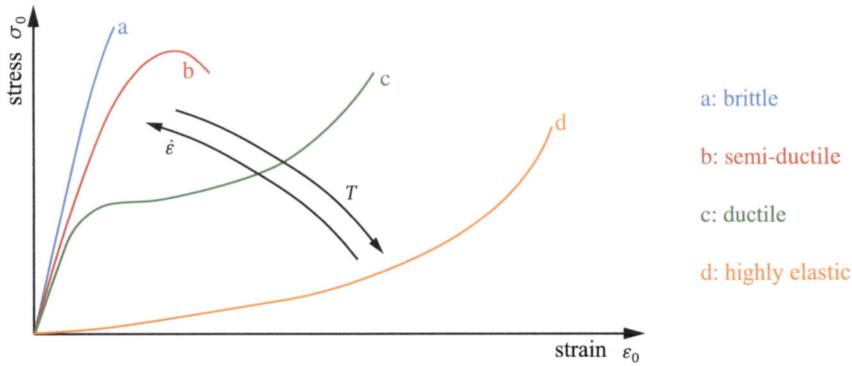

Fig. 1.16 Schematic representation of the stress-strain behaviour of polymers [26]

Figure 1.16 illustrates typical stress-strain responses for polymers obtained from uniaxial tensile tests executed at different temperatures and strain rates. A reduction in loading rate, which corresponds to a reduction in the strain rate, $\dot{\varepsilon}$, has a similar effect as an increase in temperature, T. This effect is generic for many materials and can provide a way to investigate high strain rate effects. This was actually explored for glass [48, 50, 52]. For some polymers this response follows the time-temperature-superposition principle, see e.g. Chap. 6.

A significant change in the mechanical behaviour of amorphous polymers occurs at the glass transition temperature, T_g; which separates the glassy state from the rubbery state. In the glassy state, below the glass transition temperature, the strong effect of the physical bonds between the macromolecules creates a high stiffness. The Poisson's ratio is typically between $0.3 \leq \nu \leq 0.35$ [12]. Higher temperatures lead to larger mobility of the macromolecules and thus to a small decrease in stiffness. In the range of T_g, a significant change in stiffness occurs. The thermal energy is sufficient to permit movement of macromolecules without externally applied stress [37]; in other words, Brownian motion can take place [28]. However, the transition does not occur abruptly, but within a temperature range called the glass transition range. In the subsequent rubbery state, the stiffness is generated by the effort of the macromolecules to arrange themselves into a state of higher disorder. Here, the Poisson's ratio can reach up to $\nu \approx 0.5$, and the material becomes almost incompressible [12]. Polymers are often classified according to the following criteria, which are further explained below:

- Composition of monomers: Homopolymers and Copolymers;
- Formation reactions: Chain polymerisation, polycondensation and polyaddition;
- Cross-linking degree: Linear, branched, cross-linked and networked;
- Crystallinity: Amorphous and semi-crystalline;
- Polymer structure and thermomechanical behaviour: thermoplastics, elastomers and thermosets;
- Chemical composition: polyolefins, fluorine polymers, chlorine polymers, etc.

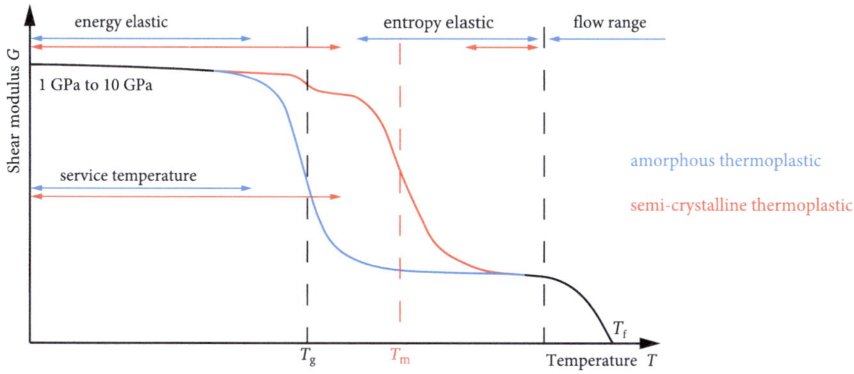

Fig. 1.19 Schematic representation of the stiffness as a function of temperature for amorphous and semi-crystalline thermoplastics

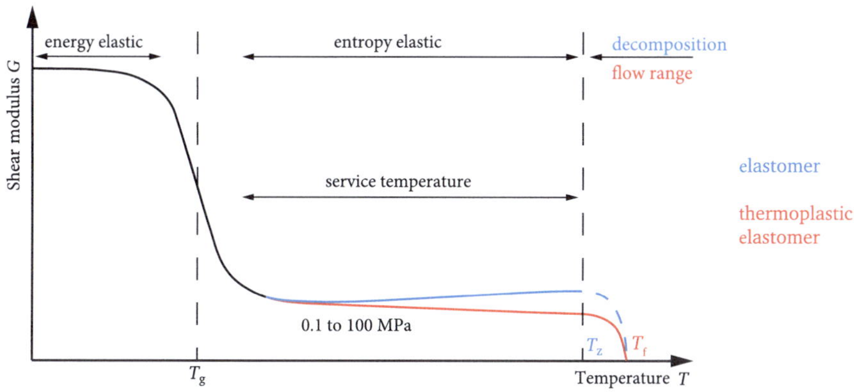

Fig. 1.20 Schematic representation of the stiffness as a function of temperature for elastomers and thermoplastic elastomers, T_z is the decomposition temperature

in a temperature range between 10 and 20 °C, and it reduces the stiffness by 3 to 4 orders of magnitude [12, 32]. At temperatures above T_g, amorphous thermoplastics are in a rubbery state until flow starts, the macro-molecules are entangled and the polymer melts at a melting temperature, T_f. In the case of *semi-crystalline thermoplastics*, the stiffness decrease at T_g, is significantly lower and a second stiffness drop caused by the melting of the crystals can be identified. The temperature that characterises this temperature range is called crystalline melting point, T_m [32].

Elastomers are chemically cross-linked polymers that form a loosely bound network structure with a rather low cross-link density and are typically amorphous. Cross-linking causes the polymers to decompose rather than melt at high temperatures. The glass transition temperature T_g is typically well below 0 °C. Elastomers are therefore typically in their rubbery state; they exhibit almost no irreversible

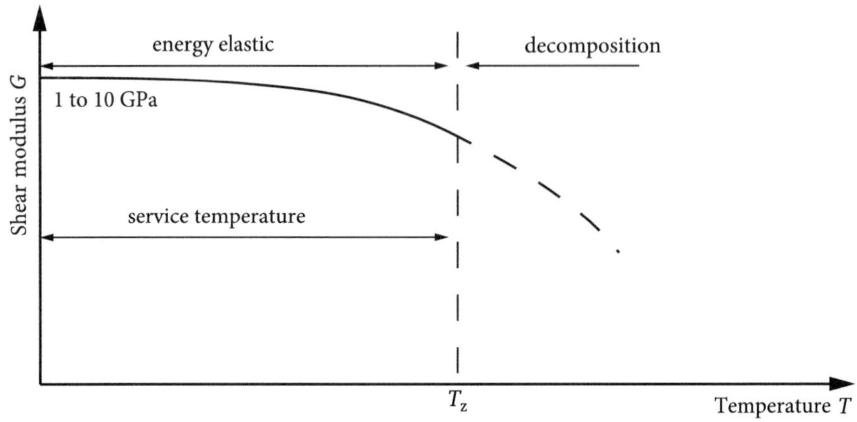

Fig. 1.21 Schematic representation of stiffness depending on the temperature of a thermoset, T_Z is the decomposition temperature

deformations and a slight increase in stiffness with increasing temperature due to the higher entropy, see Fig. 1.20.

Thermosets are closely meshed chemically cross-linked polymers and therefore amorphous and non-meltable. Due to their typical high degree of cross-linking, thermosets are also described as network polymers [22]. When heated the secondary (physical) bonds between the macromolecules "melt", causing a reduction in stiffness of the polymer. However, the chemical cross-links prevent true melting or viscous flow of the polymer. Further heating will cause the polymer to decompose [22]. The decomposition temperature is usually in the range of, or even below, the glass transition temperature. Therefore, thermosets exhibit almost solid, elastic and brittle behaviour, see Fig. 1.21.

In addition to these three main groups, there are *thermoplastic elastomers*, in which the macromolecules are physically cross-linked. They are co-polymers and consist of at least two types of monomers, one of which is in the glassy state in the temperature range for operation and the other in the rubbery state. Overall, physical cross-linking at operational temperature results in rubbery behaviour; at higher temperatures the physical cross-linking is lost and the polymers behave like ordinary thermoplastics and melt, see Fig. 1.20 [32]. The glass transition temperature is usually below 0 °C.

Finally, polymers can be classified according to their chemical composition. However, this type of classification does not provide any direct information about the thermomechanical properties and is therefore not discussed here.

1.5.2 *Test Methods*

There are several test methods for determining the mechanical properties of plastics (or polymers). The focus is on uniaxial tensile test and the Dynamic-Mechanical Thermal Analysis (DMTA). In addition to an explanation of the test principle and the evaluation, reference is made to the relevant regulations. Further test methods are then briefly explained.

1.5.2.1 Tensile Testing

The uniaxial tensile test is the basic means for determining the stress-strain behaviour of materials and the salient values derived from it. In the conventional tensile test, the specimen is clamped in a testing machine and subjected to a constant displacement rate, applied by means of a constant cross-head speed while measuring the applied displacement and the corresponding force. Special attention should be paid to the climatic conditions at which specimens are stored and tested (e.g. temperature and humidity).

The resulting stress and strain levels can be determined from the original specimen geometry, usually a dumbbell, and the force and extension from the uniaxial test. During the test, the specimen may slip at the clamps, particularly when testing smooth, low friction materials. In these instances the cross-head travel is not an accurate representation of elongation. Non-contact optical measuring methods are used instead, namely: laser extensometers; Digital Image Correlation (DIC) or video extensometers, which provide significantly better results without affecting the specimen, and are therefore preferred over clip-on extensometers.

Specimens may also experience local necking during testing. In such cases, true stress and strain can be determined by optical extensometry (e.g. DIC). In addition to the stress-strain relationship, Poisson ratio can be determined by measuring the transverse contraction using true strains:

$$\nu = -\frac{\varepsilon_{\text{transverse}}}{\varepsilon_{\text{axial}}} \tag{1.12}$$

The uniaxial tensile test on plastics is regulated in EN ISO 527-1 [35] and subsequent parts. This defines the specifications for the test procedure and characterising variables which can be derived from the test results. The US-American equivalent to the standard is ASTM D638. In addition, specific regulations exist for elastomers in DIN 53504 and ASTM D412 [25, 27].

1.5.2.2 Dynamic Mechanical Analysis

Dynamic Mechanical Analysis (DMA), also known as Dynamic-Mechanical Thermal Analysis (DMTA), is a method to determine the viscoelastic properties of a

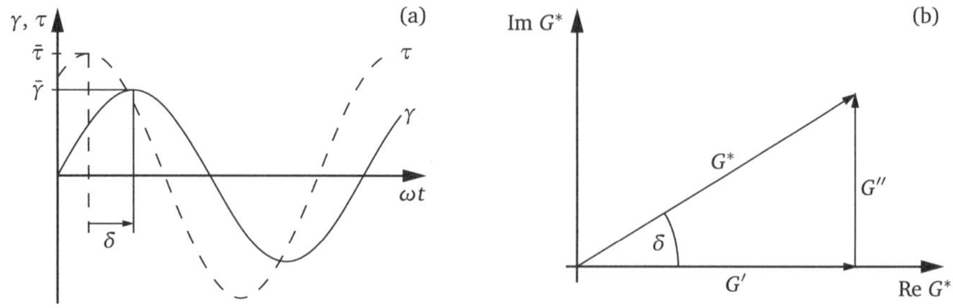

Fig. 1.22 **a** Phase shift between excitation and response; **b** definition of the complex modulus G^* transferred to the complex (Gaussian) plane

material at different temperatures and strain rates. A sinusoidal excitation is imposed on the specimen and the response is measured. A significant advantage of DMTA is that it is less laborious and time-consuming than performing the static creep or relaxation experiments at different temperatures.

The results of the DMTA in the frequency domain are given as complex numbers, see Fig. 1.22. These must then be transferred to the time domain and the real number space. The aim of the investigations is to determine the characteristic mechanical values that describe viscoelasticity such as the Prony series parameters and the Time Temperature Superposition (TTS) constants, see Sect. 4.1.5.

A DMTA can be performed in various test modes such as tension, compression, bending, double shear or torsion. Each test mode is in turn used to determine a particular mechanical characteristic such as shear modulus, Young's modulus or bulk modulus. For many polymers the viscoelastic effect originates from relaxation of the shear strain. Therfore, in a displacement-controlled experiment, the applied time-dependent shear is expressed as:

$$\gamma(t) = \bar{\gamma}\ \sin(\omega t) \tag{1.13}$$

where

$\bar{\gamma}$	Amplitude of strain;
$\omega = 2\pi f$	Angular frequency of strain;
f	Frequency of strain.

A sinusoidal stress response, $\tau(t)$, is measured, which is offset by δ in strain, see Fig. 1.22a:

$$\tau(t) = \bar{\tau}\ \sin(\omega t + \delta) = \bar{\tau}\ \sin(\omega t)\ \cos(\delta) + \bar{\tau}\ \cos(\omega t)\ \sin(\delta) \tag{1.14}$$

where

$\bar{\tau}$	Amplitude of the stress response;
δ	Phase angle.

From Fig. 1.22a, it is evident that a part of the response oscillates with the excitation and a part is shifted by 90°. The part oscillating in phase describes the elastic component; the part oscillating out of phase describes the viscous component. This results in a phase angle $\delta = 0$ for an ideal elastic material (i.e. no viscoelasticity) and $\delta = \pi/2$ for an ideal viscous material. A viscoelastic material is in between. Accordingly, a complex shear modulus G^* is introduced, which consists of an elastic component (storage modulus) and a viscous component (loss modulus) [12, 26]:

$$G^* = G' + iG'' \tag{1.15}$$

where

$$|G^*| = \frac{\bar{\tau}}{\bar{\gamma}};$$
$$G' = \frac{\bar{\tau}}{\bar{\gamma}} \cos\delta \text{ Storage modulus (elastic);}$$
$$G'' = \frac{\bar{\tau}}{\bar{\gamma}} \sin\delta \quad \text{Loss modulus (viscous).}$$

Figure 1.22b illustrates the relationship between the different moduli in the complex plane. The tangent of the phase angle, δ, is referred to as the loss factor or internal damping factor, and thus results in:

$$\tan\delta = \frac{G''}{G'}. \tag{1.16}$$

This approach is only valid if the material is linear-viscoelastic. This can be verified by a so-called amplitude sweep where the amplitude of the deformation/load is varied and the frequency is kept constant. In the range where G' and G'' is constant, the material can be assumed linear-viscoelastic.

In a temperature sweep, the excitation amplitude and frequency are kept constant and the temperature varies. These measurements are usually carried out at a test frequency of $f = 1$ Hz and are used to evaluate the temperature dependence and determine the glass transition temperature, T_g.

An isothermal experiment with varying frequency at constant excitation amplitude is called a frequency sweep. It provides information about the time dependence of the material. Master curves can be created and the TTS constants determined by performing several frequency sweeps performed at different constant temperatures.

There are several ways of converting frequency into a time range. The mathematically correct way is to quantify the Prony parameters by Fourier transform [12]. It is also possible to perform an approximate conversion from the frequency to the time domain without determination of the Prony parameters by means of simplified solutions [7, 12]. Some of them are listed below in increasing order of accuracy:

$$G(t) = G'(\omega); \tag{1.17}$$

$$G(t) = G'(\omega) - 0.4\, G''(0.5\omega); \tag{1.18}$$

$$G(t) = G'(\omega) - 0.4\, G''(0.4\omega) + 0.014\, G''(10\omega). \tag{1.19}$$

The conversion between the angular frequency and time is $\omega = 2\pi f = t^{-1}$ instead of the definition commonly used for linear frequency ($f = t^{-1}$).

The DMTA of general polymers is regulated in ASTM E1640 [20] and EN ISO 6721-1 [30].

1.5.2.3 Other Test Methods

In addition to the uniaxial tensile test, further test methods are available for investigating the stiffness of plastics up to large deformations and to determine failure criteria. These include biaxial tensile tests, shear tests, uniaxial pressure tests and triaxial (hydrostatic) pressure tests. However, the testing effort is significantly higher than with uniaxial tensile test; further multi-axial states such as hydrostatic tension are difficult to execute.

All of the above-mentioned tests can also be carried out as creep or relaxation tests, to investigate the behaviour of viscoelastic material. With different temperatures, these can be used to determine a master curve and the TTS constants, as long as the applied deformation does not exceed the linear viscoelastic range.

If damage effects such as micro-cracking or the Mullins effect are observed in the material, the aforementioned tests can also be carried out cyclically to examine initial loading and reloading separately. A large number of test cycles can lead to the failure of a material, which is referred to as fatigue testing.

In addition, further testing methods such as bending tests, shock and hardness tests are available for the characterisation of mechanical parameters. Further detailed information about the above-mentioned test methods and test methods for other physical properties of polymers is available in specialised text books such as Grellmann and Seidler provide [31].

1.6 Transparent Polymers in Construction

1.6.1 General Information

In addition to applications such as laminated glass interlayers and transparent adhesives, transparent polymers are also used as sheet or tile products in their own for façade and roof applications. One particular advantage of using polymers instead of glass is the simpler processing and a lower specific weight, which is only about half of that of soda-lime-silica glass. Other material properties can also be specifically adjusted, for example brittleness. The disadvantages are the low modulus of elasticity compared to glass, the low surface hardness resulting in a low scratch resistance, the

Fig. 1.23 **a** Entrance construction of underground station in Tokyo with multilayer fins made of thermally toughened glass and acrylic plies (Asahi Glass Ltd.; engineers: Dewhurst Macfarlane and Partners); **b** Olympic Stadium Munich, covered with acrylic panels

significant drop in strength at temperatures from approximately 70 °C and the performance in fire. It should also be noted that their strength is strongly time-dependent and that most of the polymers used become brittle due to ageing. Interesting applications in structural glass construction can also be found, for example, in the combination of acrylics and pre-stressed glass, see e.g. Fig. 1.23a.

1.6.2 Polymethyl Methacrylate (PMMA)

One of the best known transparent thermoplastics is polymethyl methacrylate (PMMA), also known as acrylic glass, which was developed by Röhm at the end of the 1920s and is distributed under brand names Plexiglas® and Perspex®, see Fig. 1.23b.

PMMA is obtained by polyaddition of methyl methacrylate monomers with the action of a peroxide as a catalyst. Usually, the monomers are heated to form a viscous mass that is poured into a mould made from two glass plates, connected by spacers that control the thickness of the resulting PMMA sheet. The complete polymerisation takes place in an oven. The glass surfaces give the polymer its smooth surface, which is excellent for cast acrylic. PMMA can also be produced by extrusion rather than casting. PMMA is therefore available as solid plates, web panels (multi-skin panels) corrugated panels, blocks, tubes or rods in various thicknesses. Solid slabs are available from 1.5 to 25 mm in thickness and multi-skin panels from 8 to 32 mm in thickness are available. The standard format is 3 m × 2 m. Colourants can be added into the mixture to provide it with different colouring [40]. The main properties of acrylic are listed in Table 1.8. An example of the application of cast acrylic in a load bearing structure is shown in Fig. 1.25 [51].

1.6.3 Polycarbonate (PC)

Polycarbonates (PC) are polyesters of carboxylic acid (H_2CO_3). In case of transparent polycarbonates, a polymer of bi-phenol acetones and carbonate is produced by polycondensation using phosgene gas ($COCl_2$). Hydrochloric acid is produced as by-product. Polycarbonates are thermoplastic polymers with a very strong network structure and a corresponding high impact, thermal and chemical resistance, and is less brittle than acrylic glass. No toxic gases are released during combustion; however, CO_2 and CO is produced.

In addition to lower UV resistance, another disadvantage compared to acrylic glass is the fact that the material is approximately 1.5 times more costly. Polycarbonate, e.g. known under the brand name Makrolon® or Lexan®, is extruded and processed into solid sheets or multi-skin sheets. It is available in thicknesses from 1 to 15 mm for solid sheets and from 4 to 40 mm for multi-skin sheets in standard formats up to 4 m × 2 m, see Fig. 1.24.

It is also possible to reinforce polycarbonates with glass fibres. These glass fibre reinforced polycarbonates are used in vehicle and container construction, but the end products are translucent rather than transparent. The salient properties of polycarbonate are shown in Table 1.8.

Fig. 1.24 Commerzbank-Arena Frankfurt am Main: Edge strips in the roof made of Makrolon® (Architects: gmp; Engineers: Schlaich Bergermann und Partner)

Table 1.8 Properties of polymethyl methacrylate (PMMA) and polycarbonate (PC) [33, 40, 41]

Properties	Polymethyl Methacrylate (PMMA)	Polycarbonate (PC)
Density ρ	1190 kgm^{-3}	1200 kgm^{-3}
Modulus of elasticity E	Approx. 3300 MPa	Approx. 2400 MPa
Poisson number ν	0.37 to 23 °C	0.38 to 23 °C
Softening temperature	Approx. 110 °C	Approx. 150 °C
Temp.-Expansion coefficient α_T	70×10^{-6} K^{-1}	65×10^{-6} K^{-1}
Tensile strength at −40 °C	100 MPa	–
Tensile strength at 23 °C	75 MPa	60 MPa
Tensile strength at 70 °C	35 MPa	–
UV resistance	Good	Can turn yellow
Fire behaviour	E (B2)	E (B2)
Conductivity coefficient $U_{g,5mm}$	5.3 W m^{-2} K^{-1}	5.2 W m^{-2} K^{-1}
Maximum weight gain after water storage	2.1%	0.30%
Mohs hardness	2–3	2–3
Specific heat capacity c_p	1.47×10^3 J $kg^{-1}K^{-1}$	1.17×10^3 J $kg^{-1}K^{-1}$
Thermal conductivity λ	0.19 W $m^{-1}K^{-1}$	0.20 W $m^{-1}K^{-1}$
Fracture toughness K_{Ic}	1.62 MPa $m^{1/2}$	–

Fig. 1.25 Skypool in London made of acrylic glass, (*credits* Simon Kennedy)

Fig. 1.26 Casting House in the Landscape Park, Duisburg: Canopy made of pneumatically supported foil cushions made of ETFE (architects: Planinghaus, Engineers: Schlaich Bergermann and Partner)

Table 1.9 Properties of ETFE films [33, 39]

Properties	ETFE
Area weight with usual thickness	86–344 gm^{-2}
Density ρ	1720 kgm^{-3}
Modulus of elasticity E	Approx. 1200 MPa
Poisson number ν	Approx. 0,44
Temperature resistance	−200 °C to 150 °C
Temp.-Expansion coefficient α_T	40×10^{-6} K^{-1}
Tensile strength at 23 °C	Approx. 50 MPa
Tensile expansion at 23 °C	450%
Further tear resistance according to DIN 53363	500 N mm^{-1}
UV resistance	Very good
Fire behaviour	B1
Conductivity coefficient $U_{g,5mm}$ (reference values)	2 Layers: Approx. 3.0 W $m^{-2}K^{-1}$ 3 Layers: Approx. 2.0 W $m^{-2}K^{-1}$ 4 Layers: Approx. 1.5 W $m^{-2}K^{-1}$
Shore hardness	D 75
Specific heat capacity c_p	0.9×10^3 J $(kg\ K)^{-1}$
Thermal conductivity λ	0.23 W $(m\ K)^{-1}$
Light transmission at 0.1 mm thickness	93%

1.6.4 *Ethylene Tetrafluoroethylene (ETFE)*

The thermoplastic copolymer Ethylene Tetrafluoroethylene (ETFE) has a content of about 75% tetrafluoroethylene and is therefore similar in properties to the better-known polytetrafluoroethylene (PTFE or Teflon®). It is highly resistant to ageing, weathering and heat. However, in contrast to PTFE, ETFE can be processed thermoplastically due to its ethylene content. This polymer is used as a membrane for foil-air cushions in extremely light roof- and façade constructions, see Fig. 1.26. They are between 0.05 and 0.20 mm thick and can be coloured by adding pigments or by depositing fluoroplastic polymers on its surface through printing. The light transmittance is higher than that of glass, which makes it ideal for greenhouse applications [19]. ETFE cushions are available in widths up to 5 m and almost any length. Table 1.9 summarises some of the physical properties of ETFE.

References

1. W.H. Zachariasen, The atomic arrangement in glass. J. Am. Chem. Soc. **54**(10), 3841–3851 (1932)
2. G. Tammann, *Der Glaszustand* (L. Voss, Leipzig, 1933)
3. R.J. Charles, W.B. Hillig, *The Kinetics of Glass Failure by Stress Corrosion* (Charleroi, Belgien, 1962)
4. S.M. Wiederhorn, Influence of water vapor on crack propagation in soda-lime glass. J. Am. Ceram. Soc. **50**(8), 407–414 (1967)
5. S.M. Wiederhorn, L.H. Bolz, Stress corrosion and static fatigue of glass. J. Am. Ceram. Soc. **53**(10), 543–548 (1970)
6. A. Petzold, W. Hinz, *Silikatchemie – Einfü hrung in die Grundlagen* (Ferdinand Enke Verlag, Stuttgart, 1979)
7. J.D. Ferry, *Viscoelastic Properties of Polymers*, 3rd edn. (Wiley, New York, 1980)
8. T.A. Michalske, S.W. Freiman, A molecular mechanism for stress corrosion in vitreous silica. J. Am. Ceram. Soc. **66**(4), 284–288 (1983)
9. D. Jaworski, *Physik-Handbuch* (Verlag Harri Deutsch, Frankfurt a. M., 1986)
10. P. Manns, R. Brückner, Non-Newtonian behaviour of a soda-lime-silicate glass at high deformation rates. Glastechnische Berichte **61**(Nr. 2), 46–56 (1988)
11. H. Scholze, *Glas – Natur, Struktur und Eigenschaften*, 3rd edn. (Springer, Berlin, 1988)
12. F.R. Schwarzl, *Polymermechanik* (Springer, Berlin, 1990)
13. C. Ullner, *Forschungsvorhaben-Nr. 43 D: Untersuchungen zum Festigkeitsverhalten und zur Rißalterung von Glas unter dem Einfluß korrosiver Umgebungsbedingungen* (1993)
14. B. Bunker, Molecular mechanisms for corrosion of silica and silicate glasses. J. Non-Cryst. Solids **179**, 300–308 (1994)
15. D. Öksoy, L.D. Pye, E.N. Boulos, Statistical analysis of viscosity-composition data in glassmaking. Glastechnische Berichte **67**(7), 189–195 (1994)
16. H.G. Pfaender, *Schott-Glaslexikon*. 5. Landsberg a. L.: mgv (1997)
17. E. Zanotto, Do cathedral glasses flow? Am. J. Phys. **66**, 392–395 (1998)
18. E. Zanotto, Do cathedral glasses flow? additional remarks. Am. J. Phys. **67**, 260–262 (1999)
19. J. Schneider, T. Hofmann, G. Reisinger, Glass foil – a new glazing possibility, in *Glass Processing Days 2001* (Tampere, 2001)
20. ASTM E1640. *Standard test method for assignment of the glass transition temperature by dynamic mechanical analysis* (2004)
21. EN 1748-1-1. *Glass in building - Special basic products - Borosilicate glasses - Part 1-1: Definitions and general physical and mechanical properties* (2004)
22. M.F. Ashby, D.R.H. Jones, *Engineering Materials 2: And Introduction to Mictrostructures, Processing and Design*, 3rd. (Elsevier, 2005), p. 352
23. F. Schneider, Ein Beitrag zum inelastischen Materialverhalten von Glas. Dissertation. Technische Universität Darmstadt (2005)

24. J.E. Shelby, *Introduction to Glass Science and Technology*. 2nd Editio. Royal Society of Chemistry (2005). eprint: arXiv:1011.1669v3
25. ASTM D412. *Standard test methods for vulcanizes rubber and thermoplastic elastomers – Tension* (2006)
26. H. Brinson, C. Brinson, *Polymer Engineering Science and Viscoelasticity* (Springer, New York, 2008)
27. DIN 53504. *Pr ü fung von Kautschuk und Elastomeren – Bestimmung von Reißfestigkeit, Zugfestigkeit, Reißdehnung und Spannungswerten im Zugversuch* (2009)
28. G. Habenicht, *Kleben: Grundlagen, Technologien, Anwendungen*. 6th ed. (Springer, 2009)
29. ASTM D638. *Standard test method for tensile properties of plastics* (2010)
30. EN ISO 6721-1. *Plastics — Determination of dynamic mechanical properties — Part 1: General principles* (2011)
31. W. Grellmann, S. Seidler, *Kunststoffprüfung*, 2nd edn. (Carl Hanser Verlag, München, 2011)
32. W. Kaiser, *Kunststoffchemie für Ingenieure*, 3rd edn. (Carl Hanser Verlag, München, 2011)
33. H. Domininghaus, *Kunststoffe – Eigenschaften und Anwendungen*, 7th edn. (Springer, Berlin, 2012)
34. EN 572-1. *Glass in building - Basic soda-lime silicate glass products - Part 1: Definitions and general physical and mechanical properties* (Beuth, 2012)
35. EN ISO 527-1. *Plastics — Determination of tensile properties — Part 1: General principles* (2012)
36. P.Y. Huang et al., Direct imaging of a two-dimensional silica glass on graphene. Nano Lett. **2**, 1081–1086 (2012)
37. J. Rösler, H. Harders, M. Bäker, *Mechanisches Verhalten der Werkstoffe*, 4th ed. (Springer Vieweg, 2012)
38. F. Schneider, Lösungen für energieeffiziente multifunktionale Verglasungen. Stahlbau **81**, 291–304 (2012)
39. 3M Dyneon. *Fluoroplastic ET 6235Z - Product Data Sheet* (2013)
40. Evonik Industries AG. *Plexiglas – Technische Informationen, Product data sheet* (2013)
41. Bayer AG. *Makrolon GP, Product data sheet* (2014)
42. Heraeus. *Base Materials Heraeus Quarzglas, Product data sheet* (2014)
43. QSIL GmbH. *Chemische / Thermische Eigenschaften Quarzglas allgemein, Product data sheet* (2014)
44. Schott AG. *Schott technical glasses – Physical and technical properties* (Mainz, 2014)
45. *ASTM C162-05 Standard Terminology of Glass and Glass Products*. ASTM (2015)
46. Schott AG, *Xensation Cover, Product data sheet* (Mainz, 2015)
47. J. Schneider, J. Hilcken, A. Aronen, R. Karvinen, J.F. Olesen, J. Nielsen, Stress relaxation in tempered glass caused by heat soak testing. Eng. Struct. **122**, 42–49 (2016)
48. S. C. Angelides, J.P. Talbot, M. Overend, The effects of high strain-rate and in-plane restraint on quasi-statically loaded laminated glass: a theoretical study with applications to blast enhancement. eng. Glas. Struct. Eng. **4**(3), 403–420 (2019)
49. J.D. Musgraves, J. Hu, L. Calvez, *Springer Handbook of Glass*, ed. by J.D. Musgraves, J. Hu, L. Calvez. Springer Handbooks (Springer, Cham, 2019), p. 1841
50. S.C. Angelides, J.P. Talbot, Blast response of laminated glass panels: a critical review of analysis and design methods. eng. in *Proceedings of the Institution of Civil Engineers: Structures and Buildings* (2021), pp. 1–13
51. G. Coult, The skypool: bringing architectural imagination to life. Glas. Struct. Eng. **6**(3), 353–373 (2021)
52. S.C. Angelides, J.P. Talbot, M. Overend, The influence of fracture pattern on the residual resistance of laminated glass at high strain-rates: an experimental investigation of the post-fracture bending moment capacity based on time-temperature mapping of interlayer yield stress. eng. Glas. Struct. Eng. **7**(4), 549–568 (2022)

Part II
Glass Products and Processing

Chapter 2
Monolithic Glass

This chapter describes the production and processing of monolithic architectural glass products. After a brief history of flat glass production in Sect. 2.1 , *several primary glass products and their production processes are presented in Sect.* 2.2. *Subsequently, Sect.* 2.3 *delineates the further processing steps that may be required to bring glass products in the desired shape, enhance their strength and/or to add further functionality to its surfaces. These production and processing steps precede possible further processing of monolithic products into multilayer products that are described in Chap.* 3.

2.1 History of Flat Glass Production

Glass is a versatile material; it can be manufactured and used in various ways. Ranging from extrusion for producing glass fibres, to moulding for producing glass containers, through to floating for architectural and automotive glass. A summary glass manufacturing processes is shown in Fig. 2.1.

The range of applications is enormous and continuously expanding. The main steps in glass production, irrespective of the final product, are: *Melting* at approximately 1800 to 1600 °C; *forming* at approximately 1600 to 800 °C; *cooling* at approximately 600 to 100 °C.

Although the earliest evidence of glass making, found in modern-day Syria, dates back to 2 000 BC. The first glass windows were made in Europe by the Romans, around the first century AD. Roman window glass was made by a primitive process of casting, resulting in one relatively rough surface. The largest known pane of approximately 1.12 m by 0.81 m was found in the bathhouses of Pompei [1]. The Romans introduced the use of glass, and eventually the manufacture of it, in the regions of the Roman Empire, in particular in the region between the river Seine and the river

J. Belis et al., *Engineering Glass Structures*,
https://doi.org/10.1007/978-3-032-03233-1_2

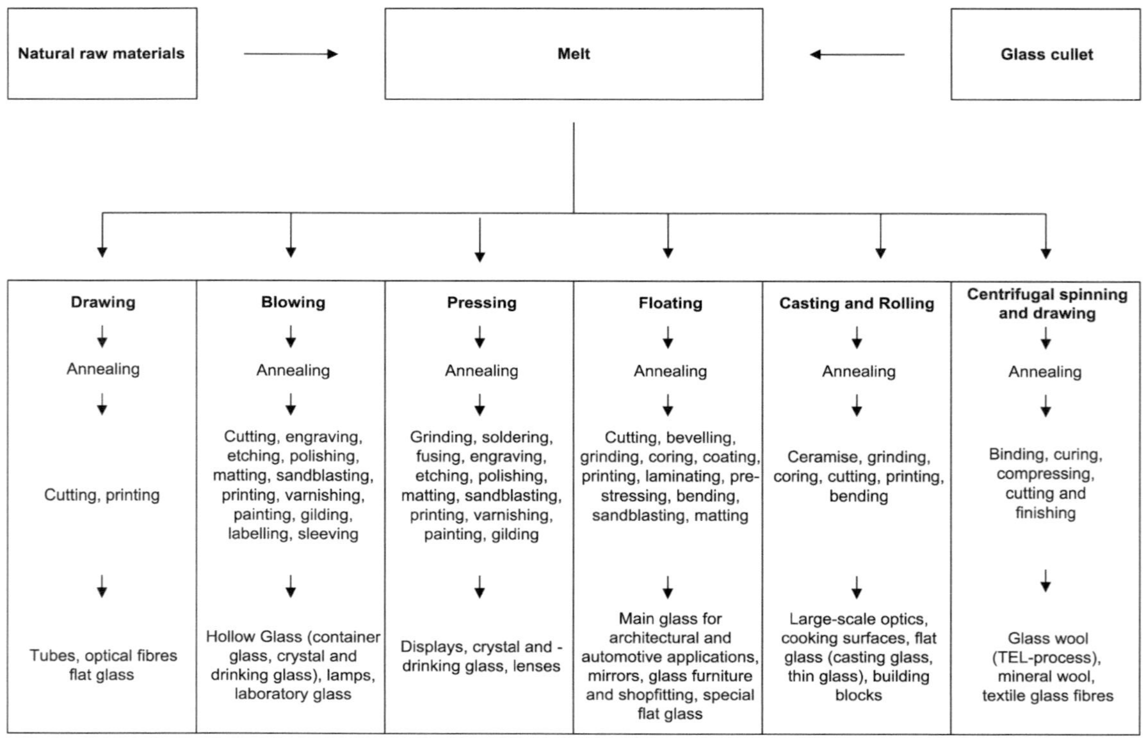

Fig. 2.1 Processes for glass production. Showing the primary manufacturing processes of glass, wherein glass types extracted directly from the glass melt are referred to as "primary glass products". "secondary glass products", such as pre-stressed glass, laminated glass or insulating glass units, can be produced from the "primary glass products" by means of subsequent processing [5]

Rhine. Glass making continued to flourish in this area, partly due to the climatic conditions and eventually due to the development of Gothic architecture which created a steady demand for glass windows. Two methods of producing window glass were prevalent in this period: the cylinder method and the crown method.

Cylinder glass (also known as cylinder blown glass) is made by elongating a blown glass sphere by swinging it back and forth, cutting off the domed ends up to the full diameter of the cylinder, cutting along its length, and unrolling the cylinder to a flat sheet in a furnace. The resulting glass sheets were up to 1.27 m long and were mainly produced in Venice and Nuremberg.

Crown glass most probably stems from the Syrian method of spinning window crowns in the eighth century. The process consisted of spinning a reheated blown glass sphere on a stick called a punty, spinning the stick on it axis thereby making use of centrifugal forces to transform the glass sphere into a larger diameter disk. Crown glass was usually much thinner than cylinder glass, more transparent and more uniform in thickness, except where the punty meets the disc, where the characteristic bull's eye is formed. This superiority of crown glass, mainly made in Normandy and later also in England, allowed it to displace other processes and it maintained its market share until the middle of the nineteenth century [1].

The next significant development in glass making was the invention of the *polished plate glass* by the Frenchman Bernard Perrot in 1687, when he developed the process to meet the high demand for large glass panes for mirrors and for glazed doors of luxurious horse-drawn carriages. In this technique the glass melt is poured out of the melting tank onto a preheated casting table and rolled out with a water-cooled metal cylinder to form a pane. In this procedure the gap between the rollers controls the pane's thickness. Subsequently, the glass panes are grinded down with quartz (in later years corundum (Al_2O_3) and carborundum (SiC)) and water and then polished using an iron oxide paste, see Fig. 2.2. The advantages of this process were the improved optical quality of the glass and previously unattainable dimensions of 1.2 m × 2.0 m. This is also how the panels got the name *grandes glaces*. However, because of the

Fig. 2.2 **a** Casting table at the end of the 19th century showing the *table rolling process*. The glass melt was poured onto a pre-heated table, (photo credits: Saint Gobain) and then **b** cooled, sanded and polished. Rolled glass production in 1908, (photo credits: Lewis Wickes Hine)

relatively time-consuming polishing work this kind of window glazing was costly and thus only afforded by the nobility. The grinding and polishing was originally done manually, but by 1800 it was performed by steam-powered machines.

Around 1900 the American John H. Lubbers developed a mechanical process for manufacturing cylinder glass. This procedure provided the basis for the *drawn glass manufacturing process*. A cylinder of regular diameter and thickness is formed out of the melting pan by applying an automated combination of pre-heated compressed air supply and drawing. In order to produce flat glass sheets, such a cylinder had to be rotated from its vertical position into a horizontal position, cut by a preheated electric wire, longitudinally split and finally flattened.

Further developments in processing came about in the early 20th century. Three processes based on the drawing method made flat glass production more economical and thus promoted the distribution of glass as a building material. In 1904, the Belgian Emile Fourcault filed his first patent for a method to pull flat glass directly from the glass melt without a reduction in width. Due to hydrostatic pressure, the hot and viscous glass mixture wells up through a rectangular slot in a fireclay float held below the surface of the glass melt, from where it is subsequently drawn out by a pulling machine. The width of glass sheets produced with a Fourcault machine range from approximately 1.9 to 2.3 m. After drawing, the glass is transported by rollers through an 8 m high annealing shaft. The slot through which the glass is drawn is typical the source of most glass defects in this process, such as: spoiled surfaces due to crystals of devitrified glass forming along the slot; and wave-like patterns caused by wear of the slot.

Similarly, the American Irving W. Colburn developed a process for the production of flat glass using the drawing process, which he patented in 1905. However, it was not until 1913 that the first glass was successfully made with this process. By then, Colburn had already sold the patent to the Toledo Glass Company. It became a well-developed process three years later. As the company changed its name to the Libbey-Owens Sheet Glass Company in 1917, this method became known as the *Libbey-Owens Process*, see Fig. 2.3b. The process differs in two essential features from the Fourcault process. First, the glass sheet is drawn from the bath without a slot, but by pairs of small rollers, thus avoiding devitrification problems. Second, the vertical draw is reheated and bent over a roller, from where it enters the annealing lehr. The continuous ribbon of glass is subsequently cut to the required lengths. The width of the glass ribbon is 2.5 m and varying the drawing speed allows for a glass thickness of between 0.6 and 20 mm. However, the surface of the glass tends to be spoiled in bending over the roller, and the horizontal processing makes it more prone to dust contamination, thereby making the glass of a poorer optical quality than Fourcault glass.

From 1928 onwards, the Pittsburgh Plate Glass Company started to use a combination of the above-mentioned Fourcault and Libbey-Owens processes.

The glass melt was drawn, in a manner similar to the Libbey-Owens process. The drawing machine then carried the ribbon into the cooling shaft up to 12 m high, such as that used in the Fourcault process.

Fig. 2.3 Historical flat glass production. **a** The *Bicheroux Process*, where molten glass is poured out of the melting pan onto a sloped surface. Two rollers immediately produce the desired glass thickness. By rolling on a firm moving plane, thin, uniformly strong sheets with a flat surface could be produced, without far fewer grinding operations. **b** The *Libbey-Owens Process*, used in the Toledo Glass Company between 1912–1914. The approximately 60 m long cooling tunnel and next to it the cutting table (photo credits: Saint Gobain)

By 1900, the development and improvement of the drawn glass process gradually replaced the polished plate glass as the process of choice. Although the quality of the drawn flat glass was sufficient for windows, thickness differences and optical defects such as streaks were unavoidable. In particular, the burgeoning automotive industry demanded improvements in flatness and transparency. In response, Max Bicheroux working at Saint Gobain in Herzogenrath between 1910 and 1914, developed a new casting process known as *Bicheroux Process*, see Fig. 2.3a. In this glass casting process, the glass melt is poured out of the melting pan in portions onto a tilted surface. Two steel rollers form a glass ribbon with the desired glass thickness. Short contact times of the steel rollers with the glass resulted in a relatively even surface, thereby greatly reducing losses in time and material during subsequent grinding operations compared to the polished plate glass process. Furthermore, in contrast to the drawn glass processes, glass could now be produced to meet the optical and thickness tolerances of the automotive industry. After rolling, the glass was cut into large glass panes up to 3.0 m $\times$ 6.0 m and cooled on moving tables in ovens. The glass produced with this process had a consistency of thickness that was far superior than other existing glass production methods at the time.

Based on the *Bicheroux Process*, the Ford Motor Company was able to produce continuous cast glass for the automotive industry in 1923. Molten glass was discharged in a continuous process via two rollers from the tank furnace and subsequently conveyed on a roller bed through an annealing lehr. Immediately after cooling, the glass is automatically grinded and polished on both sides in an operation which had already been developed by UK-based Pilkington Brothers Limited. The latter step made this production process considerably more expensive and thus limited the use for more demanding applications. Pilkington Brothers Limited further developed the Ford process to produce *plate glass* of all sizes and thicknesses, which they actually did for over thirty years. This process is still common today in the manufacturing of surface-patterned glass whereby rollers impart a surface pattern onto the glass without the need for further polishing.

The next major milestone in flat glass production was the development of the *float glass process*. It was patented by Lionel Alexander Bethune Pilkington (later Sir Alastair Pilkington) and Kenneth Bickerstaff in 1953. In this process liquid glass flows from the furnace onto a bath of molten tin; as it *floats* and spreads it takes the shape of a continuous ribbon of glass. As a result of the surface tension of the glass and the molten tin, a flat distortion-free ribbon of high optical quality is formed. The idea of using liquid tin as a carrier for flat glass goes back to the British engineer and inventor Henry Bessemer. The concept of developing a glass manufacturing process based on floating predates Pilkington. In fact in 1902 US-based William E. Heal developed Henry Bessemer's concepts and patented the manufacturing principle of continuously running the glass mixture into the tin bath thus maintaining parallel surfaces. However, this patent was never commercially exploited. It was only in 1959 that the float glass process was so well developed that from the 1960s onwards it was possible to produce on an industrial scale. In 1966 the Pilkington Brothers company started production and subsequently licensed their float line technology to other flat glass manufacturers around the world. This process resulted in glass that had a consistency of the thickness, a significant reduction of air and material inclusions and superior flatness. For the first time in the history of glass production, it was now possible to produce flat glass with excellent quality in a cost-effective manner. Consequently it quickly displaced other flat glass production methods and is currently the predominant flat glass manufacturing process.

2.2 Primary Products

2.2.1 Float Glass

Float glass generally has a soda-lime silica basis according to modern product standards (e.g. EN 572-2) and is the most widely used architectural glass today. The manufacturing process of float glass is shown schematically in Fig. 2.4. In the float glass production process, the raw materials, typically 75 to 90% virgin materials and 10 to 25% waste-glass (i.e. cullet), are mixed and subsequently heated in a furnace to approximately 1560 °C. The glass melt flows at 1100 °C, from the furnace onto a bath of molten tin, known as the float bath, which is at 600 °C. The resulting cooling of the glass melt in the float bath is sufficiently rapid to prevent crystallisation, see Sect. 1.2.1. Due to its lower specific weight, the molten glass floats on the surface of the molten tin, where it reaches an equilibrium thickness of approximately 7 mm [4].

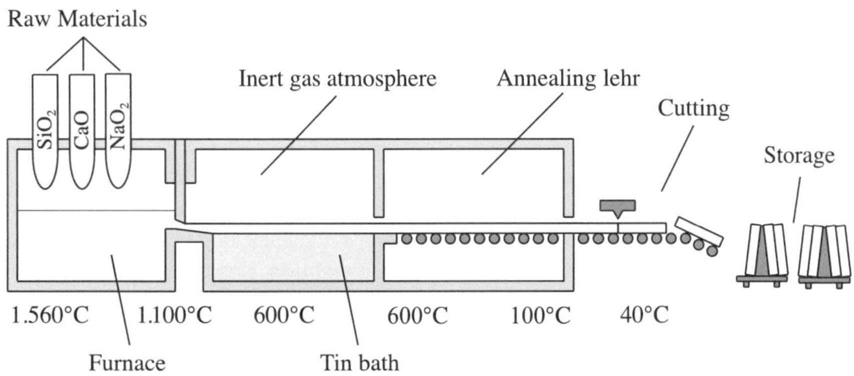

Fig. 2.4 Manufacturing process of float glass

In the float bath the glass melt flows under a protective gas atmosphere of hydrogen and nitrogen in order to avoid oxidation of the tin. The glass surface in contact with the tin is generally referred to as tin side, whereas the surface exposed to the gas atmosphere is known as the atmospheric (air) side.

The thickness of the glass can be adjusted by top rollers in the float bath. In order to produce thin glass products (down to approx. 2 mm thickness) the glass melt is pulled laterally and the draw speed is increased. In the case of thicker glass (theoretically up to approx. 35 mm thickness is possible) the glass melt is pushed laterally and the draw speed is reduced. At a thickness of 4 mm, approximately 1100 m of float glass can be produced per hour. For example, changing the thickness from 4 to 5 mm takes about 45 min, which means a loss of about 830 m of glass during the transition. Product standards such as EN 572-2 therefore allow significant production tolerances on the nominal thickness to reduce waste from such changes in thickness during production [32].

The glass is subsequently conveyed to the annealing lehr where it is slowly cooled from approximately 600 °C to approximately 100 °C. After which the glass is automatically optically checked for imperfections and cut.

During production, the float glass ribbon has a width of about 3.50 m, which is trimmed to 3.21 m to remove the edges which generally contain top roller imprints. At the end of the process, the glass is cut into panes and available for delivery at typical maximum dimensions of 3.21 m × 6.0 m. However, larger ("oversized") dimensions are also possible, up to a width of 3.6 m and a length of 20 m, but these are still rather exceptional. Standard glass thicknesses (in mm) currently are 2, 3, 4, 5, 6, 8, 10, 12, 15 and 19, although thin glass below 2 mm and ultra-thin glass below 0.5 mm can now be produced using the float process. In normal circumstances the entire process is fully automated and runs continuously for about 11 to 15 years, after which a complete overhaul of the furnace and tin bath is necessary. A float line requires significant space, particularly length; the different production zones are as follows:

- The melting furnace is usually 60 m long, 12 m wide and 1.50 m deep and is heated with an oil-fired or gas-fired burner;
- The tin bath is up to 55 m long, 7.6 m wide and 0.6 m deep;
- The annealing lehr is up to 200 m long.

Float glass is present in a wide range of applications and is for instance used in windows, façade glazing and furniture. Mirrors are also made out of high-quality float glass with a chemically applied silver layer (at the back side of the glass) that is protected by several additional coating layers. In addition, float glass can be further processed to produce other glass products e.g. pre-stressed glass, laminated glass, and insulating glass units, see Chap. 3.

The fracture resistance of soda-lime silica float glass to temperature differences across its surface is around 40 K [45]. This is relatively low compared to borosilicate glass, which due to its lower coefficient of thermal expansion, has a temperature resistance of about 100 K. This is why thermally pre-stressed borosilicate float glass is typically used as fire resistant glass, see Sect. 3.4, which is now available with maximum dimensions of 2.30 m × 3.20 m. However, borosilicate glass is much more expensive than soda-lime silica glass.

The characteristic bending strength of annealed soda-lime silica glass, $f_{g,k}$, is 45 MPa [31]. This value is determined using the coaxial double-ring test method according to EN 1288-2 or EN 1288-5 at a defined test speed of (2 ± 0.4) MPa s^{-1} and is governed by the surface strength for a degree of surface damage, see Sect. 5.4.

2.2.2 Rolled Glass

Rolled glass is produced by means of an overflow process, wherein glass is immediately taken off the rollers, annealed and cut. Nowadays it is often produced with distinct ornamentation, using rollers with a textured surface (En 572-5) [34]. Wired glass and patterned wired glass is produced by adding a wire mesh, see Sect. 2.2.4. Except for polished wired glass to EN 572-3, rolled glass, is not usually transparent, but translucent due to the surface pattern. The manufacturing process is shown schematically in Fig. 2.5.

The bending strength of rolled glass is highly dependent on the surface pattern and exhibits a relatively large scatter. Therefore, the value of 45 MPa for $f_{g,k}$ specified in EN 572-1 is not applicable. For instance, German building

Fig. 2.5 Manufacturing process of rolled glass

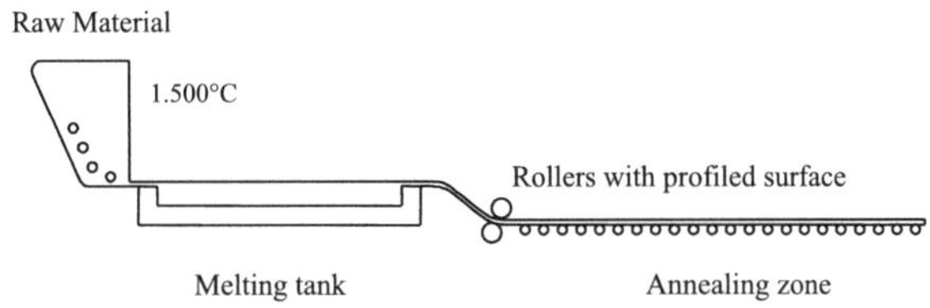

regulations require a characteristic bending strength for rolled glass, f_k of 25 MPa [42].

2.2.3 *Channel Glass*

Channel glass is a special shape of cast glass classified according to EN 572-7 [35]. It is produced in a U-shape and it is available in wall thicknesses of either 6 mm or 7 mm. Its field of application is typically in industrial construction, where it may be preferred instead of windows, but it is more recently also used in modern façade designs. Since such U-shapes provide a relatively high bending stiffness, they can span large distances without mullions. Channel glass is relatively inexpensive and is also used in double-skin façades. Table 2.1 illustrates common arrangements of channel glass.

Experimental testing of channel glass is carried out according to EN 1288-4 [8]. Studies have shown that the strength of the web and the flange differ [10, 12]. Indeed, the characteristic bending tensile strength, $f_{g,k}$, of the web is typically 25 MPa, whereas the flange has a strength of 40 MPa, see Fig. 2.6.

2.2.4 *Wired Glass and Wired Patterned Glass*

Wired glass, see Sect. 2.2.6, was already used before the development of laminated glass for preventing the glass fragments from falling out after fracture. Today, however, wired glass is no longer regarded as a safety glass [16].

In case of fire, wired glass may prevent a full transmission of flames and smoke but it does not prevent the transfer of heat by radiation. The inset wire does not increase the bending strength of glass. On the contrary: the different thermal expansions of

Table 2.1 Common arrangements of channel glass

Description	Categorisation
Single glazed wall, contiguous, flange outside	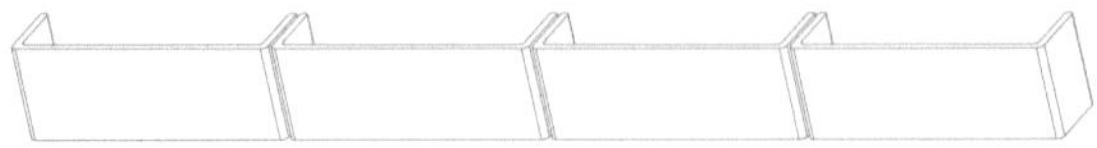
Single glazed wall, contiguous, flange inside	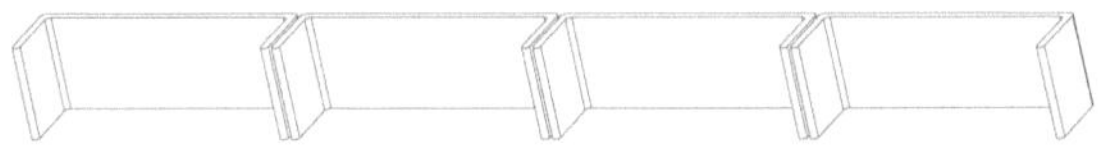
Single glazed wall, contiguous, alternating	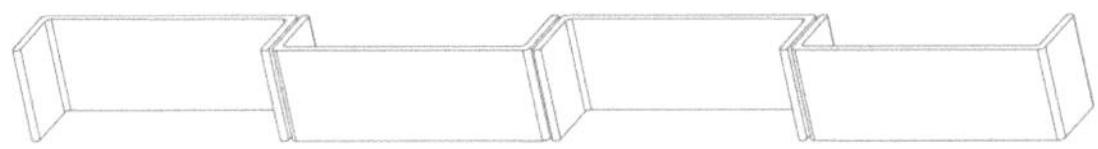
Single glazed wall, interlocking	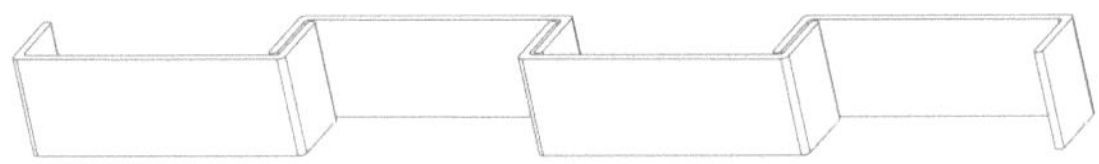
Double glazed wall, flanges of both layers joined	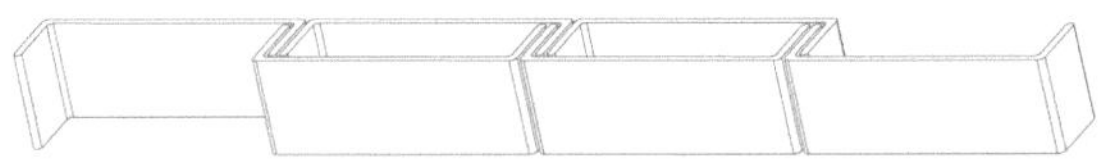
Double glazed wall, flanges of both layers shifted	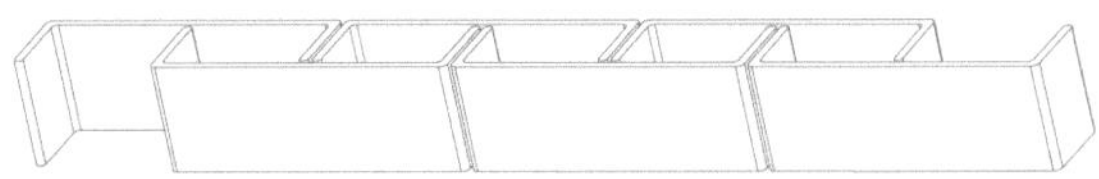

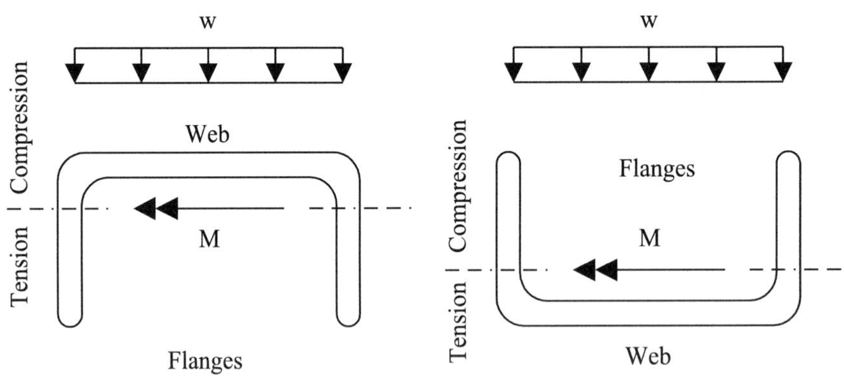

Fig. 2.6 Channel glass with different alignment of the profiles

wire and glass along with the crack inducing nature of the wire embedded in glass, typically cause fracture of the glass at lower loads than the equivalent non-wired glass. Furthermore, the edges of the glass should be protected from moisture or allowed to dry rapidly through natural convection so that the wire mesh does not corrode.

Polished wired glass is available in a thicknesses of 7 and 10 mm (EN 572-3 [33]), while ornamental wired glass comes in a thickness from 6 to 9 mm. German building regulations require wired glass to have a minimum strength, $f_{g,k}$, of 25 MPa [42].

2.2.5 Hollow Cast Glass Blocks

Hollow cast glass blocks[1] according to EN 1051-1 [11] are produced by moulding molten glass in two individual shells, which are subsequently thermally fused along their edges to compose a hollow block [39]. The result is a hermetically sealed interior cavity, which occurs due to the cooling down of the hot-shaped glass shell. This air cavity, which can optionally also be argon filled, provides the glass block a degree of thermal insulation. The glass blocks are available in different shapes (e.g., square, rectangular, or circular), dimensions (usually having edge dimensions between 19 cm and 30 cm), and their outer surfaces may be smooth or textured. The glass blocks can be coloured through body-tinting the glass or by colour-coating the inner surfaces of the glass block. Their degree of transparency varies, but the view through the blocks is typically distorted.

In architecture, hollow glass blocks are applied for interior wall partitions, exterior facades, and floors, see Fig. 2.7. These are constructed by either: (i) casting the blocks in a concrete grid; (ii) joining them by means of cement-based mortar and steel reinforcement rods or; (iii) incorporating metallic/timber/PVC extrusions in the joints between the glass blocks. Care has to be taken when designing with hollow cast glass blocks as inherent residual stresses can significantly decrease their strength. Thus, imposed loads, for example from shrinking of concrete around such blocks can lead to fracture.

2.2.6 Solid Cast Glass Blocks

Volumetric solid cast glass blocks are produced by depositing molten glass (viscosity between 1×10^1 to 1×10^3 Pa s) into a high-temperature resistant (pressed) mould and cooling it down in a controlled manner (i.e. annealing). Moulds can be either disposable, made out of low cost materials such as silica plaster, or reusable materials (e.g. steel or graphite), thereby allowing for the mass production of repetitive units. The casting process allows for considerable freedom in the size, shape, colour and

[1] This section on hollow cast glass blocks has been reprinted from [58].

Fig. 2.7 Hollow cast glass blocks in an exterior façade Maison Hermès (Renzo Piano Building Workshop, Tokyo, Japan, 2001) ©Whatsoeverhk/ Dreamstime.com

opaqueness of the resulting glass object. The size of a cast glass component is governed by the annealing time, which is in turn influenced by the coefficient of thermal expansion and mass distribution of the cast object. In principle, the thicker and larger the component, the exponentially longer the annealing time required. This has been demonstrated in the castings of the blanks of the giant telescope mirrors; for example, the solid mirror blank of the Hooker Telescope of 2.5 m in diameter and 0.32 m thickness, made of soda-lime glass, required 12 months of annealing [54]. The economic barriers imposed by the prolonged cooling process and customised production, as well as the lack of standardised strength data and building guidelines, have limited the use of cast glass to only a few real-world architectural applications [64]. The most popular examples are free-standing façades made of solid blocks, of either soda-lime or borosilicate glass and a block weights below 10 kg. Notable examples include the Atocha Memorial [19], the Crown Fountain, the Optical House [37] and the Crystal Houses [47, 54, 64], see Figs. 2.8, 2.9, 2.10 and 2.11. Cast glass is also used in larger pieces for ornamental elements or panels, such as the Youth Leading Industry (1936) at Rockfeller Centre and the Ice Falls project in New York [17]. Ongoing research

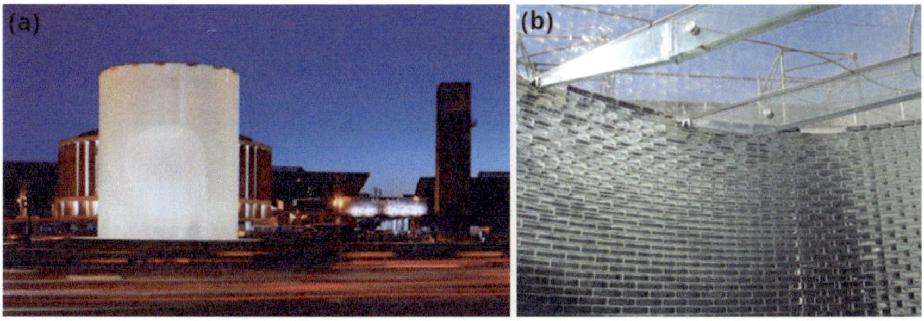

Fig. 2.8 Cast glass blocks applied in the Atocha Monument Madrid 11 March Memorial. **a** Photo by Schott/Jens Mayer, **b** photo by Christoph Paech/Schlaich Bergermann Partner [20]

Fig. 2.9 Solid cast glass blocks applied in the Crystal Houses façade in Amsterdam. From [54], published under CCBY licence

is exploring the potential of employing waste glass cullet as the raw material for solid cast glass components [56, 66]. Recent research has shown that optimising the glass mass to match the design loads while keeping a relatively homogeneous mass distribution can result in considerable reductions in the annealing time involved. Mass optimisation, in combination with cost-efficient mould technologies for complex shapes, such as 3D-printed (sand) moulds [69, 72] provide opportunities for creating large cast glass units.

2.2.7 *Extruded Glass*

Glass can be extruded to produce solid or hollow glass prismatic profiles such as (thin wall) tubes, rods or other glass elements with a constant cross-section. Extrusion is possible in glasses that have a steep viscosity curve, increased tendency to crystallize

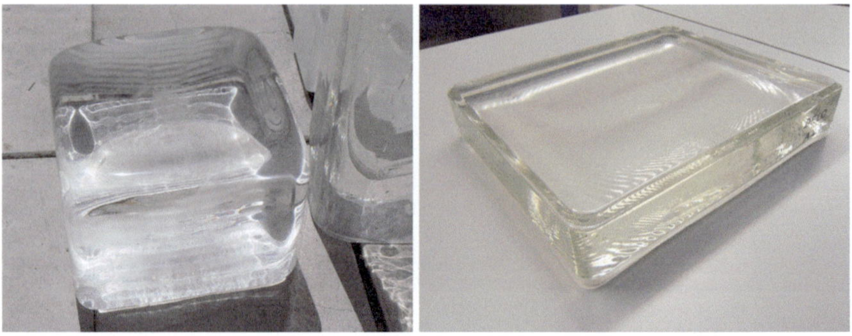

Fig. 2.10 Glass blocks made out of optical glass

Fig. 2.11 Masonry work made from glass for the *Atocha Monument Madrid 11 March Memorial*: **a** Single block and **b** finished building

and/or a considerably high softening point, such as silica glass [2]. Common fabrication processes are the Danner process and the Vello process. Extruded glass profiles are mainly used in interior architecture, art, design and lighting solutions. Extruded glass rods are not typically used in structural or cladding applications. Notable exceptions are the façade made of glass rods in the S.C. Johnson Building by Frank Lloyd Wright and the Glass Tube Field by James Carpenter Design Associates Inc [17]. The structural potential of glass rod bundles has been explored in research institutes [52] and in experimental prototype projects such as the Glass Truss Bridge [55] and the Glass Swing [65].

2.2.8 Coloured Glass

Coloured (also known as body-tinted) glass is produced by adding metal oxides to the glass melt. Colour originates from the absorption of light in the range of certain wavelengths. Standard soda-lime silica glass typically has a greenish appearance, which is primarily caused by iron oxides already present in the in raw materials.

Even relatively small quantities of raw materials can cause colouring. About 99.5% of the elements usually found in soda-lime silica glass are, in part, made out of the same colourless chemical composition; the remaining 0.5% determine the colour of the glass, see Table 1.6. This makes it very difficult to achieve an precise colour match between glass batches. This is particularly problematic when replacing individual glass panes after several years. Glass made in different glass making locations tend to differ in the degree of green colouring, largely due to the different origins of the raw materials.

Most standard coloured glass comes in green, bronze, grey and pink colours. In addition to aesthetics, the tinting reduces the total energy transmission - similar to the principle of sunglasses. However, the additional solar radiation absorbed by the glass also results in increased thermal stresses within the tinted glass and an associated increase in re-radiated energy, see Sect. 1.4.2.

2.2.9 Thin Glass

The term "thin glass" has no normative definition. Typically, glass thicknesses below 3 mm are referred to as thin, see Fig. 2.12a. Until recently glasses of this thickness could not be thermally pre-stressed, however, recent quenching technologies, see Sect. 6.2.3, make it possible to thermally pre-stress such thin glasses. More recently, the revised standard [46] defined 2 mm as the minimum thickness for thermally pre-stressed glass, which is also the current technological limit for achieving acceptable mechanical and optical properties. For heat strengthened glass this threshold is approximately 1 mm.

The way thin glass is made varies from one manufacturer to another and is driven by market, production costs and business models as well as technological limitations. In addition, thin glass making technologies are evolving rapidly. Currently, there are three principal methods for producing thin glass.

(i) The modified float process which is based on the standard float glass production method, see Sect. 2.2.1, but involves modifications to the raw materials in the batch while keeping similar furnace temperatures. This glass can be produced in thicknesses as low as 0.5 mm and widths of 3.21 m, but the maximum size is not limited by the float line, but rather by the available equipment to handle the glass. Another variation of this method is known as the micro-float process which produces glass up to 2.4 m wide and as low as 0.1 mm thick.

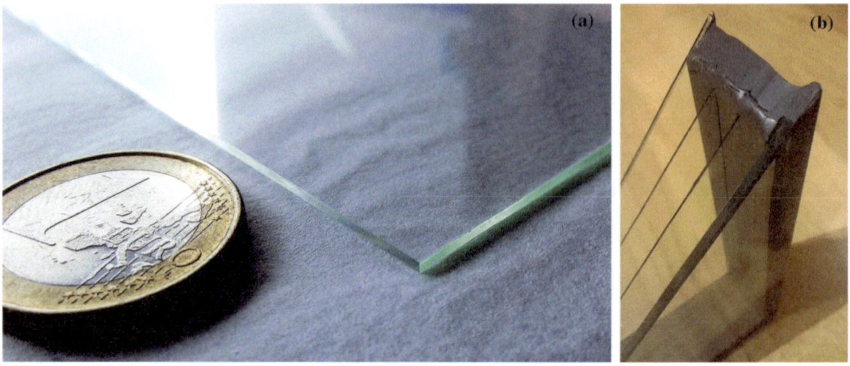

Fig. 2.12 **a** 1 mm thick thin glass compared to a 1 Euro coin; **b** application of thin glass in a quadruple insulating glass unit

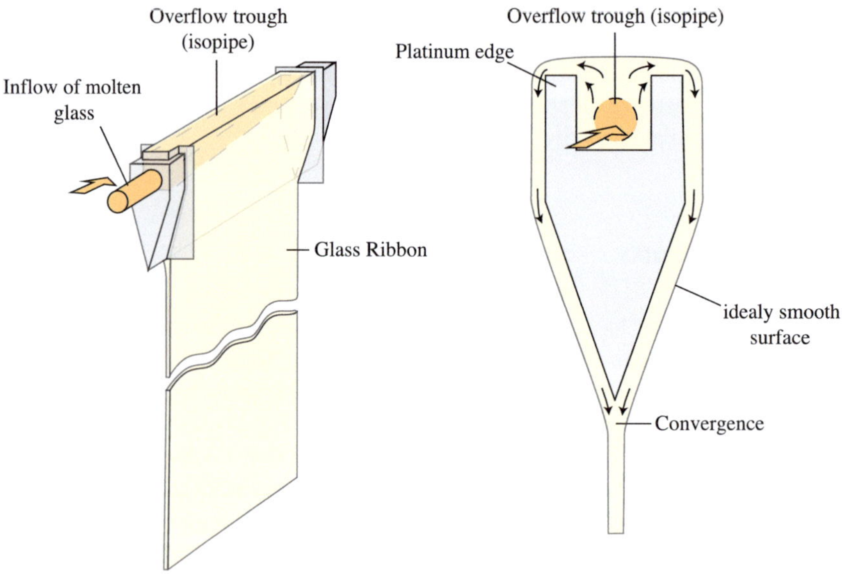

Fig. 2.13 Production of thin glass in the overflow fusion down-draw process

(ii) The down-draw process, where the glass is pulled vertically from the bottom of a tank of molten glass, thereby producing widths of up to 0.5 m and thicknesses ranging from 1.1 mm to 30 μm.

(iii) The overflow fusion down-draw process, where the glass flows over the top edges of a V-shaped trough (known as an isopipe), forming two thin streams along the outer surfaces before converging and fusing into a single sheet at the V-shaped bottom, see Fig. 2.13. The sheet is drawn vertically downwards as it lengthens and begins to cool in mid-air. This process currently produces glass sizes in the range of

3 m by 3 m and thicknesses from 0.5 mm down to 50 μm. Typically the maximum width decreases as the thickness of the manufactured glass is reduced. For example, a 100 μm thick glass can be produced up to a width of 1.2 m.

The application of thin glass in construction is currently still under development. In fact, several of these thin and ultra-thin glasses are typically used for specialised application-specific requirements such as display panels and other applications in electronics. However, an interesting potential application in the construction sector is in the middle glass panes of insulating glass units, see Fig. 2.12b, in laminated glass [44, 57, 68] and in other composite and adaptive glass products [53, 68, 70, 75].

The determination of the bending strength of thin glass (especially when governed by the surface strength) is problematic because of the large deformations before fracture. Conventional coaxial double ring testing according to [7, 9] is ill-suited for thin glass because the large deformations result in a non-uniform stress state in the internal ring area, which is a pre-requisite for this test set-up, therefore alternative testing methods for thin glass are under development [73, 74].

2.3 Processing of Glass

Once float glass has been produced, various additional processing steps may be carried out before the product is ready for use. For instance, it is common to cut the float glass into desired dimensions, to process its edges, to drill holes in it, to bend it, to coat it or to print on it. Furthermore, subtractive processes such as sandblasting and etching treatments may be applied. If thermally pre-stressed glass is to be produced, then all other processing steps usually need to be carried out before or, when possible, during the pre-stressing process.

2.3.1 Cutting

Glass is usually cut along a straight line in a two-step process. First, a cutting tool (diamond tip, hardened steel/tungsten carbide wheel) is used to score the glass surface. The glass is then broken by snapping it along the scored line (Fig. 2.14). The quality of the score can be improved by using cutting fluids (lubricants), which reduce unwanted chipping. Curved cuts are also possible, but the snapping of the glass pane can then be problematic, as the crack tends to unintentionally deviate from the score line.

Depending on the quality of the scoring, irregular and wave-like patterns can form at the glass edge which include microcracks and chips that influence the edge quality and edge strength. The glass then needs to be edge-worked, see Sect. 2.3.2 and Fig. 5.6. Such damage significantly reduces the strength of the glass in the vicinity of the edge, see Fig. 2.15.

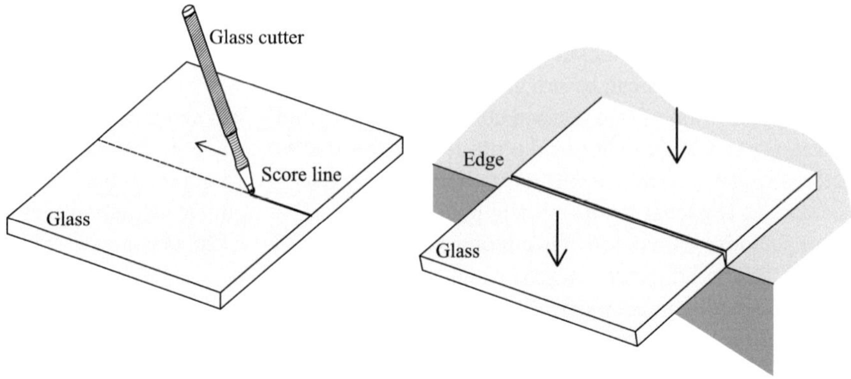

Fig. 2.14 Principles of glass cutting in two steps: by scoring with a glass cutter and through subsequent snapping by applying bending tensile stress (Picture credits: Jonas Kleuderlein)

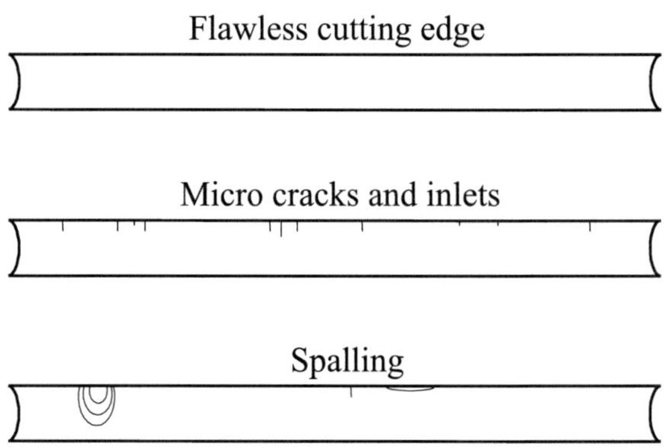

Fig. 2.15 Edges of glass after cutting

Alternatively glass can be cut by waterjet and laser cutting technologies.

In waterjet cutting, a nozzle with a diameter of approximately 0.1 to 0.6 mm funnels the water into a fine jet at a high pressure of 4 000 bar, at a speed of up to 1 000 m/s. An abrasive is added to the water to improve the quality of the cut. Almost any shape can be cut using this Computer Numerical Control (CNC) process.

Laser glass cutting technology, e.g. zero width laser cutting technology (ZWLCT™: Zero Width Laser Cutting Technology), is when a defined energy input on the surface of the glass, controlled by a computer, creates an almost invisible crack in the material. The crack is formed exactly in the desired cutting direction. With cuts of such precision, chips can almost completely be avoided. Since the cut edges have a very high surface quality, they no longer need to be post-processed. In fact, laser-cut glass components usually have a higher edge strength than mechanically cut glass.

In addition to the ZWLCT™ variant, there is the multiple laser beam absorption method (MLBA), which is also called LiSS (Laser induced stress separation method). Here, the laser and subsequent quick local cooling by an air current creates a temperature or stress state in the glass, which allows for a defined breaking of the glass. The advantage is that the number and depth of the microcracks can be minimised.

2.3.2 Edge Working

The edges of glass can be further processed in various ways after cutting. The sharp cut edge can be grinded or polished in order to reduce the risk of injury and to increase dimensional accuracy, optical quality and, in some circumstances, strength. Unprocessed glass with sharp edges is only installed in a frame where there is no risk of injury. The processing of edges in large batches is typically performed by automatic grinding and polishing machines, using corundum or diamonds abrasives. In the simplest form of edge working, (only) the sharp corners along the edge are removed to prevent injury. This is also called the *seaming* of the edge when done very lightly, sometimes even manually, or the *arrising* of the edge when done more precisely, removing more material, typically in the order of one millimetre from the edge, at a fixed angle (typically 45°) by means of a grinding machine or belt.

Grinding or *smooth grinding* (i.e. without blank spots) of the glass edges improves the quality of the cut edges as well as changing the size of the pane to the required dimension to compensate for manufacturing tolerances, e.g. those occurring during the cutting. The edge can be further improved through grinding or polishing as required; for example, by rounding or bevelling, as is often applied to mirrors or table tops, see Table 2.2.

In the case of thermally pre-stressed glass, the edges usually have to be ground and arrised, because large flaws along the cut edge typically lead to fracture during the thermal treatment process. The edge grinding also helps to evenly distribute the residual stresses in the area of the edge in a thermally pre-stressed pane. However, in principle, any edge working must be carried out before the thermal pre-stressing process. Indeed, subsequent processing of pre-stressed glass is only possible to a limited extent and the compressive zone of the glass is likely to be weakened. Drilling [22, 38, 63] and grinding [61, 62] of tempered glass has been reported in the literature. Common types of edge processing are described in [3] and shown in Table 2.2.

Table 2.2 Glass edge finishes according to DIN 1249-11

Terminology	Description	Sketch
Cut	Cut, unprocessed glass edge with sharp corners along the edge. Across the edges, light wavy lines (Wallner-lines) may be visible	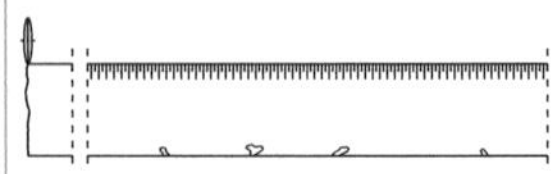
Seamed/Arrised	Cut edge with corners made blunt manually or by light machining	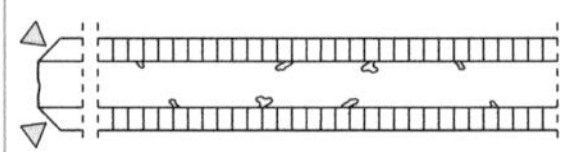
Ground (with blank spots)	Glass pane brought to the required size by grinding the edge surface, but blank spots may still be visible. The corners along the edge may be cut or seamed	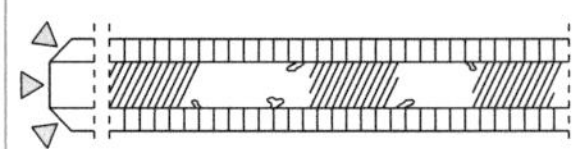
Smooth ground (no blank spots)	The edge surface is completely ground, such that no blank spots remain. The corners along the edge may be cut or seamed	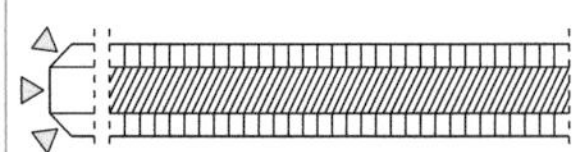
Polished	The polished edge is a refined re- ground edge. Limited polishing traces may still be visible	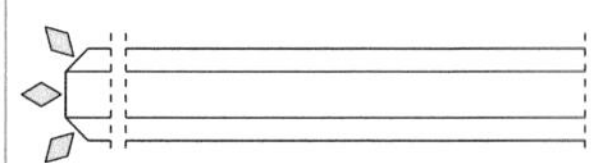
Round	The edge is, through grinding, rounded to a C-shape	 

Recent research [67] on glass edge strength shows that it does not always correlate with the type of edge working nor macroscopically with the optical quality, according to the classes defined in the glass standards [43, 45], see also Sect. 5.2.

The mean strength typically increases after seaming, arrising, grinding and polishing, but no increase is observed in the lower fractile values typically used in engineering design, see Sect. 5.4.3. This is due to the variabilities in the depth of the microcracks (flaws) resulting from scoring, breaking, grinding and polishing. These flaws are often so small and the fracture surfaces are in close optical contact that they are not visible with an optical microscope. The research to determine the glass edge strength based on the local edge damage by [41] provides a useful insight into some of the salient micro features that affect strength, although only the visible roots of the flaws were measured. The microcracks are usually decisive in determining the strength, see Fig. 5.5. It is still unclear as to which processes (e.g. cutting, snapping, edge working, etc.) have a decisive influence on the formation of microcracks.

2.3.3 *Coring*

Glass panes are occasionally connected by means of localised mechanical connections involving bolts through holes bored in the glass, see Chap. 7. Due to the high local stresses caused by bending and bearing actions, the glass intended for this application needs to be thermally pre-stressed [14]. The hole in the pane must be formed before thermal pre-stressing. Usually, purely cylindrical or combined cylindrical-conical (countersunk) holes are specified. Regardless of the core profile used, the rim of the hole should always be chamfered, see Fig. 2.16. Typically the chamfered edge is approximately 1 mm deep with a chamfer angle of 45°.

As a rule of thumb, holes in glass must have a diameter of no less than 20 mm, to enable sufficient cooling and associated effective thermal pre-stressing in the subsequent heat treatment process. Furthermore, according to EN 12150-1 the diameter of the hole should not be smaller than the thickness of the glass, and the distance between the hole and the edge should be at least twice the thickness of the glass pane. For load-bearing holes further guidance is provided in some countries; for instance DIN 18008-3 requires a minimum distance of 80 mm between the edge of the glass pane and the boundary of the hole, as well as for the minimum distance between adjacent holes [40]. The location of holes in pre-stressed glass will affect the residual stress depending on hole diameters and glass thickness [25]. The most common methods used to make holes are diamond coring and water jet cutting that typically are carried out using Computer Numerical Control (CNC) machining. Holes cut by

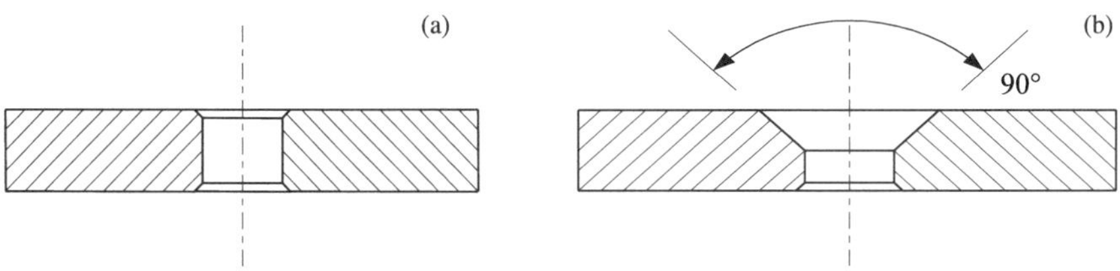

Fig. 2.16 **a** Cylindrical hole (with edge chamfer) and **b** countersunk hole (with edge chamfer)

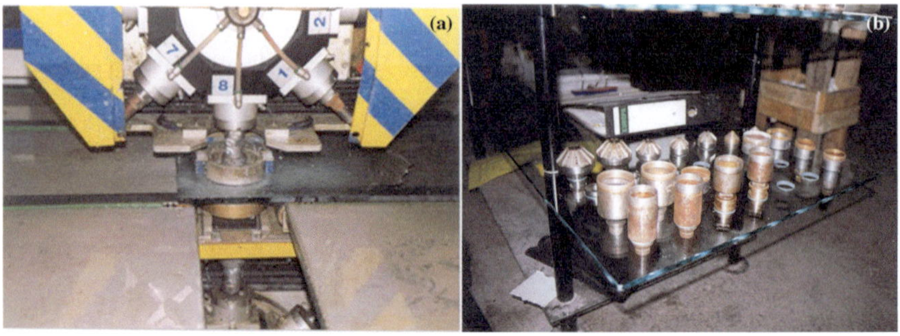

Fig. 2.17 **a** Creating a hole in glass utilising the diamond coring method; **b** Diamond cores and milling tools used to create cylindrical and countersunk holes

waterjet can be of high quality, however, they require strict quality control of the process settings and hardware [71].

In the diamond coring process, see Fig. 2.17a, coring is often carried out from both sides simultaneously, so that no glass spalling occurs at the surface. At the point in the thickness of the glass where the two cores meet, a ridge is formed that must subsequently be removed by an additional coring operation.

Countersunk holes, see Fig. 2.16b, are usually made in a two-stage process. The cylindrical hole is completed first, followed by further milling of the cylindiral hole to form the conical part. An alternative way is to work the pane from both sides simultaneously: at the top the conical part is milled out, while only the chamfer is created in the cylindrical part at the bottom. Countersunk holes enable the use of metal bolts known as point fixings that are flush with the glass surface. The depth of the conical part of the hole should not be greater than half the thickness of the glass. After coring and/or milling, the glass panes are cleaned with water and dried. Figure 2.17b shows typical heads for the diamond coring and milling process. The surface quality of these holes is similar to the that of ground edges, according to DIN 1249-11, see Sect. 2.3.2.

Finally, undercut milling and the use of undercut anchors in thermally pre-stressed glass is also possible. First, a partly penetrating cylindrical hole is made and then undercut through the inclined rotation of the milling tool, see Fig. 2.18. Undercut holes are usually carried out on 10 or 12 mm thick glass [29]. Force transfer takes place via form-locking in the undercut area, but the load-bearing capacity of undercut connections is typically lower than that of the above-mentioned through-hole alternatives.

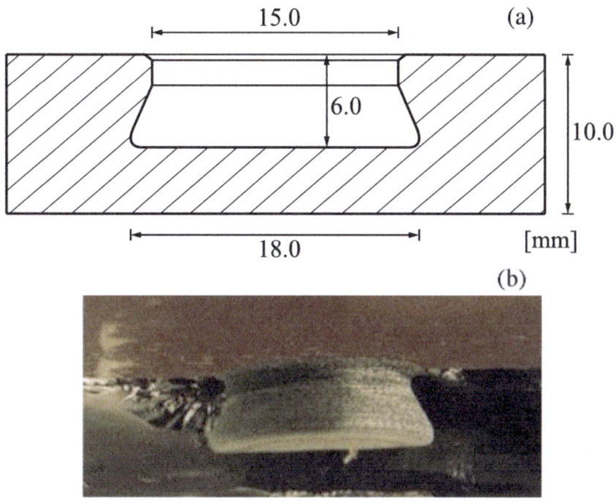

Fig. 2.18 **a** Undercut hole in glass; **b** Undercut hole in glass, with fracture mirror on the upper left edge between the glass surface and the hole

2.3.4 Pre-stressing

2.3.4.1 Thermally Pre-stressed Glass

Thermally pre-stressed glass, also known as thermally toughened glass (TTG) or fully tempered glass (FTG), contains significant residual stresses, generated by heating the glass up beyond the glass transition temperature and subsequently cooling it down rapidly, typically by air jets, see Sect. 6.2. The resulting residual stress state consists of tensile stresses in the core of the pane and compressive stresses at the glass surfaces.

Thermal pre-stressing is commonly, but not exclusively, applied to float glass. The thermal pre-stressing process starts with a thorough cleaning of the glass panes. The glass panes are then placed on a preparation table and receive a ceramic stamp (as prescribed in product standards) before entering the furnace. During the process a ceramic stamp is baked into the surface of the glass.

The characteristic bending strength, f_k, of (monolithic) thermally pre-stressed safety glass made from float glass is specified in EN 12150-1 with a value of 120 MPa (5%-fractile; 95% probability). The bending strength of enamelled thermally pre-stressed glass is lower: $f_k = 70\,\text{MPa}$. However, this value should only be used as an approximate indication, since the bending strength of enamelled or printed glass depends on the glass frit and thermal process used [50, 51].

The stress and strain profile in thermally pre-stressed glass gives rise to a elastic energy that is stored in the glass, which is released upon failure, leading to a complete fragmentation of the glass into small blunt pieces. Therefore, thermally toughened glass is also referred to as safety glass. Any major edge work, coring or cutting must therefore be executed prior to the thermal pre-stressing process.

The thermal treatment induces dimensional distortions in excess of those commonly found in annealed float glass. For example, while annealed float glass has a global bow which is limited to <1/2 000 [13] or <1/1 000 [18] of the pane's length, significantly larger values are found in thermally pre-stressed glass [26]. Due to the production process for thermally pre-stressed glass, EN 12150-1 distinguishes between global and local bow, whereby a specific local bow pattern may also be referred to as *roller waves*. The dimensional distortions must be controlled carefully because they can affect the optical aesthetic quality of the glass see Sect. 6.2.

Thermally pre-stressed glass may occasionally suffer from "spontaneous" nickel sulphide induced fracture. Therefore when monolithic thermally pre-stressed glass is installed in vertical glazing applications (e.g. as one layer of an insulating glazing units) it is commonplace to *heat soak* the glass before installation to further reduce fracture probability. Heat soak requirements vary from country to country and depend on the specific application scenario. In Europe, heat soaked glass needs to comply with EN 14179-2 [15]. Nevertheless, the required duration of the holding phase at elevated temperature is to some extent still a subject of discussion in other international markets [23].

Heat strengthened glass (HSG) is made in a manner similar to fully toughened glass but with a lower cooling rate. This generates lower residual stresses and less elastic energy in the glass, leading to glass with a tensile strength in between annealed and thermally pre-stressed glass, but with a fracture pattern that is more similar that of annealed glass i.e. with relatively large pieces, see e.g. Sect. 6.2.7.

2.3.4.2 Chemically Pre-stressed Glass

Residual surface compressive stresses can also be generated through chemical treatment of the glass, see Sect. 6.3. Chemically pre-stressed glass for the construction industry is regulated in EN 12337-1 [6]. Although the intrinsic glass strength depends substantially on the surface condition and the depth of layer (DoL), EN 12337 [6] defines a characteristic minimum bending strength, f_k, for chemically pre-stressed glass of 150.0 MPa (5%-fractile; 95% probability). However, this value indicated in the standards refers to the strength of pristine glass, while the practical strength fluctuates significantly due to unavoidable damage accumulation on the glass surface. Indeed, if surface defects penetrate beyond the compression zone, the strength level is reduced to not more than that of annealed float glass. Additionally, the longer processing time and resulting higher production costs of chemically pre-stressed glass make its application in the construction industry limited to special applications. Examples of this are hot bent, doubly-curved glass, or glass in applications with high optical requirements which cannot be achieved with thermally pre-stressed glass, such as in glazing for cruise ships, yachts or fighter jets. However, ongoing research and development on chemically pre-stressing of (ultra-)thin glass points towards future applications in light and transparent sandwich components where chemically pre-stressed thin glass is combined with suitable interlayer/core materials [48, 57, 68, 70, 75].

The chemical pre-stressing process does not induce local bow; however, global bow may occur in the process, and is typically limited to 0.002 mm/mm (1/500) [6].

2.3.5 Coating

2.3.5.1 Properties

Glass surface coatings have many purposes. They can be used for aesthetic reasons, as well as to control the energy transfer through the façade. Decorative coatings are usually applied to glass surfaces by silkscreen or digital printing processes. Functional coatings, on the other hand, are usually as transparent and colour-neutral as possible.

2.3.5.2 Effects of Coatings

Coatings on glass surfaces can fundamentally change the energy transfer properties of the glass. They have an effect on the transparency of the glass, and on the energy transmission through the glass. In uncoated soda-lime silica glass, solar light (visible to the human eye) passes relatively unhindered through the glass (visual transmission T_V). Only a small part of the solar radiation is reflected and absorbed by the glass. Transmission T, reflection R and absorption A add up to 100%, see Sect. 1.4.2 and Fig. 1.8. This applies to all wavelength ranges of electromagnetic radiation and represents the principle of universal energy conservation. The relative proportions of transmission, reflection and absorption can be controlled across a range of wavelengths. Such coatings are known as spectrally selective. The values can vary greatly depending on the spectral range considered. Glass, for example, has a very high transmittance in the spectral range of the visible light, whereas it has a high absorption in the range of long-wave infrared radiation (thermal radiation), see Fig. 1.9.

To understand the function of different glass coating, it is important to not only consider the visible light range (approximately 380 to 780 nm), but the whole solar spectrum (approx. 250 to 2500 nm). Nearly 100% of the solar energy arriving at the Earth's surface relates to this range of electromagnetic radiation. Thermal radiation, i.e., long-wave infrared radiation within the range from 5×10^3 nm to 50×10^3 nm, is not received from the sun, but occurs when solar radiation incides on an object, such as the Earth's surface. This body heats up through the absorbed solar radiation and subsequently re-radiates long-wave thermal radiation into its environment. The wavelength of this emitted electromagnetic radiation is shifted to the long-wave range, while the total energy is conserved.

There are two principal types of coatings in terms of their function: thermal performance coatings (reducing the transmission of long-wave infrared radiation to prevent heat loss) and solar control coatings (reducing the transmission of the total solar radiation with the exception of the visual range).

Thermal performance coatings are designed to prevent the transfer of thermal radiation (heat) from the indoor to the outdoor environment in heating seasons. Regular soda-lime silica glass is nearly impermeable to long-wave infrared radiation, i.e., thermal radiation T_{IR} is almost zero, see Sect. 1.4.1 and Fig. 1.9. Therefore, incident thermal radiation is first absorbed by the glass and then emitted in the form of secondary radiation (re-radiation) from the glass surfaces.

The emissivity ε of a body indicates the proportion of energy that is emitted by the body compared to ideal black-body radiation ($\varepsilon = 1$) at the same temperature. It can thus be deduced by applying Kirchhoff's law to the body in its stationery state that the absorption ratio, A, of a body is equal to its emissivity. A body with an emission level of $\varepsilon = 0.20$ thus absorbs 20% of the incoming radiation (the remaining 80% is reflected or transmitted). On heating the same body it will only radiate 20% of the energy of an ideal black-body.

Glass has a high emissivity of $\varepsilon = 0.84$ for thermal radiation since it also has a high absorption capacity within this wavelength range. To reduce the emission of thermal radiation, a thin metallic coating can be applied, which reduces the emissivity of the glass surface to a few percent, depending on the metal used. Metals such as silver ($\varepsilon = 0.02$) have a very low emissivity compared to glass. Therefore, such materials are used for thin (nanoscale) coatings that serve to reduce emissivity without significantly affecting light transmission (transparency).

The emissivity of a glass coating can be determined by measuring reflection. Since for most solids, including glass, the transmission of thermal radiation T_{IR} is almost zero, the following relationship applies:

$$1 - R = A = \varepsilon. \tag{2.1}$$

Thus, by preventing the emission of thermal radiation, the coating reflects the radiation of the glass back to the inside of a building and thereby minimising unwanted heat loss in temperate and cold climates and unwanted heat gain in temperate and hot climates. From a thermal point of view, the position of a coating in a multi-pane insulating glass unit, see Sect. 3.2, does not make a significant difference. In double glazing units and for thermal protection in heating seasons, it is usually applied to the surface on position 3, where position 1 is the outside surface of the outer pane and position 4 is the inside surface of the inner pane, see Fig. 3.9. A coating applied towards the cavity of an insulating glass unit is protected and cannot be damaged during installation or cleaning. These types of coatings are often referred to as soft *low-emissivity* or soft *low-E* coatings. Today's thermal performance coatings can have emissivity ratios as low as $\varepsilon = 0.01$ to 0.03.

Modern *low-E coatings* have made a significant contribution to improving the thermal insulation of multi-pane insulating glass units. They are nearly invisible to the human eye, i.e. they are mostly colour-neutral and have a high colour rendering index (CRI). Such thermal protection glazing is generally used in façades with low window-to-wall ratios, typical of e.g. single dwelling units. The transmissivity of solar radiation is desirable here and can be considered as heat gains in the heating season. However, for buildings with large glazed areas and in arid or hot climate zones, thermal performance coatings are often not suitable, as the transmissivity of solar radiation can lead to overheating of the building. Incoming solar radiation can pass relatively unhindered through the thermal performance coatings and heat up the interior of the building due to the absorption on floors, curtains, walls or furnishing. The radiation is absorbed and only partially reflected by the surfaces of these bodies. If the reflected radiation hits the human eye, we perceive it as the colour of the respective item. If it meets other surfaces, the radiation is again partly reflected and partially absorbed, until finally most of the solar radiation is absorbed. The absorbed radiation thus remains almost completely retained in the indoor environment of the building in the form of unwanted heat. Additionally, low-E glazing prevents emission to the outside. This greenhouse effect is desirable for façades with low window-to-wall ratios or for greenhouses. In buildings with large glass façades, however, it can lead to significant overheating problems in the cooling seasons.

Solar control coatings are used to protect buildings with larger glass surfaces from overheating in the cooling seasons. These layers usually also have a low-E function for thermal insulation in the heating seasons and are called combined solar control and thermal performance coatings. There are also solar control coatings available, which have no or only a very limited low-E function.

Solar control coatings reduce the transmission of the solar radiation and are typically applied to surface 2 in a double glazing unit, see Fig. 3.9. Total solar transmittance is reduced by reflecting or absorbing wavelengths in the range of the solar spectrum that are not visible to the human eye, thereby reducing the solar energy transmitted to the indoor environment without affecting the visible light transmittance, T_v, which is the light transmission in the visible range of the solar spectrum. These coatings are therefore called spectrally selective.

The total solar energy transmittance (also known as solar heat gain coefficient or solar factor), g, represents the proportion of total incident solar energy that is transmitted through the glass to the indoor environment. Therefore solar control coatings have a low g-value. In contrast, it is often desirable to have a high light transmittance T_v, but because visible light is a source of energy, it is not theoretical possible to simultaneously achieve a very low g and a very high T_v. In fact the maximum theoretical limit of $T_v/g = 100/53 < 2$, see Fig. 1.7. A further reduction of the g-value means that the glazing inevitably becomes darker, since the light transmission is also reduced in the visible domain. It is even possible to achieve slightly higher values of $T_v/g \geq 2$ by selectively filtering out the maxima of the visible spectrum, but this affects the colour rendering index of the glazing. Since sun protection coatings here selectively interfere with the visible spectrum of the solar radiation, colours are often perceived differently when being reflected or transmitted. In addition, the perceived colour can change with a change in angle.

In most cases, modern thermal performance coatings, as well as solar protection coatings, are known as soft coatings, because the sensitive metallic coatings must be protected against moisture and mechanical abrasion. Therefore, they are usually positioned on the cavity-facing surfaces of an insulating glass unit. Hard coatings, which may also be used on exposed glass surfaces, also exist, but rarely achieve the same levels of performance as of soft coatings. Generally, the visual transmittance T_v, the total energy transmittance g and the thermal transmittance, U, are used to characterise the light and thermal performance of glass in building envelopes (also known as façades). The typical values of double glazing units are often within the ranges of about 80% for the visual transmittance T_v, and about 60% for the total solar energy transmittance, g. Modern solar control coatings often achieve a spectral selectivity of $T_v/g \approx 2$ and are available from different manufacturers.

Other glass coatings, such as coloured screen-prints or enamels, also have an effect on the radiation properties of insulating glass units. Depending on the colour and printing coverage, the ratio of transmission, reflection and absorption can vary, so these can also be used as parameters to control the light and thermal performance of the glazing. However, these coatings have no spectral selectivity, meaning that both light and energy transmission are affected equally ($T_v/g \approx 1$). Therefore, they are mainly applied to change the aesthetic appearance of the glazing.

2.3.5.3 Coating Processes

Coatings can also be classified according to their manufacturing process, typically divided into chemical and physical processes. According to EN 1096-1 [30], chemical coating involves processes in which a chemical reaction between liquids, gases or solids and the glass surface creates (thin) deposition layers on the glass.

Chemical Deposition (CD)

Chemical deposition is a chemical reaction of the coating material with the glass surface. The procedure is often directly implemented in the glass float line. Depending on the coating material used, non-spectral selective solar coatings, thermal protection coatings, cleaning supporting layers and thin-film solar cells can be produced using this method.

Chemical processes can further be categorised into:

Chemical Vapour Deposition (CVD)

During the CVD process, the coating materials are deposited in their gaseous phase onto the glass surface. The chemical reaction is accelerated by the thermal energy of the hot float glass.

Plasma Aided CVD (PACVD)

In this process, the chemical reaction between the coating material and the glass surface is supported by additional electrical energy in the form of a gas discharge (plasma). In contrast to classical CVD procedures, this processes, take place in a vacuum rather than under ambient conditions.

Pyrolytic Process

In this process, formation of layers on the hot glass surface is carried out by a chemical reaction of the coating material with air, oxygen or water vapour. The process usually takes place on the float line in atmospheric pressure. Pyrolytic coatings are highly resistant to environmental conditions and are relatively insensitive to mechanical attacks. They are therefore also referred to as “hard coating”, and they may also be placed facing the inside of a room or the outside of the façade, thus being exposed to weathering.

Wet-chemical processes

Wet-chemical processes can be distinguished as either sol-gel processes or processes by means of a chemical reduction. In the latter, layers are deposited onto the flat glass surface by a redox (reduction-oxidation) reaction. The coating material (e.g. dissolved salts) experiences a reduction (electron uptake) and the substrate oxidises. Silver mirrors, for example, are produced using this method. The Sol-Gel method is a two-stage process in which thermal energy is also added. In the first step, the coating material (Sol) is dissolved into a liquid, applied to the glass surface converted into a gel film through the absorption of water vapour. The process takes place at room temperature under ambient conditions. In the second step, the gel film condenses onto the glass surface through the addition of thermal energy. Coatings based on chemical processes can be highly resistant to environmental conditions and relatively resistant to mechanical attacks and can therefore classify as hard coatings.

In addition to chemical processes, physical processes are also used for the production of glass coatings. Physical coating processes all take place offline, i.e. separated from the float line process. Two physical processes, i.e. Physical Vapour Deposition (PVD) and Sputter Deposition, are particularly relevant. Both procedures are carried out under vacuum conditions.

Physical Vapour Deposition (PVD)

During the vapour deposition phase, the solid or liquid coating material, possibly with the addition of thermal energy, is brought in a gaseous state under vacuum conditions and then condensed onto the surface of the substrate. This coating technology was used extensively in the early years of flat glass coatings (1960–1980).

Sputter Deposition (Plasma PVD)

During so-called sputterings or cathode atomisation, atoms from a solid body (target), are initially bombarded with energy-rich ions (plasma) and released in the vapour phase. The atomized particles are directed to the substrate by a magnetic field. An electric field is also used in the Magnetron-Sputter process to increase efficiency. The atomized coating material then is deposited on the surface of the substrate and forms a solid layer. The Magnetron-Sputter method is currently the most widely used method for thin coatings on flat glass. However, the layers are usually susceptible to corrosion and must be protected from moisture and mechanical damage. They are therefore also referred to as soft coatings, and are typically used on the cavity-facing surfaces of insulating glass units.

2.3.5.4 Structure of Coatings

Coatings can be classified as either thick film > 1 μm, e.g. applied to varnishes, colours, printing, etc. or thin film < 1 μm. Typical thin film coatings are those produced for functional coatings according to the methods described in Sect. 2.3.5.3.

Such functional coatings are available in a wide variety and from many manufacturers. Today, the magnetron-sputtering method is often used in the production of thin films. During the latter, thin, often metallic layers are applied to the glass

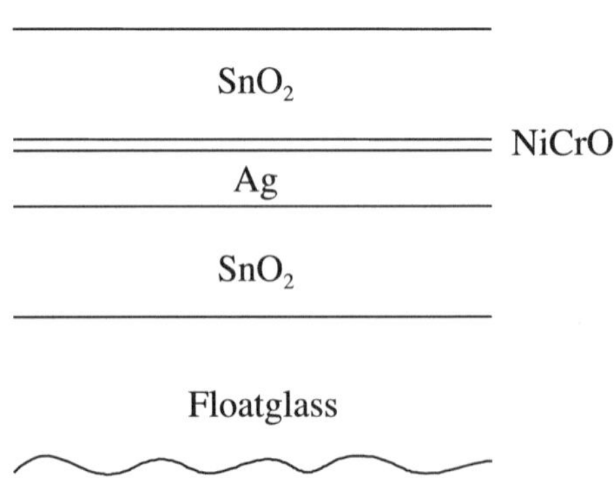

Fig. 2.19 Layering of thin coatings on glass

surface, influencing the reflection, absorption and transmittance behaviour at the different wavelengths. Thin layers can be used to create physical properties where the coating is practically invisible to the naked eye. In most cases, this effect is achieved through complex multi-layer systems, see Fig. 2.19. Different radiation properties are determined depending on the sequencing of the layers. Since the thickness of the individual layers is in the order of the wavelength, interference effects may occur. For example, weakening of the waves through destructive interference caused by the overlapping of the incoming and reflective waves. This physical principle can be used for electromagnetic shielding, radar absorption, anti-reflective layering in the visual spectrum or spectral selectivity.

The coating materials, layer thicknesses and layering order vary according to the required function. However, the overall sequence can be described as follows: first, an adhesive layer is applied onto the glass surface. This is followed by a sequence of different functional and protective layers. Lastly, a top layer serves to protect the corrosion-sensitive metallic components of the underlying layers. A layer of about 10 nm of Silver is commonly used in many layers to reduce emissivity.

2.3.6 Printing and Enamelling

In addition to functional coatings, which affect the energy transfer through the glass, coatings can also be used for aesthetic purposes. The screen printing process is widely used for decorative effects, such as dotted or dashed patterns on the glass surface. The desired decoration is first applied by perforation of a special fabric. The fabric is then stretched onto a frame and placed on the glass surface to be printed. The ink is then forced through the permeable screen with a squeegee and printed onto the glass surface through the perforations. The fabric is then removed and the ink remains as a pattern on the glass surface.

Ceramic (inorganic) inks, also known as enamels, are typically used for screen printing on glass substrates. The enamels consist of the silicates and metal oxides that produce the desired colour as well as appropriate solvents necessary for the printing process [36]. These enamels are dried after application and subsequently baked into the glass surface to create multi-coloured motifs; multiple colours are achieved by printing with different enamels before baking.

During the final baking process, the remaining solvents evaporate and the silicates are melted. In this semi-finished state, the ink or enamel is sometimes referred to as a frit. During the subsequent cooling and freezing phases, the frit forms a firm bond with the glass surface. In order to melt the enamel onto the glass surface, usually the thermal pre-stressing process can be used. However, since the quality of the glass surface is affected by the enamel, the strength of the respective glass product is also reduced, see Table 5.2. The available colour range of ceramic colours is limited. Vibrant colours often require pigments from heavy metals such as Cadmium, which are environmentally hazardous and therefore banned. On the other hand, metallic effects can be created by, for example, adding metal pigments.

In addition, the colours are weatherproof and resistant to UV radiation.

Fig. 2.20 Digital printing on glass, applied for a glass wall at *Imperial College* London (photo credits: Ali Moshiri/OKALUX GmbH)

Since the manufacturing of the printing screen is rather costly, the screen printing process is particularly suitable in the production of larger series with the same motifs. Depending on the pattern, the meshes can be reused up to 1 000 times, but the quality of the print reduces with reuse. A more cost-effective process for single multi-coloured patterns and pictures is the *digital printing method* which was developed to print photo-realistic images onto glass and allows for high resolution, cost-effective printing in small quantities, see Fig. 2.20. The digital printing method uses organic inks thereby providing a very wide range of colours. This is distinct advantage over enamelling, where the ceramic frits have a limited colour range. The organic colours used in digital printing do not have to be baked in, which means that also annealed float glass can be printed. In addition, they do not cause a reduction in glass strength, as they do not physically affect the glass surface. For this reason, however, adhesion and durability are lower than that of inorganic frits that are baked into the surface in enamlling.

Coloured coatings can also be applied to glass in other ways. Full-surface coatings can be applied to the glass by either smearing or spray painting. In the case of stained glass, individual pieces can be produced by applying paint using brushes or other painting techniques.

2.3.7 Sandblasting

The sandblasting process is commonly used to create a matt finish for the glass surface. Whereas acid etching the glass, see Sect. 2.3.8, creates a satin surface finish by means of chemical dissolution, sandblasting roughens the glass surface through a mechanically abrasive process. The resulting surface loses transparency, but remains translucent. By covering (masking) individual areas, images and patterns can be

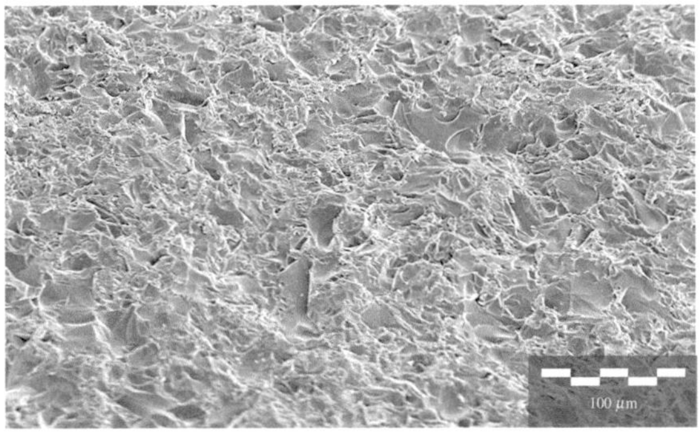

Fig. 2.21 Scanning electron microscope picture of a sandblasted glass surface [49]

applied. However, due to the mechanical removal of material from the surface, surface flaws are generated during sandblasting, see Fig. 2.21, which cause a considerable reduction in the bending strength of the glass. The tensile strength of sandblasted glass depends on various factors (grit, air pressure, etc.), typically leading to a characteristic strength in the range of $f_k = 25$ MPa for annealed sandblasted glass [49]. In the case of thermally toughened sandblasted glass, the strength for the respective product must be determined individually, taking into account the above-mentioned factors.

The strength-reducing effect of the microscopic flaws can be reduced after sand blasting by etching with hydrofluoric acid. Additionally, the sensitivity to dirt is higher for sandblasted glass than for etched glass.

2.3.8 Etching

Glass has a very high chemical resistance, only hydrofluoric (HF) acid is capable of dissolving silicates to create an immediately noticeable attack on its surface. Glass etching is in fact the oldest application of hydrofluoric acid. Hydrofluoric acid is a colourless, aqueous solution that is highly toxic; any form of skin contact must be avoided [27]. This method has always been used to clean, decorate or give a matt finish to glass surfaces. Compared to mechanical polishing, hydrofluoric acid treatment is a faster process but requires significant safety measures. In glass processing with hydrofluoric acid, a distinction is made between the following methods:

i) *Blank etching*, also referred to as acid polishing, is often used in the container glass industry to polish matt-ground surfaces;
ii) *Matt etching* is used to make the glass translucent. By depositing on the glass surface salts which are insoluble in water, the etching is carried out only in areas without these deposits;
iii) *Deep etching* is suitable for gradations or decorations. For this purpose, the entire glass surface is first covered with wax or special varnishes. This coating is removed only in the areas to be etched, which are subsequently treated with highly concentrated hydrofluoric acid.

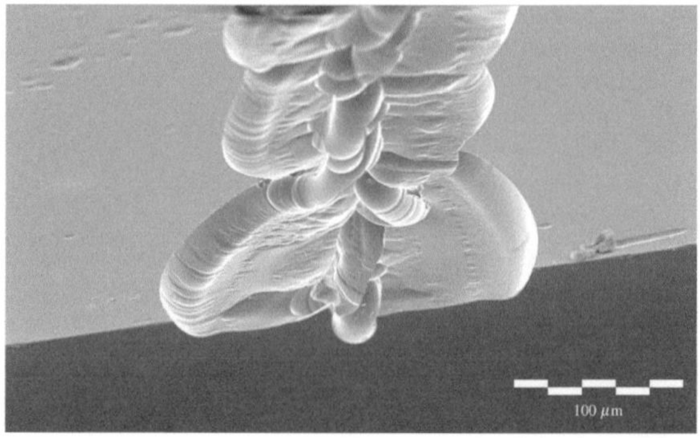

Fig. 2.22 Scanning electron microscope image taken over a period of 60 min with a 5.0% hydrofluoric acid-etched scratch on soda-lime silica glass with typically rounded crack tip in the vertical crack [49]

The optical appearance of the glass surface varies according to the method, concentration of the reactants, temperature of the etching bath and chemical composition of the glass. Matt etching, which produces satin surfaces, is the most popular method used for architectural applications. However, it is increasingly being replaced by sandblasting processes, see Sect. 2.3.7. The optical result is almost identical, but in the case of sandblasting, small surface defects occur which significantly reduce the bending strength of the glass compared to ordinary, untreated glass, see Fig. 2.21.

In addition to the usual pre-stressing methods, see Sect. 2.3.4.1, the strength of glass can be increased by the application of an acid polish. Figure 2.22 shows a scratch on soda-lime silica glass that has been treated with 5.0% hydrofluoric acid over a period of 60 min. In comparison to Fig. 5.1, it can be seen that the crack tip of the deep crack is rounded which increases the failure strength considerably compared to a sharp crack tip. In [49], the change in strength of carefully pre-damaged and subsequently etched float glass was investigated. In this study, significant strength increases were observed for annealed float glass, where $\sigma_{f,HF} > 200$ MPa or ${}^{\sigma_{f,HF}}/_{\sigma_f} = 3.68$ using a 2.5 wt% HF and ${}^{\sigma_{f,HF}}/_{\sigma_f} = 4.30$ using a 5.0 wt% HF.

2.3.9 *Hot Bending*

Curved glass is used in the building industry for instance in façades, overhead glazing, balustrades or for interior partitions. Curved glass is significantly more costly than flat glass, particularly if only a few piece of a particular curvature are needed such as in buildings. Its use has therefore been limited to iconic high-end buildings. However, recent improvements in processing have lead to cost-reductions and an increased use of curved glass in the construction sector. In contrast, curved glass is very widely used in the automotive industry. Here, multiple uses of the same size and curvature have led to mass-production processes for annealed laminated glass for windshields, monolithic thermally pre-stressed safety glass for side and rear windows.

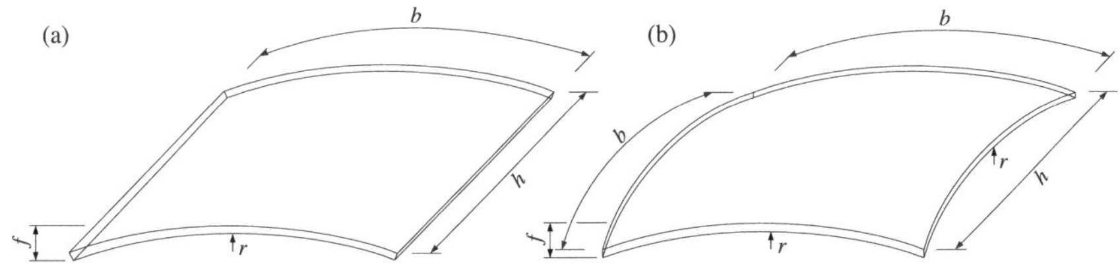

Fig. 2.23 **a** Cylindrical curved glass and **b** spherical curved glass

In addition to annealed glass, thermally pre-stressed glass, laminated glass and insulating glass units can also be curved. The production of curved, thermally pre-stressed laminated glass, with holes if necessary, is complicated by the dimensional tolerances that arise during the thermal bending processing (i.e. hot bending).

Alternatively, flat glass panes can be cold bent or permanently kept in a curved shape through bending during the lamination process (lamination bending) using a suitable interlayer material (e.g. stiff and/or cross-linking products). These curving techniques predominantly apply for multilayer glass products and are discussed in Sects. 3.3.1 and 3.3.2, respectively.

During hot bending, annealed float glass is heated to about 100 °C above the glass transition temperature T_g and then formed by means of gravity bending, press bending or mechanical bending. The most commonly used process for architectural glass is gravity bending (also known as sag bending). Here, the flat glass is heated over a mould in a furnace [21, 24]. After the glass has sagged under its own weight, the pane takes on the shape of the mould, see Figs. 2.24a and 2.28b, and the furnace temperature is subsequently reduced under controlled conditions, thereby minimising thermal residual stresses in the glass. The residual stresses can be assessed using either optical stress measurement methods or destructive tests followed by fractography, see Sects. 10.3.1 and 6.4. To avoid damage to eventual coatings, the coatings should not be put onto surfaces that come into contact with the mould during the hot bending process.

In the manufacturing of thermally curved panes a distinction is usually made between slightly curved glazing with a radius of curvature of more than 2 m and strongly curved glass with smaller radii of curvature. It is also commonplace to differentiate between single (e.g. cylindrical or conical) curved panes and doubly (e.g. spherical) curved panes, see Fig. 2.23.

The following relationships can be used to determine the geometry of cylindrically curved glass on the circular segment (see Fig. 2.24b).

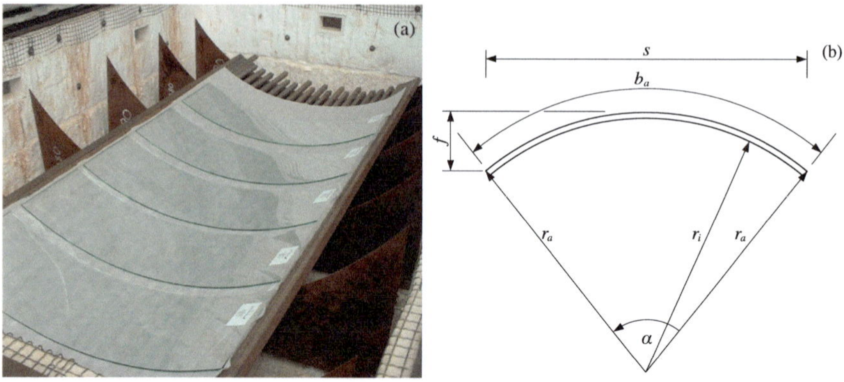

Fig. 2.24 **a** View into furnace for hot bending (here: gravity bending process) (photo credits: Semcoglas Holding GmbH); **b** determination of the geometry of single curved glass at the circular segment

$$\text{Arch Length:} \quad b_a = \pi \cdot r_a \cdot \frac{\alpha[°]}{180°} \tag{2.2}$$

$$\text{Opening angle:} \quad \alpha[°] = b_a \cdot \frac{180°}{\pi \cdot r_a} \tag{2.3}$$

$$\text{Tendon Length:} \quad s = 2 \cdot r_a \cdot \sin\left(\frac{\alpha}{2}\right) = 2 \cdot \sqrt{f \cdot (2 \cdot r_a - f)} \tag{2.4}$$

$$\text{Outer Radius:} \quad r_a = \frac{s^2 + 4 \cdot f^2}{8 \cdot f} \tag{2.5}$$

$$\text{Rise:} \quad f = r_a - \left(r_a \cdot \cos\left(\frac{\alpha}{2}\right)\right) = r_a - \sqrt{{r_a}^2 - \left(\frac{s}{2}\right)^2} \tag{2.6}$$

A standard bending process can handle panes with an arc length of about $b_a = 2500$ mm and a panel-width of 6 000 mm. A maximum rise of $a = 1\,000$ mm can be reached. The minimum bending radius is approximately 100 mm to 300 mm, with the first value being limited to glass thicknesses from 4 mm to 6 mm; a radius of no smaller than 300 mm can be achieved with glass that is 10 mm or thicker. The gravity bending process is suitable for the subsequent production of laminated glass, since the panes can be bent simultaneously in layers using special release agents between them. The minimum radius of curvature achievable with laminated glass is about 300 mm, unless thin glass is used. The production of bent laminated glass with

Fig. 2.25 **a** Doubly curved glass *Hungerburg Station—Nordkettenbahn (North Railwayline)*, Innsbruck (architect: Zaha Hadid; Engineers: Bollinger and Grohmann); **b** Doubly curved glass at the *London Eye* (Architects: David Marks and Julia Barfield; Engineer: Jane M. Wernick)

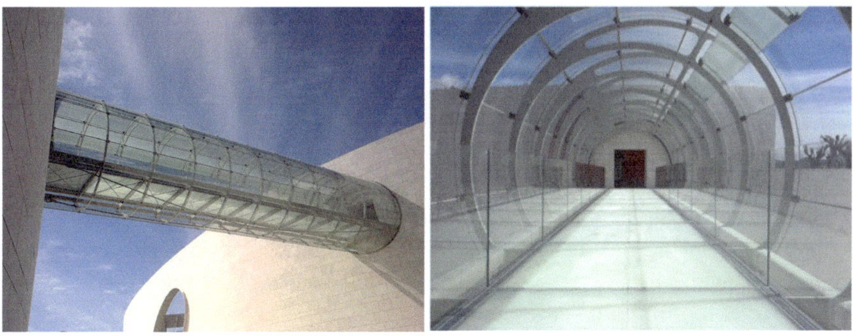

Fig. 2.26 Cylindrically curved, thermally toughened glass at the *Champalimaud Foundation*, Lisbon (Architect: Charles Correa; Engineers: Schlaich Bergermann and Partner)

holes is very complicated due to the dimensional distortions that occur at the holes during lamination.

In the (currently outdated) vertical glass bending procedure, the glass is suspended by means of clamps (also known as tongs) and then lowered or raised through the furnace. After exiting the furnace, the panes are mechanically pressed by curved moulds, but the imprints (tong marks) at the suspension points remain visible.

Doubly curved panes of defined dimensions, such as automotive windscreens, can be produced in a press bending process. The glass is conveyed on rollers through the furnace and takes on the form of the mould through a pressing process by means of a mould, which consists of a positive and a negative part. The press bending process ensures low bending tolerances and allows the desired shape to be formed with precision, see Fig. 2.25. Furthermore, doubly curved structural glass is also made with special moulds, see Fig. 2.24a, or on cushions. However, since each form must be manufactured individually, the procedure is very expensive.

Fig. 2.27 Point-supported modules with doubly curved glazing of the glass dome at *Erin Mills Center*, Mississauga/Canada (Architects: MMC Architects, Engineers: RJC Engineers)

Curved glass can also be thermally pre-stressed, see Figs. 2.26 and 2.27. One method is to bend the glass in a press using a conventional mould and to subsequently pre-stress it while it is either vertically suspended or horizontally supported in a mould. Here, too, imprints at the edge are visible where either the clamps or the mould has been in contact with the glass during the process. As a consequence of this process, the mechanical and optical quality of the glass is often compromised and only a reduced thermal pre-stress can be generated in the contact zones.

A more recent method for pre-stressing curved glass involves combining bending and thermal pre-stressing in a single process, without the need of special moulds. Here, the desired curvature is achieved by individually controlling and shifting the transport rollers of the furnace or cooling area. It is advantageous to bend the glass already in the heating area of the furnace, so that a sufficient pre-stress can be generated in the quenching area. This is, however, only possible for larger bending radii and for cylindrical curved glass because of the need of constantly moving the glass back and forth while quenching.

Since radii of no smaller than 300 mm can be produced with thermally toughened glass, curved glass with smaller radii must be pre-stressed chemically instead, see Sect. 6.3.1 and Fig. 2.28a. In particular chemical pre-stressing is currently the only way to apply a pre-stress to doubly curved glass.

The combined process of pre-stressing and curving only allows for bending glass panes individually, resulting in unavoidable small differences in their radii. Therefore, bending stresses may be induced when two curved glass plies of slightly different radii are bonded together during lamination. These stresses can also affect the fracture

Fig. 2.28 **a** Cylindrically curved, chemically pre-stressed glazing of the elevators at the *long-distance railway station*, Frankfurt a.M. (architects: BRT; Engineers: MERO); **b** Cylindrically curved (by gravity bending process) glass panes at the *Museum aan de Stroom* in Antwerp (Architects: Neutelings Riedijk; Engineers: Rob Nijsse, Bureau Bouwtechniek, ABT)

pattern, i.e. the fracture pattern of the laminated safety glass may differ significantly from the fracture pattern of the individual plies before lamination.

The bending strength and optical quality (resulting from waviness, distortions, etc.) of curved glass are inferior to those of flat glass [59, 60], see Sect. 5.4.1. The mechanical and optical qualities are highly dependent on the type of process and process parameters, especially with thermally pre-stressed, hot-bent glass. Tolerance values for bent glass are available in industry guidelines (e.g. in [28]).

References

1. R. McGrath, A.C. Frost, *Glass in Architecture and Decoration*, 2nd edn. (The Architectural Press, London, 1961), p. 712
2. E. Roeder, Extrusion of glass. eng. J. Non-Cryst. Solids **5**(5), 377–388 (1970)
3. DIN 1249-11. *Flachglas im Bauwesen; Glaskanten* (1986)
4. H. Rawson, *Glasses and Their Applications*. Institute of Metals (1991), p. 166
5. S. Glaswerke, *Glas: Ein vielseitiger Werkstoff mit außergewöhnlichen Eigenschaften* (Bundesverband Glasindustrie und Mineralfaserindustrie eV, Düsseldorf, 1996)
6. EN 12337-1. *Glass in building - Chemically strengthened soda lime silicate glass - Part 1: Definition and description* (2000)
7. EN 1288-2. *Glass in building - Determination of the bending strength of glass - Part 2: Coaxial double ring test on flat specimens with large test surface areas* (2000)
8. EN 1288-4. *Glass in building - Determination of the bending strength of glass - Part 4: Testing of channel shaped glass* (2000)
9. EN 1288-5. *Glass in building - Determination of the bending strength of glass - Part 5: Coaxial double ring test on flat specimens with small test surface areas* (2000)

10. J. Schneider, Facade structures in channel shaped glass, in *IABSE Symposium Report*. Melbourne (2002)
11. EN 1051-1. *Glass in building - Glass blocks and glass pavers - Part 1: Definitions and description* (2003)
12. F. Wellershoff, G. Sedlacek, Characteristic strength and failure probability of channel shaped glass, in *Glass Processing Days* (2003)
13. A. Luible, Stabilität von Tragelementen aus Glas. Dissertation. École Polytechnique Fédérale de Lausanne (2004)
14. J. Schneider, Glass strength in the borehole area of annealed float glass and tempered float glass. Int. J. Form. Process. **7**(4), 523–541 (2004)
15. EN 14179-2. *Glass in building - Heat soaked thermally toughened soda lime silicate safety glass - Part 2: Evaluation of conformity/Product standard* (2005)
16. GUV-SI 8027. *GUV-Informationen: Sicherheit bei Bau und Einrichtung* (2005)
17. S. Marpillero, K. Frampton, J. Schlaich, *James Carpenter: Environmental Refractions* (Princeton Architectural Press, 2006)
18. O. Englhardt, Flächentragwerke aus Glas – Tragverhalten und Stabilität. Dissertation. Universität der Bodenkultur Wien (2007)
19. H. Schober, J. Schneider, S. Justiz, J. Gugeler, C. Paech, M. Balz, Innovations with glass, steel and cables, in *Glass Performance Days* (Tampere, Finland, 2007)
20. C. Paech, K. Goeppert, Innovative Glass Joints - The 11 March Memorial in Madrid. eng. in *Challenging Glass* (2008), pp. 111–118
21. Ö. Bucak, M. Feldmann, R. Kasper, M. Bues, M. Illguth, Das Bauprodukt warm gebogenes Glas – Prüfverfahren, Festigkeiten und Qualitätssicherung. Stahlbau-Spezial (Glasbau) **78**(1), 23–28 (2009)
22. J.H. Nielsen, J.F. Olesen, H. Stang, The fracture process of tempered soda-lime-silica glass. Exp. Mech. **49**(6), 855–870 (2009)
23. O. Yousfi, Y. Brechet, P. Donnadieu, A.M. Kasper, F. Serruys, Proposal to enhance the Heat Soak Test following EN14179-1, in *Glass Performance Days 2009*. Jan 2009, pp. 584–588
24. F. Ensslen, J. Schneider, S. Schula, Produktion, Eigenschaften und Tragverhalten von thermisch gebogenen Floatgläsern für das Bauwesen – Erstprüfung und werkseigene Produktionskontrolle im Rahmen des Zulassungsverfahrens. Stahlbau-Spezial (Glasbau) **79**(1), 46–51 (2010)
25. J.H. Nielsen, J.F. Olesen, P.N. Poulsen, H. Stang, Simulation of residual stresses at holes in tempered glass: a parametric study. Mater. Struct. **43**(7), 947–961 (2010)
26. J. Belis, D. Mocibob, A. Luible, M. Vandebroek, On the size and shape of initial out-of-plane curvatures in structural glass components. Constr. Build. Mater. **25**(5), 2700–2712 (2011)
27. BGI 576. *BG RCI – Berufsgenossenschaft Rohstoffe und chemische Industrie: Merkblatt M 005: Fluorwasserstoff, Flusssäure und anorganische Fluoride* (2012)
28. Bundesverband Flachglas. *BF-Merkblatt 009/2011 – Leitfaden für thermisch gebogenes Glas im Bauwesen* (2012)
29. DIBt. *Allgemeine bauaufsichtliche Zulassung Z-70.2-122: Punktgehaltene Verglasung mit fischer Zykon Punkthaltern (FZG-G-Z)* (2012)
30. EN 1096-1. *Glass in building - Coated glass - Part 1: Definitions and classification* (2012)
31. EN 572-1. *Glass in building - Basic soda-lime silicate glass products - Part 1: Definitions and general physical and mechanical properties* (Beuth, 2012)
32. EN 572-2. *Glass in building - Basic soda lime silicate glass products - Part 2: Float glass* (Beuth, 2012)
33. EN 572-3. *Glass in building - Basic soda lime silicate glass products - Part 3: Polished wired glass* (2012)
34. EN 572-5. *Glass in building - Basic soda lime silicate glass products - Part 5: Patterned glass* (2012)
35. EN 572-7. *Glass in building - Basic soda lime silicate glass products - Part 7: Wired or unwired channel shaped glass* (2012)
36. Ferro. *Technical Information GL18: Glaze Catalogue* (2012)

37. KEN Tadashi OSHIMA, *The Architectural Review. Optical Glass House* (Hiroshima, Japan, 2012)
38. J.H. Nielsen, P.N. Poulsen, Joints in tempered glass using glass dowel discs, in *Engineered Transparency*. Sept 2012
39. C. Schittich, G. Staib, D. Balkow, M. Schuler, W. Sobek, *Glass Construction Manual*. Construction manuals (Birkhäuser, 2012)
40. DIN 18008-3. *Glas im Bauwesen – Bemessungs- und Konstruktionsregeln – Teil 3: Punktf ö rmig gelagerte Verglasungen* (Beuth, 2013)
41. M. Lindqvist, Structural glass strength prediction based on edge flaw characterization. Dissertation. École polytechnique fédérale de Lausanne (2013)
42. DIBt. *Allgemeine bauaufsichtliche Zulassung Z-70.3-197: Verbund-Sicherheitsglas mit einer Verbundfolie der Produktfamilie EVASAFE mit Ansatz eines Schubverbunds* (2014)
43. J. Kleuderlein, F. Ensslen, J. Schneider, Investigation of edge strength dependent on different types of edge processing, in *Engineered Transparency 2014*. Düsseldorf (2014), pp. 259–268
44. M. Overend, C. Butchart, H. Lambert, M. Prassas, The mechanical performance of laminated hybrid-glass units. Compos. Struct. **110**(1), 163–173 (2014)
45. M. Vandebroek, Thermal Fracture of Glass. PhD thesis. Ghent University (2014)
46. EN 12150-1. *Glass in building - Thermally toughened soda lime silicate safety glass - Part 1: Definition and description* (2015)
47. F. Oikonomopoulou, F. Veer, R. Nijsse, K. Baardolf, A completely transparent, adhesively bonded soda-lime glass block masonry system. J. Facade Des. Eng. **2**(3-4), 201–221 (2015)
48. J. Schneider, Thin glasses – a future envelope? in *The Future Envelope 9* (Delft, 2015)
49. S. Schula, Charakterisierung der Kratzanfälligkeit von Gläsern im Bauwesen. Dissertation. Technische Universität Darmstadt (2015)
50. M. Bergers, K. Natividad, S.M. Morse, H.S. Norville, Full scale tests of heat strengthened glass with ceramic frit. Glas. Struct. Eng. **1**(1), 261–276 (2016)
51. I. Maniatis, M. Elstner, Investigations on the mechanical strength of enamelled glass. Glas. Struct. Eng. **1**(1), 277–288 (2016)
52. F. Oikonomopoulou, E.A. van den Broek, T. Bristogianni, F.A. Veer, R. Nijsse, Design and experimental testing of the bundled glass column. Glas. Struct. Eng. **2**(2), 183–200 (2017)
53. C. Louter et al., Adaptive and composite thin glass concepts for architectural applications. HERON **63**(1) (2018)
54. F. Oikonomopoulou, T. Bristogianni, F.A. Veer, R. Nijsse, The construction of the Crystal Houses façade: challenges and innovations. Glas. Struct. Eng. **3**(1), 87–108 (2018)
55. A. Snijder, R. Nijsse, C. Louter, The glass truss bridge. Heron **63**(1-2), 139–157 (2018)
56. T. Bristogianni, F. Oikonomopoulou, C.J. de Limaa, F. Veer, R. Nijsse, Structural cast glass components manufactured from waste glass: diverting everyday discarded glass from the landfill to the building industry. Heron **63**(1) (2018)
57. T. Weimar, S.A. López, Development of thin glass-polycarbonate composite panels. ce/papers **2**(5-6), 1–17 (2018)
58. J. Belis, C. Louter, J.H. Nielsen, J. Schneider, Architectural glass, in *Springer Handbook of Glass*, ed. by J.D. Musgraves, J. Hu, L. Calvez. Springer Handbooks (Springer, Cham, 2019), p. 1841
59. P. Bukieda, M. Engelmann, B. Weller, M. Elstner, M. Broich, T. Teichmann, J. Kohn, R. Gruber, J. Wittwer, Vierpunkt-Biegeversuch für thermisch gebogenes Glas - Part I. Bautechnik **96**(5), 380–389 (2019)
60. M. Engelmann, P. Bukieda, B. Weller, M. Elstner, M. Broich, T. Teichmann, J. Kohn, R. Gruber, J. Wittwer, Vierpunkt-Biegeversuch für thermisch gebogenes Glas - Part II. Bautechnik **96**(5), 390–399 (2019)
61. K. Lohr, Thermisch vorgespanntes Glas mit nachgeschliffenen Kanten. PhD thesis. TU Dresden (2019)
62. K. Lohr, B. Weller, Residual stress distribution in tempered glass with reground edges. Glas. Struct. Eng. **4**(1), 99–115 (2019)

63. J.H. Nielsen, M.J. Meyland, B.E. Thorup, A. Zugravu, J.F. Olesen, Investigating the strength effects of drilling in tempered glass: a first step towards a novel dowel joint. Glas. Struct. Eng. 2009 (2019)
64. F. Oikonomopoulou, Unveiling the third dimension of glass. PhD thesis. Nov 2019
65. A.H. Snijder, L.P.L. van der Linden, C. Goulas, C. Louter, R. Nijsse, The glass swing: a vector active structure made of glass struts and 3D-printed steel nodes. Glas. Struct. Eng. **5**(1), 99–116 (2019)
66. T. Bristogianni, F. Oikonomopoulou, R. Yu, F.A. Veer, R. Nijsse, Investigating the flexural strength of recycled cast glass. Glas. Struct. Eng. **5**(3), 445–487 (2020)
67. P. Bukieda, K. Lohr, J. Meiberg, B. Weller, Study on the optical quality and strength of glass edges after the grinding and polishing process. Glas. Struct. Eng. 1–18 (2020)
68. J. Hänig, B. Weller, Load-bearing behaviour of innovative lightweight glass–plastic-composite panels. Glas. Struct. Eng. **5**(1), 83–97 (2020)
69. F. Oikonomopoulou, I.S. Bhatia, W. Damen, F.V.D. Weijst, T. Bristogianni, Rethinking the cast glass mould, in *Challenging Glass Conference Proceedings*, vol. 7. 1. Challenging Glass Conference, Sept 2020
70. J. Hänig, B. Weller, Experimental investigations and numerical simulations of innovative lightweight glass-plastic-composite panels made of thin glass and PMMA. Glas. Struct. Eng. **6**(2), 249–271 (2021)
71. K. Sanders, F. Bos, E. ten Brincke, J. Belis, Edge strength of core drilled and waterjet cut holes in architectural glass. Glas. Struct. Eng. **6**(2), 131–145 (2021)
72. R. Giesecke, B. Dillenburger, Three-dimensionally (3D) printed sand molds for custom glass parts. Glas. Struct. Eng. **7**(2), 231–251 (2022)
73. S. Nategh, M. Zaccaria, J. Missinne, J. Belis, Experimental strength characterisation of thin chemically pre-stressed glass based on laser-induced flaws. Glas. Struct. Eng. **7**(3), 471–486 (2022)
74. M. Zaccaria, T. Peters, J. Ebert, N. Lucca, J. Schneider, C. Louter, The clamp bender: a new testing equipment for thin glass. Glas. Struct. Eng. **7**(2), 173–186 (2022)
75. D. Pfarr, C. Louter, Prototyping of digitally manufactured thin glass composite façade panels. Arch. Struct. Constr. (2023)

Chapter 3
Multilayer Glass

Glass is often assembled into multilayer products for architectural applications. This chapter describes these products, which essentially consist of a combination of the monolithic glass products that were previously discussed in Chap. 2. *Compared to monolithic (primary) glass products, multilayer glass products provide additional functionality in terms of post-fracture safety, insulation performance or curvature, as discussed in Sects.* 3.1, 3.2 *and* 3.3, *respectively. Whereas fire, burglary, bullet and blast resistance functionalities of mutilayer glass products are discussed in Sects.* 3.4 *and* 3.5. *Finally, multilayer glazing products that relate to energy harvesting are described in Sects.* 3.6 *and* 3.7.

3.1 Laminated Glass

This section deals with the materials and processes used in the production of laminated glass products. The engineering design of laminated glass is described in Sect. 9.7.

3.1.1 Definition

Laminated glass is defined in EN ISO 12543-1 [1] as an assembly consisting of one ply of glass with one or more plies of glass and/or plastic sheet material joined together with one or more interlayers. Laminated glass generally consists of two layers of glass as illustrated in Fig. 3.1 but multi-ply laminated glass with 3 or more glass plies, is also possible. Durability is an essential requirement for laminated glass and is assessed through tests, such as those specified in EN ISO 12543-3 [64], involving exposure to high temperatures (exposure to 100 °C for 16 h or for two

J. Belis et al., *Engineering Glass Structures*,
https://doi.org/10.1007/978-3-032-03233-1_3

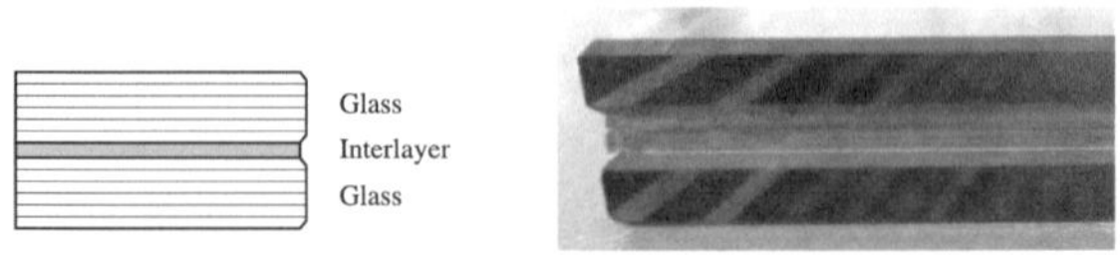

Fig. 3.1 Layout of laminated glass

hours in boiling water), humidity (two-week exposure at 50 °C and 100%RH) and radiation (2 000-hour exposure to artificial solar radiation at 45 °C). The detailed test methods are defined in EN ISO 12543-4 [65]. In practice, particular measures are taken to protect the interlayer from the environment at the edges of a laminated glass pane, thereby improving its durability.

One for the most important features of laminated safety glass is that it provides a better performance than monolithic glass after fracture by reducing the risk of injury from glass fragments[1] and providing a degree of residual load-bearing capacity. Laminated safety glass with an interlayer of Polyvinyl Butyral (PVB), ionomer or Ethylene Vinyl Acetate (EVA) have received offical approval as products for use in the building industry in Europe. Other interlayers are possible, but they may require testing and classification as laminated safety glass before use.

A detailed description of the interlayer materials and their properties is given in Sect. 3.1.3. A product standard for the evaluation of conformity of laminated safety glass is available as EN 14449 [2].

In most countries, laminated safety glass is mandatory for applications that require residual load-bearing capacity or the residual stiffness, for instance, to prevent people from falling through glass floors in the event of glass fracture. Laminated glass is also increasingly being used in façades. Four-edge supported vertical glazing provides a good residual capacity even when thermally pre-stressed glass plies are used, because in contrast to horizontal glazing, the interlayer is usually able to hold the small fragments together after fracture.

3.1.2 Production

This section describes the manufacturing process of laminated glass with a PVB interlayer. Differences arising from the use of alternative interlayer materials are subsequently explained.

The constituent layering materials are the float glass plies and the transparent or coloured PVB films of thickness 0.38 mm or multiple films thereof (up to 2.28 mm). After carefully stacking the individual layers (glass and PVB) in a clean room, but before the main laminating process, the laminate is typically prepared by pre-pressing (pre-nipping) and de-airing. For instance, for the majority of (simple) glass laminates, a nip-roll process is used. This process consists of a brief exposure to heating elements

[1] While the term "safety glass" on its own, often refers to thermally pre-stressed glass, "laminated safety glass" is not necessarily composed of thermally pre-stressed glass.

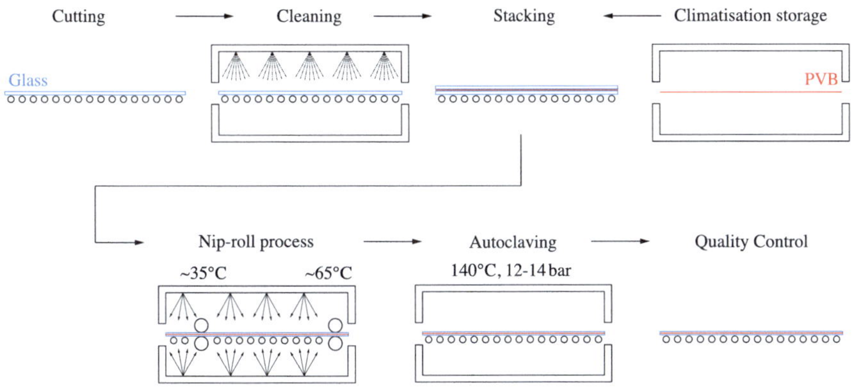

Fig. 3.2 Typical manufacturing cycle of laminated glass

and pressure rollers to de-air and pre-bond the laminate. After the nip-roll process, the interlayer may appear milky translucent as numerous small air inclusions are still entrapped in the laminate. Alternatively, for more demanding and/or high-value laminates, which require a high success level of lamination, de-airing is achieved by means of a vacuum either by placing the laminate in a vacuum bag, or by providing vacuum rings around the perimeter of the laminate. The vacuum is also applied and maintained during the subsequent principal laminating process. The vacuum process is more costly, but it generally provides a better quality of laminated glass, where haze, air inclusions or (local) delaminations occur less frequently.

The subsequent principal laminating process creates a permanent bond between the glass and the interlayer film. This can be achieved in an autoclave at a pressure of about 12 to 14 bar and a temperature of about 125 to 140 °C, see Fig. 3.2. The PVB binds the entrapped air molecules and the resulting laminated glass is ideally completely transparent. Depending on the total thickness of the laminate and the mass and shape of the glass, this process may take 30 min to 12 h.

An alternative to the autoclave process, is the use of a vertical or horizontal hotbox (also known as vacuum oven). Lamination in a hotbox always requires vacuum, typically provided using either vacuum rings or (silicone or nylon) vacuum bags. Vacuum bags should be used with caution, in particular to ensure that pressure they exert on the glass edges, does not cause excessive elastic bending of the glass plies which can remain permanently bent or draw in ambient air after the vacuum is removed. This can be prevented e.g. by providing an incompressible spacer around the perimeter of the glass inside the silicone vacuum bag, so the glass edges are not inadvertently bent during the vacuum bagging process.

A similar procedure is followed when laminating with ionomer interlayers (SentryGlas®), but the control parameters (i.e. temperature and pressure) differ from those used with PVB. The glass plies should preferably be positioned with the tin side facing towards the foil to achieve optimum adhesion through ion bonding, see Sect. 3.1.3.3.

The production of EVA glass laminates does not require autoclaving; they are usually heated in a two-stage vacuum process in a furnace, wherein cross-linking of the copolymers occurs in the second stage, see Sect. 3.1.3.4. However, cross-linking in an autoclave process is also possible.

Cast in place resin (CIP) (also refereed to as cold-poured resin) does not come as film material but as a liquid which is poured into the cavity between the glass plies. CIP is available as a single component material which cures when exposed to UV light, or as a self-catalysing multi-component material. Single-component materials are more easily processed, as they do not require dosing and mixing equipment.

Laminated glass can be produced in a curved form by hot bending. In this process, the monolithic glass plies are stacked on top of each other over a curved mould, separated by a temporary silicone film spacer and subjected to gravity bending as one package in a furnace, see Sect. 2.3.9. This results in a smooth curvature fit of all the plies, which can subsequently be laminated in the normal way by replacing the temporary film with desired interlayer. However, this 2-stage process of sag bending followed by lamination is only suitable for annealed float glass. For hot-bent laminates with thermally toughened glass, the bending and pre-stressing process of the glass plies must be carried out for each ply individually, which introduces discrepancies in the curvatures when the glass is subsequently stacked to form a laminate. The imperfections are often compensated by using thick interlayers which accommodate differences in curvature. Nevertheless, the resulting product is more likely to have inferior optical quality and is more prone to delamination.

3.1.3 Interlayers for Laminated Glass

3.1.3.1 General Information

Polymer interlayers are typically transparent, but can also be dyed or printed with patterns. This section describes the commonly used types and corresponding properties of interlayers.

3.1.3.2 Polyvinyl Butyral (PVB)

Polyvinyl butyral interlayers are the most commonly used interlayer for laminated glass in the construction industry.

The polymer film consists of three main components: PVB resin, plasticizer and additives. The PVB resin is a compound of vinyl alcohol, vinyl acetate and butyral, see Fig. 3.3. In addition, small quantities of additives (alkali and alkaline salts) are added during production of the PVB to control adhesion while while maintaining the mechanical properties.

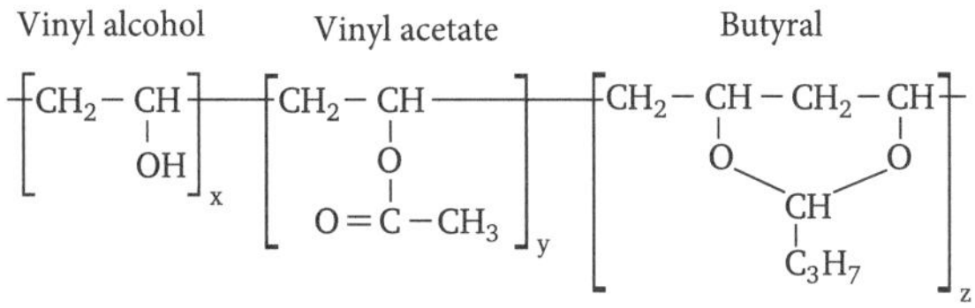

Fig. 3.3 Chemical composition of PVB resin [3]

Properties such as sheet adhesion, colouring, UV transmission, can be controlled by using additional additives. However, the additives have no significant influence on the mechanical stiffness of the sheets.

PVB resin, plasticizer and additives are made into a film by extrusion. During the process, particular attention is paid to the surface roughness, the film moisture and the avoidance of residual stresses. The film is prevented from sticking on the roll, by keeping it below ambient temperature or by wrapping it with polyethylene separating sheets. Wrapping also prevents moisture from reaching the sheet surface [3].

The adhesion of PVB sheets to glass is primarily based on the formation of hydrogen bonds (physical bonding). The influencing factors, in addition to the type of sheet used, are the quality of the washing water and the surface of the glass (air side or tin side, printing). In addition, sheet moisture has a significant influence on adhesion. Standard products for the construction industry should be laminated with a film having a moisture content between 0.4 and 0.5% to achieve Pummel values of at least 3 and thus sufficient adhesive strength, see Sect. 3.1.4. If a film becomes moistened it can be dried again through appropriate storage.

PVB exists in a variety of different grades. Because normal PVB films are almost impermeable to UV radiation, they are sometimes referred to as UV protection films. Special PVB films for sound insulation represent alternatives to cast resin composites. Stiff, plasticizer-reduced PVB films are used where increased demands on shear transfer or residual load capacity is needed.

The polyvinyl, as well as the PVB film, are classified as amorphous thermoplastics. The glass transition temperature of a film is approximately at ambient room temperature, whereby the mechanical behaviour is entropy-elastic at low strain rates. As a result, the material behaves almost perfectly elastically after release, but time-delayed (viscoelastic), which is commonly known as linear-viscoelastic behaviour. At large deformations, the stress and strain relationship becomes non-linear, which can be described as being hyper-elastic, see e.g. Sect. 4.1.3, as long as the deformation rate is constant or negligible.

3.1.3.3 Ionomer

An ionomer is a specific classification of polymers based on the ionised linking of macromolecules. Where in addition to the usual secondary valence bonds, there are also ion bonds between the macromolecular strands [4]. A well known example is SentryGlas® produced by Kuraray (formerly DuPont de Nemours), which was originally developed in the 1990s as an interlayer for hurricane-protection glazing. This type of interlayer is also known as "Ionoplast", but the term "Ionomer" is preferred because it refers to the actual molecular structure.

The ion bonds produce a significantly higher stiffness at room temperature, therefore Ionomer interlayers are especially used in glass applications where a high shear transfer is required between the glass panes or a to achieve a high level of out-of-plane post-fracture stiffness. In addition, the product often has good adhesion to metals, such as stainless steel or titanium, and thus enables the reinforcement of glass composites or the attaching of metal inserts, commonly used at joints [5–7].

SentryGlas® is a modified polyethylene or an ethylene copolymer, which is classified under semi-crystalline thermoplastics.

Apart from a UV absorber, no additives, such as plasticizers or adhesion agents, are added [8]. The manufacturer asserts that it also outperforms PVB in durability and edge stability.

The adhesion to glass is due to the hydrogen bridge bonding via the polar OH-groups analogous to PVB. In addition, an ionic bond is formed, particularly with the ions on the tin side of the glass.

Due to its high stiffness, the material is produced and supplied as sheet material in a thickness of 0.89, 1.52 and 2.28 mm. A thickness of 0.89 mm is also available on rolls. In contrast to PVB, it is very difficult to desiccate moist SentryGlas® film.

3.1.3.4 Ethylene Vinyl Acetate Copolymer (EVA)

Polymer films made of acetate (EVA or EVAC) are popular in solar panel applications to encapsulate the solar cells in glass-glass or glass-film modules (current market share above 95%). Broader applications in the construction industry have increased in recent years, yet they still represent a niche product (market share of less than 5%) for special applications compared to PVB films. PVB films are typically more costly than solar-grade EVA films, but cheaper than construction-grade EVA films. The special feature of the EVA films is the three-dimensional crosslinking of the molecular chains during the lamination process. As a result, EVA films have high resistance to heat and ageing and a certain minimum stiffness, even at very high temperatures and long loading time.

EVA is a copolymer produced through chain polymerization consisting of ethylene and Vinyl Acetate (VAC) monomers. The properties of the polymer are largely dependent on the percentage composition of the two monomers (VAC content). The VAC is used to disturb the chain structure of the polyethylene. Increasing VAC content reduces the crystallinity, hardness, stiffness, glass transition and melting temperature, but increases the transparency, density, elongation at break, adhesion, weather stability, cross-link ability and gas permeability. The tensile strength reaches a maximum at about 25 wt. − % VAC. Consequnetly, polymers made of EVA are classified according to their VAC content as follows and assigned to application areas. Below 30 wt.% to 40 wt.%, they are in the energy-elastic range at room temperature, above that they are in the entropy elastic range [4]. Films for the solar cell market have a VAC content of about 33 wt. – % [9], while in films for construction applications, the VAC content is about 25 wt. – %. Various additives are used in conjunction with the EVA base component resin to achieve the desired properties in the final EVA film including: UV absorbers, UV stabilizers, antioxidants, cross-linkers and adhesion agents [10]. Up to this point in the process, no crosslinking of the macro-molecules has taken place, which means that the polymer is still meltable. In fact, crosslinking in EVA films takes place only during the lamination process.

Adhesion to the glass is achieved by means of Silanes (silicon-organic compounds) in the EVA that enable a covalent bond to the glass.

Due to the sparsely dispersed network after lamination, the polymer can be categorized as an elastomer. However, there are also EVA films without cross-linking additives (typically peroxides with organic residues), which can accordingly be classified as a thermoplastic.

3.1.3.5 Thermoplastic Polyurethane (TPU)

Films made of thermoplastic polyurethane (TPU or TPE-U) are used in particular when the laminate includes one or more polycarbonate plies, typically used in protective glazing applications, see Sect. 3.5. The high flexibility of the TPU films is such that it accommodates the differential thermal expansions of glass and polycarbonate. In addition, the film has good adhesion with both glass and polycarbonate, and lamination is also possible at relatively low temperatures.

This is a useful, but relatively niche application, which explains the small share of TPU in the laminated glass market. The distinguishing feature of TPU films is the physical cross-linking of the molecular chains in the thermoplastic elastomer material, see Sect. 1.5.

A TPU is a linear macromolecule consisting of hard and soft segments. At ambient temperature, the hard segments are in the energy-elastic range while the soft segments are in the entropy-elastic range. The hard segments of different macromolecules form crystalline areas via hydrogen bridge bonding, and thus physical cross-linking points [11], while the soft segments represent a chemically bound plasticiser [4]. In addition to physical reticulation, reversible chemical cross-linking points can also arise [4].

Intermediate thermoplastic elastomer layers of TPU can be re-melted at any time. However, at ambient temperature, their mechanical response is similar to an elastomer in the entropy-elastic area. The mechanical behaviour can be controlled by the composition of hard and soft segments.

In addition to TPU resin, additives can be added to the film to control the processing and the properties of the final product. These are, for example, adhesion agents, UV stabilisers and additional plasticisers [12].

3.1.3.6 Cast in Place (CIP) Resin

Cast in place resins (also known as cold poured resins) are a special form of interlayers in laminated glass, as they are produced in liquid form rather than as a film material, and are therefore poured between the glass plies to form a laminate. CIPs can either be of the single or multicomponent type. While single component resins are cross-linked by UV radiation, multicomponent resins are mixed shortly before they are poured into the glass. The linking then takes place independently.

The resins are typically acrylate or methacrylate polymers, which are classified as thermosets and include other additives such as fillers, adhesion promoters, photoinitiators for one-component casting resins, plasticisers and stabilisers [13].

CIPs outperform PVB films in sound insulation applications and in gap filling (e.g. to compensate for high tolerances between plies). Due to cross-linking, casting is also suitable for cold-bent laminated glass, see Sect. 2.3.9. In particular cases, CIP can also be used for embedded connections in glass [14, 15].

3.1.4 Laminated Glass Test Methods

Several tests are carried out on laminated glass. These include the tests to determine the resistance of laminated glass according to EN ISO 12543-3 [64] and the determination of interlayer humidity (e.g. by means of infrared radiation methods) as well as other tests to ascertain compliance with specific performance requirements (e.g. post fracture performance) or for quality control need. The test specimens are typically made up of laminates with two glass plies and an interlayer. With the exception of the pendulum impact test and the ball-drop test, several of these test methods are not embedded in international standards. This means that alternative methods may be used for ascertaining a particular performance.

Pummel Test

In this adhesion test the laminate, made up of two float glass plies, each with a maximum thickness of 4 mm, is placed onto a tilted metal table and subsequently systematically hammered. The glass is thereby damaged and then visually assessed. The level of adhesion is classified according to Pummel values from 0 to 10 for each surface of exposed film or of adhering glass splinters, see Fig. 3.4.

When using standard films made from PVB, the specimens are cooled to −18 °C before testing to ensure a consistent assessment. Both glass plies are tested to detect the possible influence of the tin side or air side on the adhesion quality. For PVB laminated glass, the tin side usually provides slightly lower Pummel values than the air side. The sample dimensions are usually about 80 mm × 200 mm [3]. In contrast, ionomer specimens are tested at ambient temperature. This test is seldom used on EVA-laminated glass because high Pummel values are always reached, making the test rather meaningless. Pummel tests are not standardised and their execution and

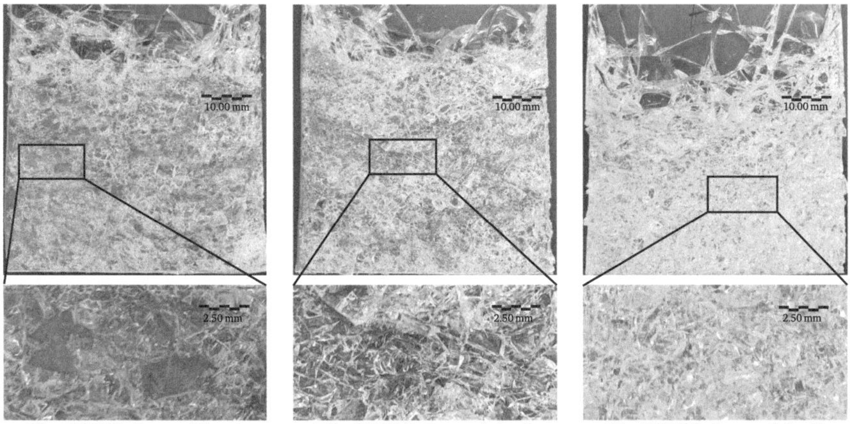

Film BG R10: Pummel 3 to 4 | Film BG R15: Pummel 4 to 6 | Film BG R20: Pummel 8 to 9

Fig. 3.4 Different levels of adhesion of PVB and corresponding values of Pummel test

evaluation is rather dependent on the operator, although recently some improvements have been proposed to reduce this variability [16].

Compressive Shear Test (CST)

In a CST, a square laminated glass specimen with dimensions 25 mm × 25 mm is tested at an angle of 45° so that the interlayer experiences both compression and shear. The test is carried out at ambient temperature. The engineering stress (applied force divided by the film surface) is used as a measure of adhesion [3].

Torsion Test

In this process, circular laminated glass specimens with a diameter of 22 mm are subjected to a turning shear load until the interlayer is detached from the glass, at 20 °C and 50% RH. The results correlate predominantly with those of the compression shear test [3].

Peel Test

Test specimens cannot be taken from normal production, because a separating film needs to be inserted before the lamination, so that the interlayer is bonded to only one glass ply. Often additionally a tear resistant film is laminated in to prevent high elongation of the interlayer during the test. The interlayer must extend beyond the glass so that it can be properly mounted into a test machine. The force (constant) needed to peel the interlayer from the glass is used as a measure of adhesion. In principle different loading angles can be tested, however, $\theta = 90\,^{\circ}$, is often used as this leaves no horizontal forces and eliminates the need for horizontal displacement control, see Fig. 3.5 and [17].

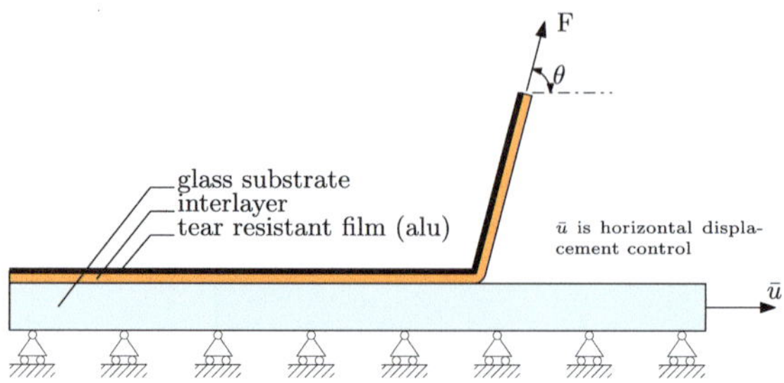

Fig. 3.5 Sketch of the peel test

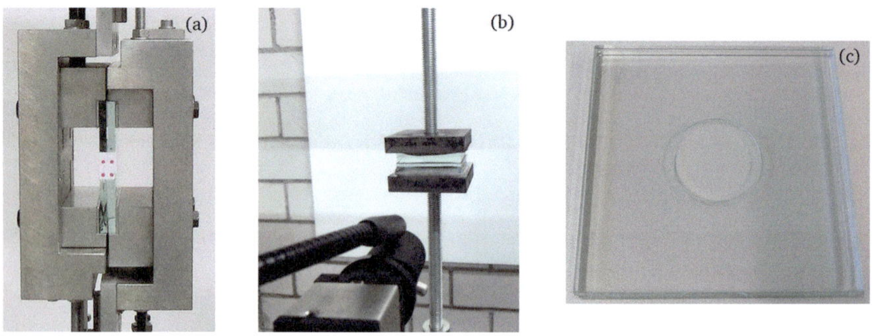

Fig. 3.6 **a** Shear test, **b** pull test and **c** test specimen with circular PVB film

Shear Test

In this test, laminated glass specimens are subjected to pure shear, see Fig. 3.6a. The engineering shear stress (shear force divided by the film surface) is often considered a measure of the adhesion, although this approach may underestimate local stress peaks. The use of circular samples provides a more homogeneous stress distribution. Such samples may be created using a circular piece of interlayer bonded to a larger rectangular glass, see e.g. Fig. 3.6c.

Pull Test

Here, laminated glass samples are perpendicularly loaded in tension, until a loss of adhesion of the film to the glass occurs, see Fig. 3.6b. The maximum tensile force divided by the film surface is used as a measure for adhesion in tension. To enable a proper load introduction, the test specimens are bonded onto steel pieces, which makes the experiment laborious. Regarding the geometry of the specimens, it is common practice to use rectangular cut-out laminated glass, although circular specimens can also be used to avoid edge effects.

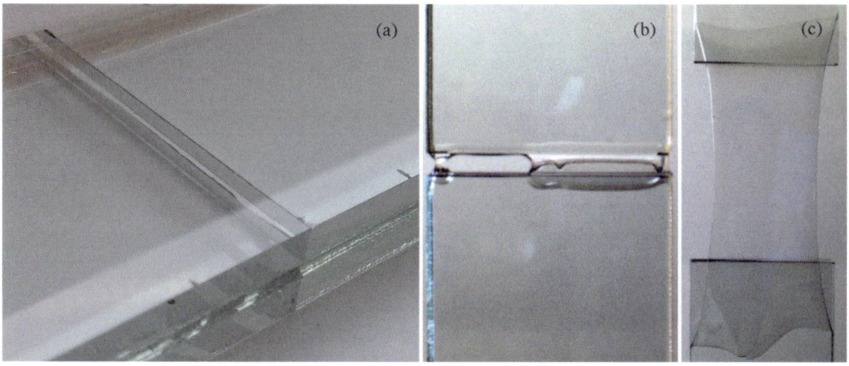

Fig. 3.7 TCT test: **a** Prepared test specimen before testing; **b** High-adhesion PVB laminated specimen during testing; **c** Low-adhesion PVB laminated specimen during testing

Through Cracked Tensile (TCT) Test

A crack is induced at mid length of both glass plies of a laminated glass and then subjected to an in-plane tensile test, see Fig. 3.7. Consequently, in the mid-length zone the tensile load is completely transferred through the interlayer. In addition to the qualitative and quantitative assessment of the local delamination behaviour, a determination of the energy release rate is hereby also possible [18–20].

Pendulum Impact Test

According to EN 12600 the method and classification of glazing is described in terms of its safety on impact. The impact resulting from the collision of a person with a glass panel is simulated by a double tyre impactor from various heights, where each height represents a different speed and energy of impact. According to EN ISO 12543-2 [63], a *laminated safety glass* must obtain a classification 3(B)3 according to EN 12600. This means that the glazing must either remain intact or must have no significant openings/holes after being impacted at the lowest height of the pendulum (190 mm). In contrast, a *laminated glass* does not have to meet any requirements regarding the pendulum impact. More details can be found in Sect. 9.5.2.

Ball Drop Test

EN 14449 [2] Annex C describes a ball drop test that is intended for laminated glass producers to check the quality of the laminate. In this test, a steel ball (diameter 63.5 mm, mass 1030 g) is dropped from a certain height onto the laminated glass sample (length and width: 500 ± 5mm), which is enclosed in a standard frame, see Fig. 3.8. The ball should fall within a circle of 100 mm diameter in the centre of the sample and should not penetrate the test sample within 5 s after impact. In the evaluation of a similar ball drop test that is described in DIN 52338 [52] a number of penetrated samples, fractured samples and the mass of the detached glass splinters are recorded. In general, a lower adhesion between interlayer and glass usually leads

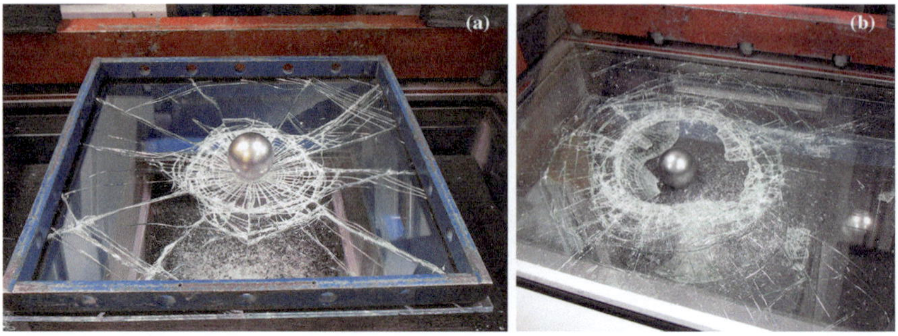

Fig. 3.8 Drop ball tests

to a better result in these ball drop tests, because a local delamination allows the interlayer to stretch more before tearing in the vicinity of the cracks in the glass. Ball drop tests that are related to burglary resistant glazing are described in Sect. 3.5.2.

3.2 Insulating Glass Units (IGU)

3.2.1 General Information

Insulating glass units (IGU) have a lower thermal conductivity and therefore provide a better thermal performance than traditional glazing. The development of early insulating glass units, consisting of two glass panes separated by a sealed gas (air) cavity, dates back to the American Thomas D. Stetson, who secured a patent in 1865 for this construction (US pat. No 49167). For a long time, however, it was not technologically possible to maintain a hermetically sealed glazing unit.

The first edge solutions were relatively rigid, as both glass panes, for example, were directly fused at their adjacent edges or soldered to metal spacers. These systems produce a good hermetic seal, but microcracks resulting from movements within the construction often led to failure. The more flexible edge system, which is still in use today, solves this problem and is based on an invention by Alfred Arnold in 1959. Indeed, contemporary insulating glass units consist of at least two glass panes, separated by a spacer filled with a desiccant to absorb entrapped residual moisture, see Fig. 3.9. The cavity between the glass panes is filled with either air or a noble gas. A two-pane unit (with one cavity) is referred to as a double glazing unit (DGU), whereas a three-pane unit (with two cavities) is referred to as a triple glazing unit (TGU).

As a rule, today, two barriers are used to seal the edges of IGUs and to obtain a hermetically sealed system, see Fig. 3.9 and Sect. 8.2.

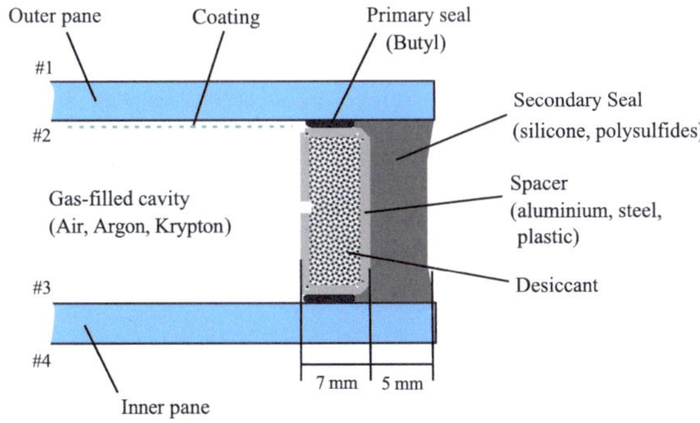

Fig. 3.9 Typical structure of double glazing unit with two-stage edge seal systems

Regardless of the type of gas in the cavity, the elastic two edge seals are not completely diffusion-proof. In fact, molecular migration takes place continuously through the sealing materials. This basically causes two problems.

Firstly, water molecules that penetrate the seal into the cavity can be absorbed by the desiccant until its absorption capacity is exhausted. Beyond that stage, water vapour can accumulate in the cavity leading to corrosion of the metallic functional coatings on the glass surfaces facing the cavity, and condensation may form on the cavity-facing surfaces of the glass. As a consequence, the pane can lose its transparency over time.

Secondly, when the cavity is filled with noble gas, this tends to leak from the cavity over time. This gradual gas loss results in a slight but continuous deterioration of the thermal insulation. In contrast to the above-mentioned corrosion and condensation problems, this deterioration is not visible with the naked eye. However, both phenomena are physically inevitable. Therefore, limit values are given in the product standard EN 1279-1 [53] for insulating glass units.

This standard regulates the requirements, in particular the durability, of insulating glass units.

The manufacturer of insulating glass units has to prove, by means of an initial type test (ITT), the ability to produce insulating glass according to the requirements defined by the product standards. The manufacturer must define the products and the components used, for example the spacer type and the components of the edge seal. However, according to common practice, individual components from other IGU producers or component manufacturers may be used as long as they are from ITT-proven IGUs.

The standard EN 1279-1 provides limit values for:

- moisture penetration through the edge seals;
- gas leakage rate and gas concentration for gas-filled cavities;
- unacceptable condensation (fogging) on the glass surfaces facing the cavity.

Defined test specimens are artificially aged in accelerated climatic tests and then assessed with respect to the above-mentioned properties. It should be noted that the requirements and limit values only apply for the defined test specimen geometry and the specified ageing cycles. However, in reality, the effects on the insulating glass unit and thus also its response may deviate from these standardised conditions. Nonetheless, contemporary IGUs have a service life of up to 25 years [21].

3.2.2 Energy Transfer Characteristics

The thermal transmittance (also known as U-value)

$$U = \frac{Q}{A \cdot \Delta T \cdot t}, \quad [U] = \frac{\mathrm{W}}{\mathrm{m}^2 \cdot \mathrm{K}} \tag{3.1}$$

is defined as the heat quantity, Q, transmitted through a unit area, A, of an object within a unit time, t, driven by a temperature difference, ΔT, between the opposite faces of the object. The U-value is thus an inverse measure of the thermal insulation of a component. The lower the U-value, the better the thermal insulation. The reciprocal of the U-value is the R-value, which is known as thermal resistance.

In the case of window systems it is essential to distinguish between the different U-values. The overall U-value of the complete window (U_w, *window*) is made up of the glazing (U_g, *glass*) and the frame (U_f, *frame*), where the latter can significantly contribute to the window's overall thermal performance. The calculation of the heat transfer coefficient is described in EN 673 [61] (U_g) and EN ISO 10077-1 [59] (U_w). An experimental determination of the U-value is described in EN 674 [62] (U_g) and EN ISO 12567-1 [60] (U_w).

In the case of a conventional double glazing unit the heat transfer depends on the internal heat transfer h_i, the external heat transfer h_e and the heat transfer in the cavity h_t and can be calculated according to Eq. 3.2:

$$\frac{1}{U_\mathrm{g}} = \frac{1}{h_\mathrm{i}} + \frac{1}{h_\mathrm{t}} + \frac{1}{h_\mathrm{e}}. \tag{3.2}$$

Thermal insulation is of major importance in many regions of the world, even in temperate climatic zones. Further improvements to the thermal performance of the IGUs are nowadays achieved by low-E coatings (see Sect. 2.3.5) and corresponding gas fillings, such that insulating glass units today achieve thermal transmittances of $U_\mathrm{g} = 1.0\ \mathrm{W/(m^2K)}$ for double glazing units and $U_\mathrm{g} = 0.5\ \mathrm{W/(m^2K)}$ for triple glazing units. The optimal width of the cavity is governed by the respective gas used (air, Argon or Krypton), more so than the glass thickness which has an insignificant influence on the U_g-value. In the heating season, the heat losses through an IGU are governed by the radiation rather than the conduction and convection, see Fig. 3.10. The main improvement of thermal insulation is therefore achieved by coatings that reflect the heat (long-wave IR), whereas the contribution of the cavity filling gas to the U_g-value is relatively modest. This is also why a gas leakage does not directly lead to a very significant loss in thermal perfromance of the insulating glass unit.

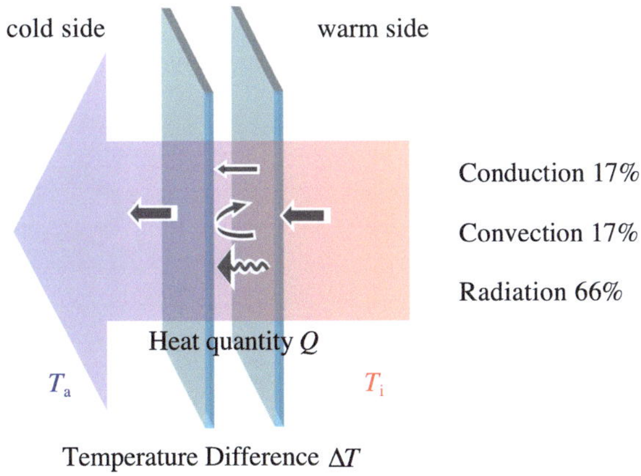

Fig. 3.10 Heat transfer in insulating glass

Table 3.1 Comparison of U_g-values

Glass	Heat transfer coefficient, U_g W/(m^{2}K)
Single glazing, 6 mm	5.7
Double-glazed insulating glass, cavity 16 mm	2.7
Double-glazed insulating glass, with air and low-E coating ($\varepsilon = 0.03$), cavity 16 mm	1.4
Double-glazed insulating glass, with Argon and low-E coating ($\varepsilon = 0.03$), cavity 16 mm	1.1
Double-glazed insulating glass, with Krypton and low-E coating ($\varepsilon = 0.03$), cavity 16 mm	1.0
Triple-glazed insulating glass, with 2 × Krypton and 2 × low-E coating ($\varepsilon = 0.03$), cavity 2 × 12 mm	0.5

U_g-values for some typical IGU compositions are listed in Table 3.1. However, these are not only dependent on the coating, the filling gas and the cavity width, but also on the inclintion angle of the installed IGU. According to EN 673 or EN 674, U_g-values can be determined for different inclinations. However, the manufacturers typically only specify the value for vertical glazing. In this position, the warm gas in the cavity will rise, thereby causing a relatively large, slow flow with relatively little gas movement. However, if the glazing is tilted towards the horizontal, convection in the cavity will increase. Indeed, the ascending warm gas reaches the cold side significantly faster and several small convection vortices are formed, see Fig. 3.11. The proportion of convection in the heat flux $\dot{Q}$ increases and the U_g-value increases accordingly. This leads to an increase in the U_g-value of up to 50% for roof glazing. If this is not taken into account, unexpected heat losses may occur. This increase is less pronounced in a triple glazing units with two separated cavities since the difference in temperature between the inner and middle panes or between the middle and outer

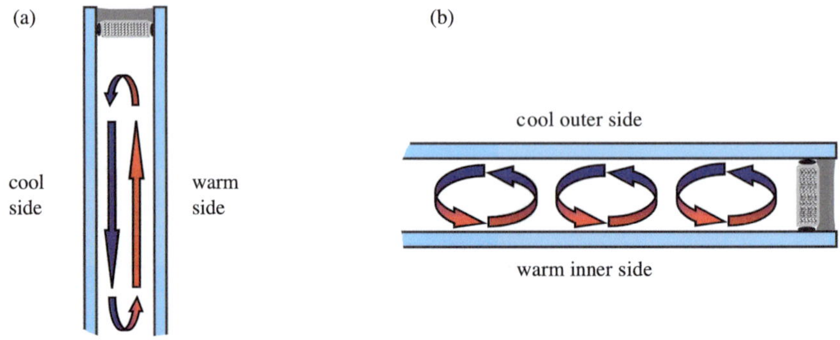

Fig. 3.11 Convection in the cavity in case of **a** vertical and **b** horizontal mounting

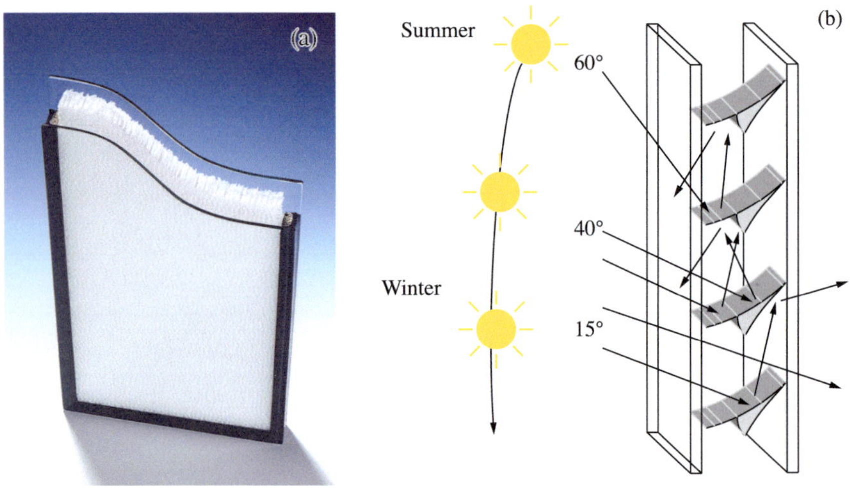

Fig. 3.12 **a** Translucent thermal insulation in the cavity of an IGU; **b** operation of an optically controlled sun protection with light lamellas in the cavity

panes is correspondingly lower. However, convection can also be prevented by means of a suitable filling of the cavities e.g. with a translucent thermal insulation such as aerogel, see Fig. 3.12a. In this case the U_g-value is independent of the inclination of the IGU.

For large glazed areas, in most climatic zones, the protection against solar heat gain has a larger impact on energy demand in buildings than preventing heat loss in winter.

Energy Considerations and Daylight Usage

The intensity of solar radiation at the perimeter of our atmosphere is 1367 W/m^2 (solar constant). However, on its way to the earth's surface, a part of the radiation intensity is lost through reflection and absorption in the atmosphere. Depending on the latitude and in the absence of cloud cover, up to 800 to 1 000 W/m^2 solar radiation arrives at the Earth's surface. Depending on the position of the sun and the orientation of the façade or roof, this solar energy can very quickly lead to overheating of buildings. The ratio of the solar energy transmitted (through a window) is known as the solar gain and denoted as the g-value. Through a glazed roof with a g-value of 0.5 and exposed to perpendicular irradiation, 400 to 500 W/m^2 of solar radiation may still enter the interior, which converts into heat when hitting a surface in the indoor environment.

On the other hand, in the heating season, at a temperature gradient of 30 K between indoor and outdoor temperature and for a glazing with $U_g = 1.2\,W/(m^2K)$, only $30\,K \cdot 1.2W/(m^2K) = 36\,W/m^2$ is lost.

An additional factor to consider is that the necessary energy consumption for cooling an indoor environment by a set temperature with air conditioning units is about twice as that required energy for heating that environment by the same temperature by means of conventional combustion technology. Consequently, in most cases low g-values are required for larger glass units. However, a low g-value has its disadvantages. Solar energy gains are reduced all year round, including in the heating season when they would be beneficial and less natural daylight is transmitted into the building leading to increased energy demand from artificial lighting.

The efficiency of natural sunlight becomes apparent when comparing to artificial light sources. Humans can perceive 53% of the total solar radiation with the naked eye as light, see Fig. 1.7. By using spectral-selective coatings, this efficiency can theoretically be increased to 100%, as all the components of solar radiation that cannot be perceived with the human eye are filtered out. An incandescent light bulb, on the other hand, converts only about 3% of the electrical energy into light which is why these are now banned in many countries. The rest of the energy is delivered directly as heat to the room. With fluorescent tubes, the efficiency of 17 to 18% is still slightly higher and sodium vapour lamps can even reach up to 30%. LED (Light Emitting Diode) lighting achieves a level of efficiency similar to natural daylight. However, if the transmission losses are taken into account, the real efficiency is significantly lower. In addition, the ratio of usable light to radiated energy and the associated heat development is better than for most artificial light sources. Overall, the use of the freely available daylight is therefore preferred from a lighting energy point of view.

Basically, the energy transfer through glazing is governed by the U-value, g-value and light transmission. These values should therefore be specified on project by project basis and the optimal values should be selected for each building depending on the particular boundary conditions. While most attempts are made, especially with large glass surfaces, to minimise the U-value and g-value and at the same time to maximise light transmission, this does not always lead to the lowest energy demand (for heating, cooling and lighting) in buildings. In cold climates, a lower U-value and a higher g-value are often the more optimal solution, as solar gains are beneficial. In hot regions, the U-value plays a smaller role, since the energy transmitted due to solar radiation are many times greater than those that result, even with low thermal insulation, from conduction. Here it is essential to achieve a low g-value.

3.2.3 Cavity Between the Panes of an IGU

The hermetically sealed cavity of an insulating glass unit (IGU) is filled with either air or inert gases. The internal cavity pressure corresponds to the atmospheric conditions at the moment of production, or another pressure can be set, for example taking into account the targeted installation altitude. Desiccated air, Argon, Krypton or Xenon are commonly used as fill gases. The noble gases improve the thermal insulation of the insulating glass units, although the extra costs, especially for Krypton and Xenon, are significant. As these fill gases are heavier than air, they move slower and thereby reduce the heat exchange between inner and outer panes due to convection. In addition, they have infrared-absorbing properties so together with a low-E coating they further reduce the heat transfer. However, compared to a low-E coating the influence of the filling gas on the thermal insulation is relatively low, which is why a slow, continuous loss of gas does not lead to a significant reduction of the thermal performance of the insulating glass unit. The optimal cavity width depends on the gas used and also the thermal conductivity changes depending on the cavity width. Usually, the cavity is between 12 and 16 mm wide, although widths between 4 and 30 mm are also possible.

In the past, the gas Sulfurhexafluoride (SF_6) was also used to improve sound insulation. However, due to its environmentally harmful properties, the use of SF_6 has been banned in the European Union since 2008.

The cavity can also accommodate other components in a space-saving manner and protect them against environmental influences. For example, built-in window dividers are popular for subdividing the transparent surface. However, at the same time they can cause a certain level of thermal bridging.

Systems which provide light scattering or daylight control can be integrated into the cavity. Moreover, sensitive materials such as wood and metals can be durably preserved to keep their original appearance. Another advantage lies in the simple installation, which is usually not different from the installation of other insulating glass units. Also, additional structures, such as an external sun protection, can be avoided. As such, the function of an integrated system is relatively independent of wind and other external influences. Contamination also does not occur in the cavity; cleaning of the elements is the same as with any other glazing.

A widespread application is the integration of translucent thermal insulation in the cavity of an IGU, see Fig. 3.12a. As such, the sensitive translucent materials are protected from dirt and mechanical damage. Their function is to improve thermal insulation and uniform light scattering towards the interior. The materials used are, for example, capillary panels made of transparent UV-stable plastics, mats made of glass fibres or granular silica-aero gels. The latter not only prevent convection in the cavity, but also largely stop direct heat conduction due to their nano-porous structure. Today, translucent components with U_g-values less than 0.3 $W/(m^2K)$ can be achieved.

Another way to integrate additional functionality into the cavity is to use angle-selective systems, e.g. mirror lamellas. In addition to conventional external sun protection, lamella systems can also be integrated in the cavity of an insulating glass unit. However, since the space between the panes is limited, the shape and size of the lamella profiles must be adjusted accordingly. Today, different fixed or movable lamella systems are available for use in the cavity, which either are retro-reflective or specifically direct daylight into the interior. In this case, in contrast to daylight systems based on translucent insulation material, the light is directly reflected into the interior using highly reflective lamellas. However, their primary function is to reduce the amount of irradiated solar energy. In contrast to solar protective coatings, lamellas offer a variable g-value and a variable light transmission depending on the position of the sun, see Fig. 3.12b. In the heating season, at low solar angles, more radiation reaches the building, enabling the active use of solar energy. In contrast, in the cooling season, direct irradiation is prevented and the building is protected from overheating.

3.2.4 Edge Seal

The edge seal plays a key role in the durability of an IGU. Today, a two barrier edge seal is typically used, see Fig. 3.9. The primary seal consists of permanently soft butyl (Polyisobutylene (PIB), see Sect. 8.2.8) and serves to protect the cavity from penetrating moisture and to prevent the premature loss of filling gas. The materials used for the secondary seal have a significantly lower diffusion density than butyl. Silicones, Polysulfides or Polyurethanes are used for this purpose, see Sect. 8.2. They serve to permanently connect the individual glass panes of the IGU. Since both the ambient air pressure and, as a result of temperature changes, the internal pressure in the cavity can vary thereby exerting tensile or compressive loads on the edge seal, see Fig. 3.13. The butyl alone cannot absorb these loads, which would cause it to migrate into the cavity. The secondary sealing material therefore also has the task of relieving the butyl of these loads. This way, the secondary sealant makes an indirect contribution to the systems tightness.

Compared to polysulphides or polyurethanes, silicones are extremely open to diffusion. Openings in the primary seal result in a high gas loss and the penetration of water. However, in contrast to polysulphide or polyurethane, the inorganic silicone is UV-resistant, because its chemical bonds are very

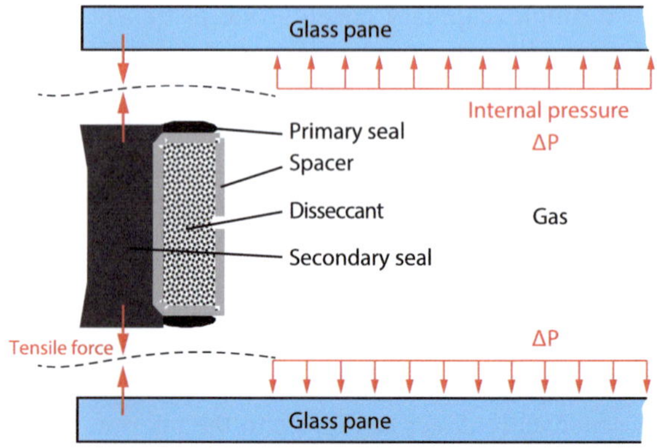

Fig. 3.13 Free-body diagram with a cut through the insulating glass edge seal

stable. In case the edge is exposed to direct sunlight, silicones have to be used as secondary seal, e.g. for structural glazing, see Sect. 8.5, ETAG 002 [22]. For several years now, manufacturers have also been offering what is known as gas tight silicones. These silicones, however, usually do not have a higher density, but a higher stiffness. Due to the higher stiffness, deformations in the area of the edge seal are reduced and the primary seal is exposed to smaller movements. This indirectly leads to improved durability.

The load-bearing behaviour of the edge seal has as yet not been mechanically sufficiently explained. The rheology of the sealant used is complex. The materials behave in part hyper-elastic and are extremely strain-rate and temperature dependent. Numerical solutions mostly fail due to the numerous unknown or inaccurate boundary conditions. Therefore, relatively simple engineering formulas are often used to design the required width (commonly known as bite) of the secondary seal [23]. This usually assumes simply supported boundary conditions of the glass, without accounting for the clamping effect and compliance of the edge seal. In particular for uniformly distributed surface loads, it is then possible to resort to an analytical solution of the plate equation (see Sect. 9.6). The necessary bite of the secondary edge composite can thus be estimated relatively easily.

In the case of common insulating glazing, the statically effective bite of the secondary seal is often only 3 mm to 5 mm, see Fig. 3.9. In general, it should be noted that, due to their larger internal climatic loads, small, thus stiff IGUs will require larger edge joint bites than large, thus more flexible IGUs, see Sect. 9.8.1. Also curved IGUs may behave very rigidly, depending on the bending radius. In addition, larger cavity widths or a higher absorption will lead to increased cavity pressure and thus to larger edge seal bites.

If the cavity is used for the integration of additional functions, it is usually necessary to reinforce the edge seal. The additional weight of the insert must be added and all the fixtures have an increased absorption compared to the gas, thus leading to higher temperatures in the cavity and larger tensile stresses in the edge seal due to the climatic loads, see Sect. 9.8.1. Therefore, the bite of the edge seal can be up to 30 mm in specific cases.

This should be considered at an early stage, since an edge seal that is not UV-stable must be completely covered with framing profiles. In polysulphide or polyurethane, UV radiation very quickly leads to a loss of mechanical properties and consequently to a malfunction of the IGU. When installing the IGU into a window frame or in a façade system, the usual glazing guidelines or the installation regulations of the glass manufacturer must be followed. To ensure a permanent function of the insulating glass unit, the rebate must be sufficiently ventilated and condensation must be drained, as permanent moisture may damage the secondary seal. A common cause of damage is also the loss of mechanical properties due to material incompatibilities. The materials of the profiles used, the setting blocks, the weather sealants etc. must be compatible with the edge seal components, see also Sect. 8.4. Otherwise it may lead to the migration of volatile constituents such as silicone emulsions or plasticizers. Depending on whether these constituents are absorbed or delivered, an undesirable softening or brittleness may occur, see Sect. 8.4.

3.2.5 Production

Insulating glass units may be manufactured in a horizontal or vertical process. In the horizontal procedure, the assembly is performed in a horizontal position and usually manually. The vertical process is the more common production line in which the glass is moved along an edge, slightly tilted towards the back wall, see Fig. 3.15. Today, production is largely automated. The insulating glass production line, which is up to 100 m long, consists of several stations where different processing steps are carried out, see Fig. 3.14. The arrangement can differ slightly from one manufacturer to another, but typically consists of the following steps.

(a) Insertion Station and Edge Coating Removal.

Production starts when the cut to size single pane is positioned on the lower conveyor belt with its uncoated side facing towards the rear panel, see Fig. 3.14a. From here, the glass is first transported to the edge coating removal station. In this first processing station, the functional coating, if present, is removed from the front glass surface in the area where later the secondary edge seal is applied. This is done by mechanical grinding along the edges of the glass. The aim is to remove the coating layer, which inhibits subsequent adhesion.

(b) Edge Machining.

A mechanical edge treatment is provided, see Fig. 3.14b, whereby an arris is applied to remove the sharp edges of the glass, see e.g. Sect. 2.3.2.

(c) Cleaning and Scan.

After the mechanical processing stations, a washing machine cleans and dries the glass, see Fig. 3.14c. Behind the washing machine a scanner detecting scratches, coating defects or remaining contaminations might be installed.

(d) Spacer Application.

At the next station, the spacer is positioned on the glass surface, see Fig. 3.14d. This is done either manually or in an automated manner. In case a classical hollow metal spacer is used and mounted manually, the production of the spacer is done separately from the insulating glass production line. The spacer can either be bent into the required geometry or composed of pieces cut at the required length and joined together at the corners. Generally, the spacer is filled with a desiccant in advance, but it is also possible to do the filling afterwards. The primary seal of butyl is then extruded onto the two flanks of the spacer profile. At this stage, the spacer, tailored to the glass size, enters the production and is manually set onto the glass. Here also a visual quality control of the glass and the evaluation of the optical scans takes place.

More recently, production lines place the spacer directly into the insulating glass production line using extrusion. These so-called thermoplastic spacers (TPS) consist of an extrudable seal with an integrated desiccant. The process was developed to increase productivity and reduce the impact of climatic loads on the durability of the edge seal.

(e) Press, Gas Filling and Secondary Seal.

The glass pane with the spacer is then transported further into a hydraulic press, where it is joined with the counterpane, see Fig. 3.14e. The counterpane is first sucked into the press and pulled a few centimetres forward by a vacuum to provide space for the incoming pane with the spacer. The latter moves behind the counterpane and is automatically positioned.

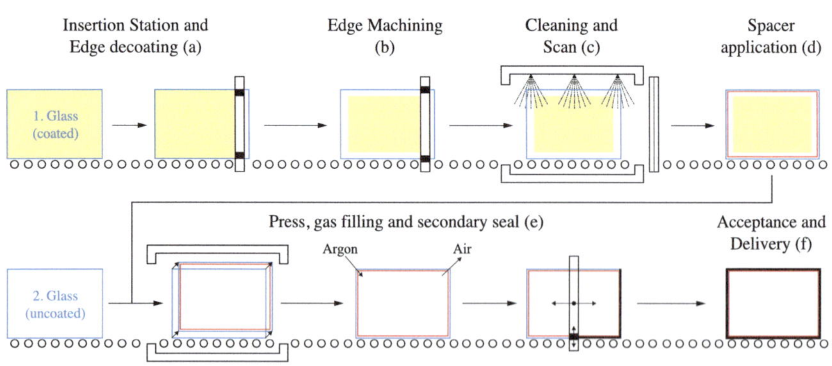

Fig. 3.14 Vertical production line for insulating glass units (insulating glass lines)

Fig. 3.15 Manufacturing of insulating glass units

Various methods can be used for gas-filling the insulating glazing. Modern IGU manufacturing lines contain a closed "gas filling press" chamber. After the glass panes have entered in upright position, the gas filling press is closed and subsequently filled with a carefully pre-defined volume of noble gas. In this condition the glass panes are then pressed together, entrapping the gas in the cavity. After pressing, the butyl must form a continuous line with a constant width. At this point, the glass panes are only connected to each other by the permanently viscous butyl on both sides of the spacer. The spacer is placed such that sufficient width for the secondary seal is left at the perimeter of the glass. The glass is somewhat cantilevering over the spacer to form a small, continuous channel along the external perimeter of the IGU. This channel is filled in the last station with the respective secondary sealant to permanently bond the individual panes and form the final IGU. The materials for the secondary sealant are usually two components that cure chemically.

Fig. 3.16 Filling of a cavity using a filling nozzle

An alternative method is to first close the cavity in ambient conditions, allowing it to be filled with air, and only fill it with gas subsequently. This is done using a filling nozzle, see Fig. 3.16, through which the gas enters through a small hole drilled in the spacer at the bottom of the upright IGU. At the same time a measuring device is used at the top of the upright IGU to check whether the cavity actually has been filled completely. After filling, the holes for the nozzle and the probe are hermetically sealed again.

These processes are repeated in the manufacturing of triple insulating glass units (TGU). It is important to mention that gas filling is a laminar flow process. This influences the formation of the nozzle geometry and the filling time.

(f) Acceptance and Delivery.

Finally the finished insulating glass unit can be packed for transport and delivery, see Fig. 3.14f.

3.2.6 Vacuum Insulating Glass

Intensive research has been carried out since the 1980s on IGUs with an evacuated cavity, and to date only a few systems are available on the market. The pressures in the cavity of VIGs is close to a vacuum, typically in the order of 10×10^{-6} bar. The high pressure difference from ambient atmospheric pressure outside the VIG leads to high deflections in the glass panes that are prevented by tiny *spacers* (also called pillars) that are arranged in a regular grid in the cavity between the glass panes. The spherical or cylindrical spacers are usually made out of glass, ceramics or metal. In order to maintain a permanent vacuum in the interior, extremely high demands are to be placed on the tightness of the edge seal. Polymeric adhesive bonds are unsuitable for this application; therefore often glass fusing or soldering techniques are used instead, see Fig. 3.17.

Thermal insulation for vacuum insulating glass (VIG), also known as Vacuum Glass Unit (VGU), is in the range of $U_g \approx 0.4\,W/(m^2K)$ to $0.7\,W/(m^2K)$ and this

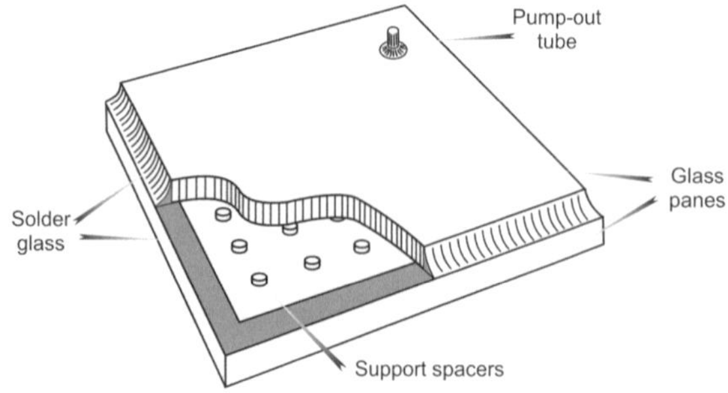

Fig. 3.17 Schematic depiction of vacuum insulating glass [24]. Also VIGs without pump-out tube are currently commercially available

for a very limited glazing thickness. The cavity width is usually ≤ 1.0 mm. The vacuum is obtained using a vacuum pump and a small pump-out tube that can be located in the corner of a glass or at the evacuation gap. The vacuum is beneficial to minimize the heat transfer as heat is transmitted only through radiation and the conduction through the small pillars and the edge seal. However, as the edge seal and the spacers (pillars) act as thermal bridges they need to be minimized. In addition, the pillars induce a complex contact stress state in the glass, which must be taken into account when designing the VIG [24–26].

The thermal performance of a vacuum insulating glass is comparable to that of triple insulating glass units, but a VIG is considerably thinner and lighter [27, 28]. Therefore, vacuum insulating glass could play an important role in the renovations of historic buildings as a high thermal performance can be achieved in a small overall thickness and with relatively thin (and lightweight) glass.

3.3 Curved Multilayer Glass

There are three methods for making curved glass: hot bending; cold bending and bending by lamination. Hot bending is performed on a monolithic glass panel, as described in Sect. 2.3.9, which may be subsequently laminated. Cold bending and bending by lamination typically refer to multilayer glass products (such as laminated glass and IGU) and are discussed in the following subsections.

3.3.1 *Cold Bending*

Cold bending of glass[2] is gaining popularity in architectural applications. It comprises the elastic bending of a glass panel at ambient temperature and subsequent fixing of the panel, usually along its perimeter so that it remains in the curved shape. Cold bending has several benefits compared to hot bending techniques. Cold bending does not require high costs for moulding or heating energy, and it typically results in fewer optical distortions than hot bent glass. Also, since cold bending can be done on site, transport of the glass can be done more efficiently in flat conditions as opposed to the impractical transport of curved hot bent glass panes. Furthermore, eventual replacements are less expensive compared to hot bent glass replacements. However, the degree of curvature of cold bent glass is more restricted than of hot bent glass, since cold bending curvatures must remain within the elastic bending limits of the glass. Therefore, its use is typically restricted to small curvatures of developable surfaces, such as cylindrical and conical surfaces.

Cold bent glass panels remain permanently stressed due to the imposed curvature. To avoid static fatigue effects, see Sect. 5.3, which emerge in the presence of permanent tensile stress, cold bending typically implies the use of thermally pre-stressed

[2] This Section on cold bending of glass has been reprinted from [29].

glasses that contain an inherent compressive residual stress at the surface, see Sect. 6.2. Although cold bending is most easily done with thin monolithic glass panes, their application is usually not acceptable for safety reasons. Cold bent laminated glass panes are, therefore, used more often, but attention should be paid to the shear interaction provided by the interlayer. For multi-ply laminated glass panes, the cold bending process can be assisted by softening the interlayer through modest heating of the panel. Cold bending of insulating glass units which consist of glass plies separated by edge spacers requires specific attention because failure of the edge seal will cause leakage and related loss of thermal performance of the insulating glass unit, see Sect. 3.2.

3.3.2 *Bending by Lamination*

In addition to cold bending and hot bending, see Sect. 2.3.9, glass bending can also be achieved simultaneously with the lamination process, see Fig. 3.18. For this purpose, a stack of individual glass panes and interlayer films which are not yet bonded is forced into a mould in the autoclave. The plies are then laminated in this shape. After the laminate is made, the mould is removed. This causes a limited instantaneous elastic spring-back and a temporally delayed creep, which is governed by the visco-elastic properties of the interlayer. For this reason, interlayers used in this bending process must have sufficient stiffness to maintain the desired shape until the end of the lifetime of the laminated glass. These can be either cross-linked polymers such as EVA films or very stiff plastics such as ionomers, or PVB with reduced plasticiser content.

Fig. 3.18 Glass bent by lamination: **a** Manufacturing of a large uniaxially curved glass panel; **b** biaxially curved walkable glass at the Städel Museum in Frankfurt am Main

3.4 Fire Resistant Glazing

Two aspects need to be differentiated in fire protection, namely the *reaction to fire* of building materials and the *resistance to fire* of building components. The reaction of construction products to fire is classified according to the European standard EN 13501-1 [55]. This standard defines seven classes (A1, A2, B to F) and provides information on certain classes regarding smoke production and flaming droplets or particles of burning materials.

Following the introduction of the CWFT (classified without further testing) list in the context of European harmonisation [30], glass is classified as a non-combustible building material, but laminated glass is not, due to the presence of the polymeric interlayers. Fire tests show that, for laminated glass with PVB films for example, can achieve a Class B rating according to EN 13501-1. This is important e.g. in case it is required that external walls must consist of non-combustible construction materials. Fuelled by recent fire incidents some authorities have restricted the use of laminated glass because of combustibility of the interlayer.

3.4.1 Fire Resistance

In case of fire, float and wired glass are likely to fracture after a short exposure, resulting in a considerable risk of flashover and fire spread after the glass fragments have fallen out. The European standards EN 13501-2 [56] and EN 1634-1 [66], respectively, regulate the classification and the testing of fire resistance of building components. For fire protection glazing, EN 357 [54] summarises the classification and tests. Accordingly, for glazing the following relevant classes are distinguished:

- R: Load-bearing capacity of a supporting component;
- E: Integrity of an enclosure. Prevention of flames and gases from entering;
- W: Reduction of radiation. Maximum thermal radiation allowed is $15\,\mathrm{kW}/^2\mathrm{m}$;
- I: Thermal insulation. Maximum surface heating of the side not exposed to the fire is 140 K on average.

High fire resistance classes for E are usually achieved by special thermal pre-stressing processes or with wired glass, (thermally pre-stressed) borosilicate glass, glass ceramics or glass blocks. Due to the lower thermal expansion coefficients and associated higher resistance to thermal shocks and the higher glass transition temperature, borosilicate glass is superior to soda-lime silica glass in fire scenarios. Consequently, it is often possible to reach a fire resistance duration of 120 min (class E 120) with borosilicate glass.

The prevention of heat transfer (classes EW and EI) is achieved by the use of fire-protective films in laminated glass. Thin glass panes thereby enclose layers of sodium silicate glass gel (also known as hydrogel or water glass). In the event of fire, these intumescent layers start to foam while forming water in an endothermic

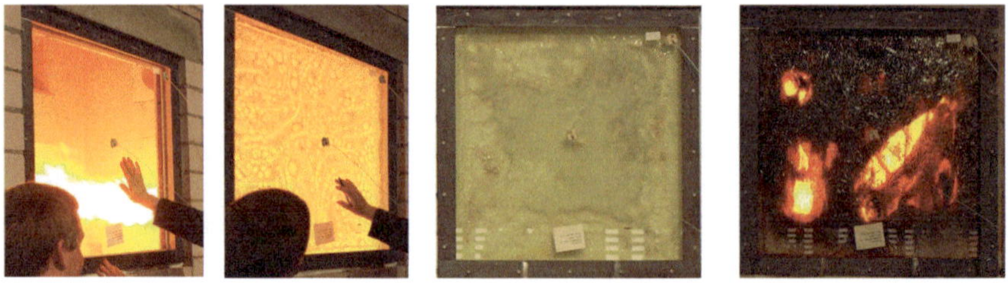

Fig. 3.19 Fire test on an EI glazing at *Schott AG*

reaction, cool the laminate, become opaque as a result of salt formation and thus hinders the transfer of radiated heat, see Fig. 3.19. The use of multiple layers results in a longer fire resistance duration. In 52 mm thick glass the class EI 120 is reached [31]. In applications with direct sunlight, PVB films are usually applied as outer interlayers of the laminate because hydrogel layers are generally not completely UV-stable. This makes it possible to classify them as laminated safety glass; however, the overall resistance against impact must usually be proven separately.

Since fire protection glazing is a non-regulated construction product, a technical approval or project-specific testing is usually required. The latter regulates the entire system, including the frame design. Therefore, manufacturers of fire protection glazing usually offer system solutions of which properties are technically approved and documented in the certifications.

3.5 Protective Glazing

3.5.1 General Information

The term *protective glazing* covers the aspects related to the protection against burglary, explosions and bullets. Detailed descriptions are given in the following Sections. A normative distinction is often made between the glazing as a building product and the overall construction.

In addition, some countries have requirements for safety in use for ball impacts on sporting hall glazing. In Germany for instance, this is tested according to DIN 18032-3 [32] with a ball-throwing machine with which two types of balls (handball and hockey ball) are projected on the glazing at an angle of 45° or 90°. Thermally pre-stressed glass of thickness 8 mm and laminated safety glass with 2 mm × 4 mm float glass with 0.38 mm PVB would meet the "no-break" requirement.

Alarm glass is glazing which reports a glass fracture to the alarm system. Thermally pre-stressed glazing is usually used in these applications whereby a conductive circuit is applied onto the glass. For insulating glass units, usually combinations with laminated panes are chosen to achieve increased prevention against burglaries. In a different variant, electrically conductive silver wires (thickness 0.1 mm) are embedded in the interlayer at intervals of 20 to 35 mm in a meandering form. When the wire is torn, the alarm is triggered.

3.5.2 Burglary Protection

The resistance of glazing against manual attack, also called burglary protection, is regulated in Europe by two standards. EN 356 [33] describes the requirements for glazing, while the standard EN 1627 [34] to EN 1630 [35] regulates the requirements for windows as a type of building element.

Drop tests with a steel ball are used to obtain certification for the first five classes (P1A to P5A) of EN 356. Here, a 4.1 kg steel ball is dropped three times, at three different locations, onto a 1100 mm × 900 mm, four-side supported pane. The ball must not penetrate the glazing within 5 s after impact; neither must the sheet slip from its supports. The other three classes (P6B to P8B) are tested with an axe Table 3.2. In these tests the number of axe impacts required to make an opening of 400 mm × 400 mm in the glazing are counted. Three test specimens must pass the test in order to receive certification. The testing of the relatively small standard specimens, however, refers only to a qualitative comparison of various glazing subjected to a specific attack and always with the same boundary conditions. They therefore provide only a limited assessment of actual burglary resistance particularly if the boundary conditions and the mounting depth of the glazing in the test does not correspond to real-world installation.

High classifications of anti-burglary glazing are only achieved with thick glass laminates, see Fig. 3.20. Laminates consisting of multiple glass plies provide better results than laminates composed of only two glass plies of the same overall thickness and type. It is possible to use stiff polymer sheets such as polycarbonate in the core, as this results in a thinner and lighter glazing unit for the same classification. However, the application possibilities of glass/polycarbonate composites are limited due to

Table 3.2 Resistance of security glazing against manual attack classes according to EN 356

DIN EN 356	Drop height or number of axe impacts
P1A	1.5 m (3 drops)
P2A	3.0 m (3 drops)
P3A	6.0 m (3 drops)
P4A	9.0 m (3 drops)
P5A	9.0 m (9 drops)
P6B	30–50 axe impacts
P7B	50–70 axe impacts
P8B	>70 axe impacts

Fig. 3.20 Manual testing of a burglary-resistant glazing with axe and chisel

the disparity (factor of 6) in the coefficient of thermal expansion, which is why only small dimensions are possible. In addition, the durability is worse than with glass/glass laminates. Therefore alternatives such as rigid interlayers (e.g. ionomers or stiff PVB) can be an option.

The classification P8B according to EN 356 is for instance achieved with a multiple laminate of 25 mm total thickness [36]. When using polycarbonate, this can be reduced to 22 mm [37].

EN 1627 regulates the procedure for the classification of burglary-resistant window and façade constructions. The components can be classified in six resistance classes (RC1 to RC6). The first three classes are intended to withstand attacks by casual offenders; the classes RC4 to RC6 the attacks by professional perpetrators. The minimum requirements for the glazing to be used according to EN 356 are regulated here. The following parts of the standard series describe the three related test methods:

- a static load test in particular for frame construction, see e.g. EN 1628 [38];
- a dynamic load test (pendulum test according to EN 12600 [39]) for classes RC1 to RC3 EN 1629 [40];
- a manual intrusion test with different tool sets, see EN 1630 [35].

3.5.3 Bullet Resistance

The resistance of glazing against ballistic attack, see Fig. 3.21a, also referred to as bullet resistance, is regulated in Europe by two standard series, analogous to the resistance against manual attack. EN 1063 [41] describes the requirements for glazing as a construction product, whereas EN 1522 [42] and EN 1523 [43] regulate the requirements for windows in buildings. In addition to the civil standards, military standards exist in the area of bullet-proofing, such as the NATO Standardization Agreement STANAG 4569, which differs in particular in the type of weapons and ammunition used.

Table 3.3 Resistance classes against bullets according to EN 1063

Classification	Caliber	Bullet type	Bullet weight [g]	Speed [m s^{-1}]	Number of hits	Shooting distance [m]
BR1	.22 LR	L/RN	2.6	360	3	10
BR2	9 mm Luger	FJ/RN/SC	8.0	400	3	5
BR3	.357 Magnum	FJ/CB/SC	10.2	430	3	5
BR4	.44 Rem. Magnum	FJ/FN/SC	15.6	440	3	5
BR5	5.56 × 45	FJ/PB/SCP1	4.0	950	3	10
BR6	7.62 × 51	FJ/PB/SC	9.5	830	3	10
BR7	7.62 × 51	FJ/PB/HC1	9.8	820	3	10
SG1	12/70	Solid lead slug	31.0	420	1	10
SG2	12/70	Solid lead slug	31.0	420	3	10
HC1:	Steel Hard Core					

EN 1063 uses seven classes (BR1 to BR7) to define resistance against handguns and rifles and two additional classes (SG1 and SG2) for the resistance against shotguns. The glass test specimens are 500 mm square. Three test specimens must be tested for obtaining a classification. The glazing must withstand three shots, carried out depending on the resistance class with different ammunition from a defined distance in an equilateral triangular field with an edge length of 120 mm, see Table 3.3. If glass fragments penetrate the splinter indicator on the back, the classification is marked with an additional S label. If the splinter indicator remains intact, the classification is marked with NS.

As with anti-burglary glazing, high classifications are only achievable using thick glass laminates, typically consisting of multi-layered asymmetric laminated glass with different film thicknesses. For example, the classification BR6 NS according to EN 1063 is achieved with a multi-laminate of 73 mm total thickness [36]. Using polycarbonate could reduce this to 42 mm [37].

EN 1522, together with EN 1523, defines the basic procedure for the classification of windows with similarities to EN 1063.

3.5.4 Blast Resistance

The resistance of glazing against explosions, also known as blast resistance, is regulated in Europe according to EN 13541 (product standard) as well as EN 13123 [44] in conjunction with EN 13124 [45].

EN 13541 defines four classes: ER 1 to ER 4. The certification is done according to the maximum of the positively reflected overpressure, p_R, the positive impulse, i_+, and the duration of the positive pressure phase, t_+, see Table 3.4 and Sect. 9.5.4. The test is carried out in a shock tube setup. To acquire certification, three test specimens with dimensions 900 mm × 1100 mm have to pass the test. The tested glazing must not have any continuous holes and must have no openings between the tensioning frame and the edge of the glazing. The resistance class can have an additional S or NS label, depending on whether glass splintering has occurred on the back of the glazing. Certification according to EN 13541 is only to a limited extent useful to evaluated real life situations due to the small dimensions of the test specimens and the unrealistic supporting conditions.

The aim of a blast resistant glazing is to prevent a room, and its occupants in particular, from the effects of the shock wave and from glass splinters. Laminated glass is essential in blast resistant glazing because monolithic glazing produces unacceptable splintering when the glass fractures, see Fig. 3.21b. However, fracture of the glass in a laminate is often permissible and even desirable, because the blast energy is partially dissipated in the elasto-plastic deformation of the fractured laminate and thus the substructure is subjected to smaller peak stresses (*balanced design*). This is in contrast to burglary and fire-retardant glazing, where thick stiff glass laminates are desirable. In addition to the glass splinters, which must also be retained by the interlayer in case of glass fracture, it is important to ensure that the glazing as a whole is not detached from its supports. An effective measure against this is to bond the glass into the frame or to the clamping brackets by means of structural silicone adhesive [46].

It should be noted that novel façade components have been developed which have the ability to upgrade conventional or lightly enhanced glass façade systems to higher blast protection levels [47].

An ER1 certification can, for example, be achieved with two 6 mm thick float glass panes and a 1.52 mm thick PVB film. In addition to the optimisation of the glazing, a significantly better performance can be achieved through optimisation of the frame and sub-construction.

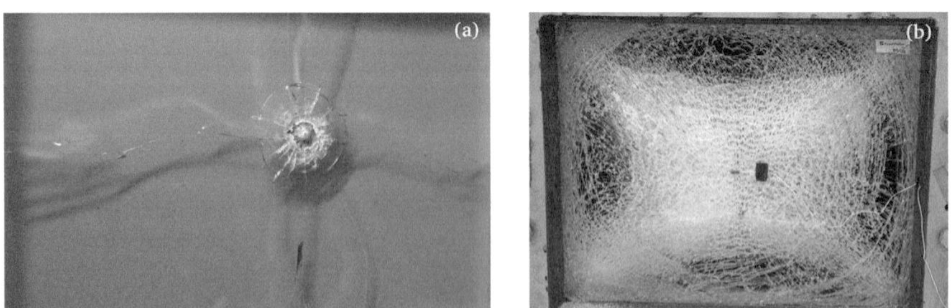

Fig. 3.21 **a** Pattern of a bullet resistant glazing and **b** test specimen of an explosion-resistant glazing after a shock tube test

Table 3.4 Classification of explosion-resistant glazing according to EN 13541 and [48]

Classification	Maximum overpressure of the reflected blast wave p_R [kPa]	Positive specific impulse i_+ [kPa ms]	Duration of the overpressure phase t_+ [ms]	Outside Exp. load [kg TNT]	Exp. distance [m]
ER 1	50–100	370–900	$\geq$ 20	80	30
ER 2	100–150	900–1500	$\geq$ 20	350	35
ER 3	150–200	1500–2200	$\geq$ 20	900	40
ER 4	200–250	2200–3200	$\geq$ 20	1900	45
GSA C	$\geq$28	$\geq$192	$\geq$ 14	30	30
GSA D	$\geq$69	$\geq$614	$\geq$ 18	200	35

In design, EN 13123 [44] regulates requirements and classification, whereas EN 13124 [45] deals with the test procedure. The standards are divided into two parts. The first deals with shock tube testing and the second with the outdoor range test. The classification according to the shock tube test, which is to represent a large detonation at a larger distance (plane pressure wave), is largely identical to EN 13541 [49]. For the outdoor experiment, which represents a close detonation, the mass of the explosive charge and the distance to the test specimen are defined as classification characteristics instead of maximum pressure and impulse. Companies that specialise in explosion resistant windows and doors usually offer standard tested and classified system solutions which combine glazing and frame.

In addition, the US GSA Security Criteria [48] has been established worldwide as a classification standard for blast resistance, see Table 3.4.

3.6 Photovoltaic Glazing

3.6.1 General Information

The use of glass for photovoltaic panels has been growing since the beginning of this millennium due to declining module prices, feed-in tariffs and the trend towards glass-glass modules instead of glass-film modules, see Sect. 3.6.2. However, there are significant differences between desired safety levels needed in the building industry and the practices of the solar industry. In the next Section the different types of modules, support systems and the loading particular to photovoltaic glass are described.

3.6.2 Module Types

The module types can be distinguished in terms of the solar cells used. Thick layer modules consist of mono or polycrystalline silicon cells that are about 200 μm thick, whereas thin-film modules consist of cell thicknesses in the order of a few μm. Thin film cells are directly applied on the glass substrate by means of different technologies, namely: Amorphous Silicon (ASI) cells, Cadmium Telluride (CdTe) cells and Copper Indium Diselenid (CIS). The latter is also often supplemented with gallium (CIGS).

The cells are encapsulated essentially in glass-glass or glass-film modules, embedded in intermediate films made of Ethylene Vinyl Acetate (EVA). Polyvinyl Butyral (PVB) films or cast-in-place resins are also possible, but less common. The glass is predominantly thermally pre-stressed, as this leads to thinner panes. Indeed, thin glass results in lighter structures and, in the case of the cover glass, to higher efficiency levels. The thermal pre-stressing level is rarely defined in the product data sheets, complicating the derivation of a design stress considerably. In the case of thin-film modules, a thermal toughening of the substrate glass is not possible since the application of the thin film is carried out in a specially designed thermal process, in which the residual stresses in the glass partially relax again, see e.g. Sect. 6.2. In all module types the cover glass is usually textured and/or provided with an anti-reflective coating to minimise reflection losses.

Thick-film modules are usually supported by a circumferential frame, whereas thin-film modules are often frameless. The sizes of the rectangular modules usually range between 0.7 and 1.7 m^2 having widths of 0.6 to 1.1 m.

3.6.3 Support Systems

The fields of application of photovoltaic modules are very diverse: on or in sloping roofs (e.g. solar tiles), on shed roofs, on or in flat roofs, as façades or as free-standing systems.

However, the support systems share some similarities. The main distinction is the type of support, in this case, simply supported edges, discrete clamps or backrail-bearing modules, see Fig. 3.22. For the structural analysis of the glass, the edge frame may be assumed to provide a continuous support along the edges, irrespective of the actual boundary conditions of the frame.

3.6.4 Special Loads

If the loads on photovoltaic glazing are to be determined by means of the regulations from the construction industry, two questions are fundamentally asked:

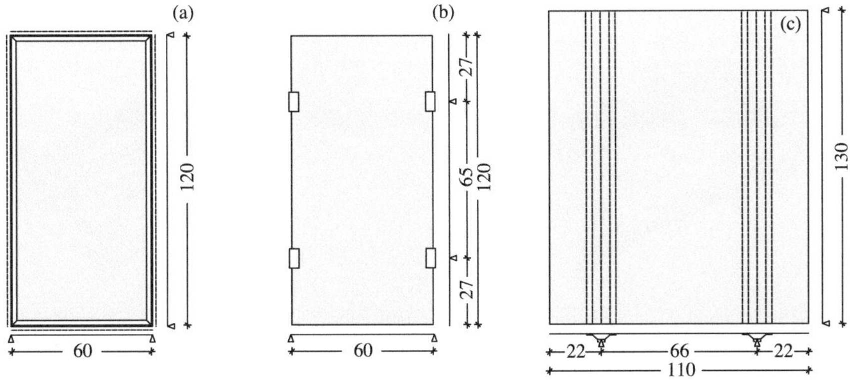

Fig. 3.22 Typical supports of glass-glass modules: **a** Framed, **b** locally clamped and **c** backrail-supported; dimensions in cm

- Which partial factors are to be used to determine the wind and snow loads?
- How are the high thermal actions due to solar absorption and the associated induced stresses taken into account?

The first question arises only in the case of on-roof solutions and elevated systems. Here, either specific research results (e.g. [50] for snow loads in elevated systems) have to be used, wind tunnel tests are carried out or conservative engineering assumptions are made. The loads can be found in e.g. the European standards EN 1991-1-3 [57] and EN 1991-1-4 [58].

The second question can only be answered at present by considering the most extreme conditions, since there are currently no design standards that provide the thermal loads, which are specific for the unique module configuration. However, the solar radiation on dark cells with high solar absorption can lead to temperatures of up to 80 °C in temperate climates. The maximum thermal stress in the glass might occur for example in the case of sudden cooling of the glass, e.g. by rain.

Finite element software can be used to calculate the thermal stresses from any temperature distribution, but the temperatures themselves are not regulated in standards. For example, in the case of the module shown in Fig. 3.22a a linear temperature gradient over the thickness with a temperature difference between the thin film and the glass surface of $\Delta T = 20\,\text{K}$, results in principal tensile stresses in the glass of $\sigma_1 \approx 10\,\text{MPa}$ [51].

3.7 Glazing for Solar Thermal Applications

Solar thermal systems convert heat energy from solar radiation into usable thermal energy. Such systems are popular in small domestic installations as a supply for hot

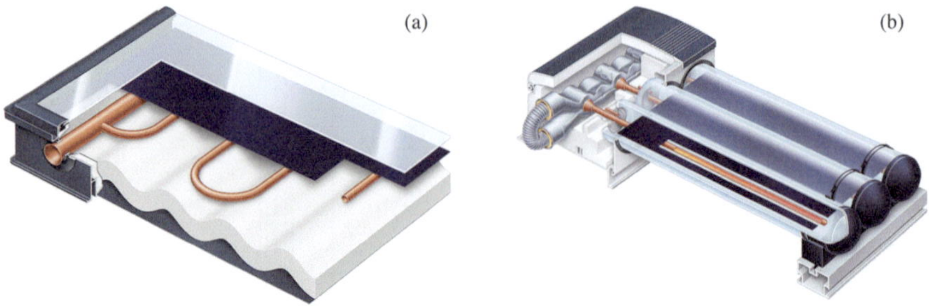

Fig. 3.23 Schematic representation of a a flat panel collector and b a tube collector (Picture: Viessmann)

water or heating of buildings. Conversely, a solar thermal system can also be used for cooling.

In principle, there are two types of collectors: flat and tubular, see Fig. 3.23. Flat collectors consist of a flat dark absorber plate placed over the heating circuit. Solar irradiation heats the absorber plate, which in turn releases the heat energy to the fluid in the heating circuit. The temperature in the heating circuit is usually between 30 and 110 °C. A glazing cover is placed above the absorber to retain the heat in the solar thermal module. Tubular collectors consist of evacuated glass tubes, in which inner tubes are arranged for the heating circuit. An absorber is located on the inner tubes similarly to the flat collectors. The temperature range of the heating circuit is usually between 50 and 150 °C.

The load capacity of the glazing in solar collectors is determined in the same way as for the photovoltaic glazing by a mechanical test of the entire collector with a load of 2.4 kN/m^2 (EN ISO 9806 [67]).

However, it is recommended to take into account the specific building site conditions and load combinations in the design of the collectors.

Fig. 3.24 **a** Photomontage of a solar chimney power plant (image credits: Schlaich Bergermann Solar); **b** Parabolic dish with Stirling engine (photo credits: sbp sonne GmbH)

Fig. 3.25 Parabolic trough power plant (Picture by: sbp sonne GmbH)

Solar energy can be converted to electrical energy by means of turbines (e.g. steam trubines) or engines (e.g. Stirling engines), but this is a rather exceptional solar thermal application. These can be classified in non-focusing systems (e.g. updraft power plants, Fig. 3.24a), point focusing systems (e.g. solar tower power plant or Dish-Stirling plants, Fig. 3.24b) and linear focusing systems (e.g. parabolic through power plant, Fig. 3.25). In the case of linear focused systems, very high demands are placed on the dimensional accuracy of the mirror lenses, since even slight deviations from the target geometry may lead to high energy losses. In parabolic trough power plants, 4 mm thick curved float glass is typically used, which is connected to the substructure by means of adhesively bonded point fixings. The fracture of a glass can also be critical for the adjacent glasses due to the expansion on fragmentation. In the case of solar updraft wind power plants, flat glass is used as a collector roof. Such solar power plants have been implemented in e.g. Spain and the USA.

References

1. EN ISO 12543-1. *Glass in building - Laminated glass and laminated safety glass - Part 1: Vocabulary and description of component parts* (2011)
2. EN 14449. *Glass in building - Laminated glass and laminated safety glass - Evaluation of conformity/Product standard* (2005)
3. Trosifol. *TROSIFOL Manual*. Kuraray Europe (2012)
4. H. Domininghaus, *Kunststoffe - Eigenschaften und Anwendungen*, 7th edn. (Springer, Berlin, 2012)
5. S. Feirabend. "Steigerung der Resttragfähigkeit von Verbundsicherheitsglas mittels Bewehrung in der Zwischenschicht". Dissertation. Universität Stuttgart (2010)
6. C. Louter, J. Belis, F. Bos, D. Callewaert, F. Veer, Experimental investigation of the temperature effect on the structural response of SG-laminated reinforced glass beams. Eng. Struct. **32**(6), 1590–1599 (2010)
7. K. Puller, "Untersuchung des Tragverhaltens von in die Zwischenschicht von Verbundglas integrierten Lasteinleitungselementen". Dissertation. Universität Stuttgart (2012)
8. DuPont. *SentryGlas von DuPont: Verarbeitungshinweise* (2010)

9. M.D. Kempe, G.J. Jorgensen, K.M. Terwilliger, T.J. McMahon, C.E. Kennedy, T.T. Borek, Acetic acid production and glass transition concerns with ethylene-vinyl acetate used in photovoltaic devices. Sol. Energy Mater. Sol. Cells **91**(4), 315–329 (2007)
10. A. Czanderna, F. Pern, Encapsulation of PV modules using ethylene vinyl acetate copolymer as a pottant: a critical review. Sol. Energy Mater. Sol. Cells **43**(2), 101–181 (1996)
11. G. Habenicht, *Kleben: Grundlagen, Technologien, Anwendungen*, 6th ed. (Springer, 2009)
12. BASF. *Thermoplastische Polyurethan Elastomere (TPU): Elastollan – Materialeigenschaften, Product data sheet* (2011)
13. A.P. Galac, G. Griebel, *Verfahren zur Herstellung von Verbundglas, Europäisches Patent 1321290B1* (2002)
14. E. Volakos, C. Davis, M. Teich, P. Lenk, M. Overend, Structural performance of a novel liquid-laminated embedded connection for glass. Glas. Struct. Eng. **6**(4), 487–510 (2021)
15. J. Hänig, B. Weller, Integrated connections for glass-plastic-composite panels: an experimental study under tensile loading at +23, +40 and +60 °C and different glass build-ups. Glas. Struct. Eng. **7**(2), 211–229 (2022)
16. M. Schuster, J. Schneider, T.A. Nguyen, Investigations on the execution and evaluation of the Pummel test for polyvinyl butyral based interlayers. Glas. Struct. Eng. **5**(3), 371–396 (2020)
17. C. Butchart, M. Overend, Influence of Moisture on Post-fracture Performance, in *Glass Processing Days* (2013)
18. D. Callewaert, D. Delincé, D. Sonck, J. Belis, R. Van Impe, Experimental investigation of the influence of temperature on local bridging behaviour in laminated glass elements in post-breakage state. eng., in *proceedings of 10th Asia-Pacific conference on AEPA 2010*, ed. by Y. Huang (World Scientific, 2011)
19. J. Franz, Untersuchungen zur Resttragfähigkeit von gebrochenen Verglasungen. Dissertation. Technische Universität Darmstadt (2015)
20. C. Butchart, M. Overend, Influence of aging on post-fracture performance, in *Glass Processing Days* (2017)
21. BBR. *Bundesamt für Bauwesen und Raumordnung: Leitfaden Nachhaltiges Bauen – Anlage 6: Bewertung der Nachhaltigkeit von Gebäuden und Liegenschaften* (2001)
22. ETAG 002. *Guideline for european technical approval for structural sealant glazing kits (SSGK) – Part 1: Supported and unsupported systems, EOTA* (2012)
23. P. Descamps, V. Hayez, Silicone joint dimensioning calculation methods. en., in *Challenging Glass Conference Proceedings* (2018), Vol. 6 (2018): Challenging Glass 6
24. C. Kocer, The thermal and mechanical performance of a vacuum insulating glazing, in *Glass Performance Days 2015* (Tampere, 2015), pp. 7–10
25. R. Collins, T. Simko, Current status of the science and technology of vacuum glazing. Solar Energy **62**(3), 189–213 (1998)
26. VDMA. *Vakuum-Isolierglas (VIG) Abschlussbericht* (2007)
27. C. Kocer, A. Aronen, R. Collins, O. Asano, Y. Ogiso, Measurement of heat flow through vacuum insulating glass part 1: measurement area in direct thermal contact with specimen. Glas. Struct. Eng. (2022)
28. C. Kocer, A. Aronen, R. Collins, O. Asano, Y. Ogiso, Measurement of heat flow through vacuum insulating glass part 2: measurement area separated from glass sheets with buffer plates. Glas. Struct. Eng. (2022)
29. J. Belis, C. Louter, J.H. Nielsen, J. Schneider, Architectural glass, in *Springer Handbook of Glass*, ed. by J.D. Musgraves, J. Hu, L. Calvez. Springer Handbooks (Springer, Cham, 2019), p. 1841
30. KEG. *Festlegung eines Verzeichnisses von Produkten, die in die Kategorien A gemäß der Entscheidung 94/911/EG zur Durchführung von Artikel 20 der Richtlinie 89/106/EWG des Rates über Bauprodukte einzustufen sind, Die Kommission der Europäischen Gemeinschaften* (1996)
31. Schott AG, *PYRAN, PYRANOVA, NOVOLAY secure, PYRANOVA secure, Product data sheet* (Mainz, 2014)

32. DIN 18032-3. *Sporthallen – Hallen für Turnen und Spielen und Mehrzwecknutzung – Teil 3: Prüfung der Ballwurfsicherheit* (1997)
33. EN 356. *Glass in building - Security glazing - Testing and classification of resistance against manual attack* (2000)
34. EN 1627. *Pedestrian doorsets, windows, curtain walling, grilles and shutters - Burglar resistance - Requirements and classification* (2011)
35. EN 1630. *Pedestrian doorsets, windows, curtain walling, grilles and shutters - Burglar resistance - Test method for the determination of resistance to manual burglary attempts* (2011)
36. Saint Gobain Glass. *Memento Glashandbuch* (Stolberg, 2005)
37. Sicurtec. *sicurLITE: highly resistant glass, Product data sheet* (2011)
38. EN 1628. *Pedestrian doorsets, windows, curtain walling, grilles and shutters - Burglar resistance - Test method for the determination of resistance under static loading* (2011)
39. EN 12600. *Glass in building - Pendulum tests - Impact test method and classification for flat glass* (2003)
40. EN 1629. *Pedestrian doorsets, windows, curtain walling, grilles and shutters - Burglar resistance - Test method for the determination of resistance under dynamic loading* (2011)
41. EN 1063. *Glass in building - Security glazing - Testing and classification of resistance against bullet attack* (2000)
42. EN 1522. *Windows, doors, shutters and blinds - Bullet resistance - Requirements and classification* (1999)
43. EN 1523. *Windows, doors, shutters and blinds - Bullet resistance - Test method* (1999)
44. EN 13123. *Windows and doors and shutters - Explosion resistance - Requirements and classification*. eng. (2001)
45. EN 13124. *Windows, doors and shutters - Explosion resistance - Test method*. eng. (2001)
46. J. Schneider, J. Kuntsche, Rückhaltesysteme für geklemmte Verbundglasplatten bei Explosionsbeanspruchung, in *Baustatik-Baupraxis* (München, 2014), pp. 605–614
47. G. Lori, C. Morison, M. Larcher, J. Belis, Sustainable facade design for glazed buildings in a blast resilient urban environment. Glas. Struct. Eng. **4**(2), 145–173 (2019)
48. GSA. *Security criteria* (1997)
49. EN 13541. *Glass in building - Security glazing - Testing and classification of resistance against explosion pressure* (2012)
50. N. Grammou, Ermittlung von geometrischen Formbeiwerten für Flachdächer mit aufgeständerten Photovoltaikanlagen anhand eines physikalischen Analogiemodells. Dissertation. Technische Universität Darmstadt (2015)
51. J. Schneider, J. Kuntsche, J. Kleuderlein, Load-bearing capacity of thin film photovoltaic modules, in *Advanced Building Skins* (Graz, 2012)
52. DIN 52338. *Prüfverfahren für Flachglas im Bauwesen – Kugelfallversuch für Verbundglas* (1985)
53. EN 1279-1. *Glass in Building - Insulating glass units - Part 1: Generalities, system description, rules for substitution, tolerances and visual quality* (2004)
54. EN 357. *Glass in building - Fire resistant glazed elements with transparent or translucent glass products - Classification of fire resistance* (2005)
55. EN 13501-1. *Fire classification of construction products and building elements - Part 1: Classification using data from reaction to fire tests* (2010)
56. EN 13501-2. *Fire classification of construction products and building elements - Part 2: Classification using data from fire resistance tests, excluding ventilation services* (2010)
57. EN 1991-1-3. *National Annex - Nationally determined parameters - Eurocode 1: Actions on structures - Part 1-3: General actions - Snow loads* (2010)
58. EN 1991-1-4. *National Annex - Nationally determined parameters - Eurocode 1: Actions on structures - Part 1-4: General actions - Wind actions* (2010)
59. EN ISO 10077-1. *Thermal performance of windows, doors and shutters - Calculation of thermal transmittance - Part 1: General* (2010)
60. EN ISO 12567-1. *Thermal performance of windows and doors - Determination of thermal transmittance by the hot-box method - Part 1: Complete windows and doors* (2010)

61. EN 673. *Glass in building - Determination of thermal transmittance (U value) - Calculation method* (2011)
62. EN 674. *Glass in building - Determination of thermal transmittance (U value) - Guarded hot plate method* (2011)
63. EN ISO 12543-2. *Glass in building - Laminated glass and laminated safety glass - Part 2: Laminated safety glass* (2011)
64. EN ISO 12543-3. *Glass in building - Laminated glass and laminated safety glass - Part 3: Laminated glass* (2011)
65. EN ISO 12543-4. *Glass in building - Laminated glass and laminated safety glass - Part 4: Test methods for durability* (2011)
66. EN 1634-1. *Fire resistance and smoke control tests for door and shutter assemblies, openable windows and elements of building hardware - Part 1: Fire resistance test for door and shutter assemblies and openable windows* (2014)
67. EN ISO 9806. *Solar energy — Solar thermal collectors — Test methods* (2014)

Part III
Glass Engineering Fundamentals

Chapter 4
Mechanical Fundamentals for Glass Engineering

This chapter explains the fundamental mechanical principles that underpin glass engineering design. It covers glass itself as well as the necessary background for other materials commonly used in glass structures, starting with the commonly-used constitutive models for polymers in Sect. 4.1, *followed by the different failure criteria for glass in Sect.* 4.2.

4.1 Mechanical Performance of Polymers

4.1.1 General Information

The mechanical response of many materials including polymers is generally non-linear and may depend on time and temperature, see e.g. Fig. 1.16. However, the experimental determination of all necessary material parameters is difficult and usually uneconomical. Furthermore, the description of non-linear material behaviour in a computational algorithm is complex and commercial finite element software do not always include the required combinations of material non-linearities. Consequently, it is commonplace to establish and use simplified material models. These simplified models often have implicit assumptions and limits of applicability. It is therefore important that the user is aware of the limits of these material models. Salient material models are listed below and provide an overview, for the modelling and sizing of the polymeric components that are often used in glass structures. More detailed information can be found in literature, e.g. [1–7].

In the context of this book, a material model typically provides the relationship between stress and strain (deformation). Thus, the term stress is the generic term used for tensile, compressive and shear stresses. The related strain variables are elongation, compression and shear, respectively. The distinction between engineering (nominal) strain and true strain is clarified using the example of the uniaxial tensile test, which

J. Belis et al., *Engineering Glass Structures*,
https://doi.org/10.1007/978-3-032-03233-1_4

Fig. 4.1 Tensile test on a transparent polymer (PVB film). **a** initial (unloaded) situation; **b** current (loaded) situation; **c** test specimen geometry

is illustrated in Fig. 4.1. The general three-dimensional case is available in literature, e.g. [2–4, 8, 9].

- Engineering strain (also known as *nominal strain*) is the change of length, Δl, relative to the initial length, l_0:

$$\varepsilon_0 = \frac{\Delta l}{l_0} = \frac{l - l_0}{l_0} = \frac{l}{l_0} - 1$$

- True strain (also known as logarithmic strain) is the integration of the incremental change of length, dl, relative to the current length, l:

$$\mathrm{d}\varepsilon = \frac{\mathrm{d}l}{l} \quad \Rightarrow \quad \varepsilon = \int_{l_0}^{l} \frac{\mathrm{d}l}{l} = \ln\left(\frac{l}{l_0}\right) = \ln\left(1 + \varepsilon_0\right)$$

- Engineering stress is the force, F, relative to the initial area, A_0:

$$\sigma_0 = \frac{F}{A_0}$$

- True stress is the force, F, relative to the current area (current configuration), A:

$$\sigma = \frac{F}{A} = \sigma_0 \frac{l}{l_0} = \sigma_0 \left(1 + \varepsilon_0\right)$$

The shown conversions assume a constant volume ($A_0 l_0 = Al$). For small strains, engineering and true are equal.

4.1.2 *Linear Elasticity*

The simplest material model is linear elastic, i.e. where the stress-strain relation is linear, see e.g. Fig. 4.2a.

If the material is also isotropic and homogeneous, it can be completely described with two elastic constants, for example, Young's modulus, E, and Poisson's ratio,

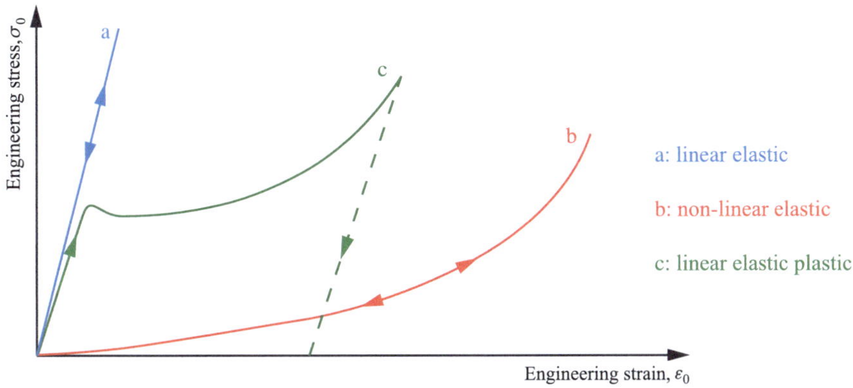

Fig. 4.2 Schematic representation of **a** linear elasticity, **b** non-linear elasticity and **c** plasticity

ν. All other elastic characteristics, namely the shear modulus, G, and the bulk modulus, K, can be derived from the first two. This linear elastic model is a very good representation of glass below its transition temperature T_g and it is therefore used for glass engineering applications, where the temperature is typically well below T_g of glass.

Hooke's Law describes the relation between stress and strain in a linear elastic material. In the uniaxial case, it can be written as:

$$\sigma = E \cdot \varepsilon. \tag{4.1}$$

In a rheological model, linear elasticity is commonly represented as a spring. In the case of small strains, linear elasticity can often be used as a suitable model for many polymers and other materials, as long as the influence of temperature and load duration on the material behaviour can be neglected for the given problem. If the material is incompressible, Poisson's ratio is $\nu = 0.5$, which can cause problems in numerical modelling due to zero terms in the denominators of the material stiffness matrix.

4.1.3 Non-linear Elasticity

Figure 4.2b shows a non-linear stress-strain relationship of a hyper-elastic material. The stress response is independent of the stress history and the material is completely elastic. No hysteresis can be detected from viscous effects or irreversible deformations. This is typical of many polymers in the entropy-elastic range, especially elastomers. Rheological models for non-linear elasticity also consist of non-linear springs.

The mechanical description of hyperelasticity involves the definition of a strain energy density, which is associated with this nonlinear stress response. The derivative of the strain energy density with respect to the strain results in the stresses. If the material is incompressible, the hydro-static pressure component of the strain energy can be neglected. This is the case for many polymers in the entropy-elastic range.

There are various hyper-elastic material models used to represent the polymers commonly found in glass structures e.g. Neo-Hooke, Mooney-Rivlin, Blatz-Ko, Ogden, Arruda-Boyce and Yeoh. Further details can be found in literature, e.g. [10, 11]. In addition, hyperelasticity can be modelled without explicitly setting up the strain energy density, by using experimental data to directly determine the derivatives according to the invariants [12, 13].

If the linear or non-linear elasticity are dependent on the strain rate, this is referred to as viscoelasticity, which is described in Sect. 4.1.5.

4.1.4 Plasticity

In classical time-independent plasticity, non-reversible strains occur when the stress exceeds the yield limit and the material changes stiffness, see Fig. 4.2c. In a uni-axial tensile test, plastic behaviour is usually manifested as a 'yield plateau'. In a multi-axial state of stress, the von Mises yield criterion known from material modelling of steel is often used, although this is not strictly correct for polymers, see Sect. 4.1.6.

Plastic deformations are particularly pronounced in thermoplastics, as a result of the slipping of polymer chains or by crazing which is an internal fracture that does not change the surface texture, but may be the precursor to macroscopic cracking. Crazing is usually visible as a white discoloration. Thermoplastic elastomers (TPE) can also be plastically deformed through the addition of heat. On the other hand elastomers and thermosets show almost no plastic deformations, due to the cross-linking of the molecular chains.

If the plastic behaviour is dependent on the strain rate, this is referred to as viscoplasticity. Further information on plasticity can be found in [2, 4, 5, 7].

4.1.5 Viscoelasticity

4.1.5.1 Linear Viscoelasticity

The term viscoelasticity is a combination of the terms elasticity and viscosity. It describes a material behaviour in which the stress resulting from a constant applied strain decays over time through creep. The stress (and strain) state is therefore a function of time and the load history is an important parameter in viscoelastic material models.

An increase in the strain rate, which represents the load velocity, results in a stiffer material response for viscoelastic materials. Furthermore the loading and unloading

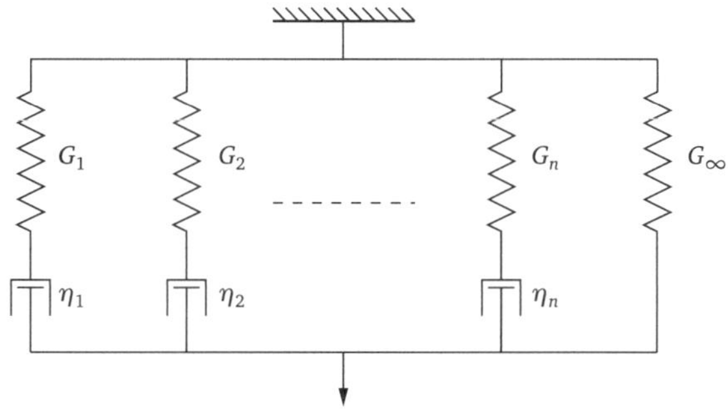

Fig. 4.3 Generalised Maxwell model

stress-strain paths differ i.e. hysteresis occurs and the area enclosed by the loading and unloading curves represents the energy converted into heat.

Viscoelasticity may be either linear or non-linear. Non-linear viscoelasticity may refer to non-linearities in the elastic components or the viscous component, or both. Linear viscoelasticty means that the time-dependent deformation is proportional to the applied stress and the BOLTZMANN superposition principle is valid.

The deformation behaviour of linear-viscoelastic materials can be described by means of rheological models. The *Generalised Maxwell Model* shown in Fig. 4.3 is often used for this purpose. In this model, the time dependent shear modulus, $G(t)$, which is also known as a relaxation function, is expressed as a so-called Prony series:

$$G(t) = G_\infty + \sum_{i=1}^{n} G_i \cdot e^{\frac{-t}{\tau_i}} = G_0 \left(g_\infty + \sum_{i=1}^{n} g_i \cdot e^{\frac{-t}{\tau_i}} \right). \tag{4.2}$$

The relaxation time, $\tau_i = \eta_i / G_i$, is a measure of the duration of the relaxation process of the corresponding Maxwell element (spring + dashpot pair) in the generalised Maxwell material model. At time $t = \tau_i$ the relaxation process is completed to $1 - e^{-1} \approx 63\%$. A Prony series element consisting of a spring + dashpot pair is required to represent one time decade.

An extension from the simple two-dimensional shear test to the general three-dimensional stress state is straightforward for homogeneous and isotropic materials. This results in two independent material parameters, for example, the shear modulus, G, and the bulk modulus, K. In the linear-viscoelastic case, these are no longer constant but time-dependent and can be represented by two Prony series. If the viscoelastic material is subjected to (almost) pure shear the bulk modulus can be assumed constant, $K(t) = K_0$.

At low strains linear viscoelasticity is applicable to many polymers. The transition to non-linear viscoelasticity is known as the linearity limit. For polymers in the glassy state, a minimum linearity limit of 1% strain can be assumed [14]. However, in the rubbery state, this transition generally occurs at significantly larger strains [1].

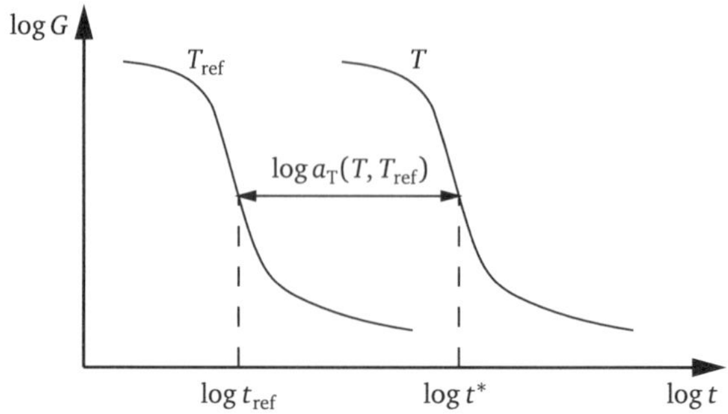

Fig. 4.4 Time-temperature superposition principle

4.1.5.2 Time-Temperature Superposition Principle (TTS)

As described in Sect. 1.5, a reduction in stiffness caused by a temperature increase is proportional to the reduction in stiffness caused by the effects of a longer load duration. This is caused by the thermal activation of the molecular movement and transfer processes in the polymer, that occur faster at higher temperatures. If the variations in time and temperature are proportional, the TTS principle applies.

A temperature increase therefore corresponds to a reduction in the relaxation times of the viscoelastic material, and thus a horizontal shift of the relaxation function, $G(t)$, on the time axis. This is usually plotted logarithmically, see Fig. 4.4. Consequently, a (time-) shift factor, a_T (T, T_{ref}), can be written as:

$$\log_{10} a_T(T, T_{ref}) = \log_{10} t^* - \log_{10} t_{ref} = \log_{10} \frac{t^*}{t_{ref}} \tag{4.3}$$

where t^* is known as the reduced (or shifted) time.

The TTS approach is an approximation of the real material behaviour that is often sufficiently accurate for engineering applications [1]. If the material is *thermorheologically simple*, it is possible to apply the TTS. A *master curve* can be created from measurements at different temperatures to predict the material stiffness at long load durations, which would otherwise require excessively long testing, see Sect. 1.5.2.2. The shift factor, $a_T(T, T_{ref})$, depends on the considered temperature, T, and also on the reference temperature, T_{ref}, which can be arbitrarily selected. At different temperature ranges, approaches can be used to describe the shift factor, $a_T(T, T_{ref})$, mathematically.

The empirical shift factor by William, Landel and Ferry (WLF) is given as:

$$\log_{10} a_T(T, T_{ref}) = -\frac{c_1 \, (T - T_{ref})}{c_2 + T - T_{ref}} \tag{4.4}$$

and is commonly used for polymers [15]. The constants c_1 [-] and c_2 [°C] are to be determined and depend on the reference temperature. The approach often finds good agreement with test results within the glass transition range and in the rubbery state [1].

The Arrhenius shift function is derived through the thermal activation of secondary dispersion phenomena (lateral chain movements, local movements in the main chain of macro-molecules or melting crystalline phases), and thus predominantly describes the glassy state [1]. It is often used for the modelling of glass in its transition range:

$$\log_{10} a_T(T, T_{ref}) = -k_A \left(\frac{1}{T} - \frac{1}{T_{ref}} \right) \tag{4.5}$$

where:

$k_A = \log_{10}(e) \frac{E_A}{R_m T} = 0.434 \frac{E_A}{R_m T}$	Arrhenius Constant
E_A	Material and process-specific activation energy [J/mol]
$R_m = 8.314\,J/(molK)$	Universal gas constant
T	Temperature [K].

Further details on both time-temperature shift approaches are available in literature. In addition, there are alternative time-temperature shift approaches, which are less widespread such as Narayanaswamy-Tool for glass modelling [13] or the Arcus tangent approach [16].

4.1.6 Polymer Failure Criteria

In polymers, failure is not always clearly defined, because there is no clear limit of proportionality or yield point and complete failure is often accompanied by very large strains. This is particularly pronounced in thermoplastics, where failure is often defined as reaching a specified strain limit. However, actual failure occurs only at rupture. Both states can be described by appropriate failure criteria. An attempt is made to define a comparative value, which indicates the failure for all stress conditions occurring in the material. This can be based on stress, strain or even energy density, whereas engineers typically prefer to use stresses. For example, in the case of glass, the principal tensile stress is used as a failure criterion at macroscopic scale; for steel, the von Mises equivalent stress is used as a yield criterion. All failure criterias have in common that they have to be invariant.

In order to determine a suitable failure criteria, material failures have to be induced and compared under different stress states. Uniaxial tension is simple to execute, see e.g. Sect. 1.5.2.1. Biaxial tension, uniaxial compression and pure shear stress, are more complex to perform especially when testing involves high strains to failure. Other stress states (hydrostatic tension and compression) are difficult to realise experimentally.

If material failure is governed by hydrostatic compression, which is often the case with polymers, classical failure hypotheses that are independent of hydrostatic effects, such as *Tresca* or *von Mises* cannot be directly applied. Suitable failure

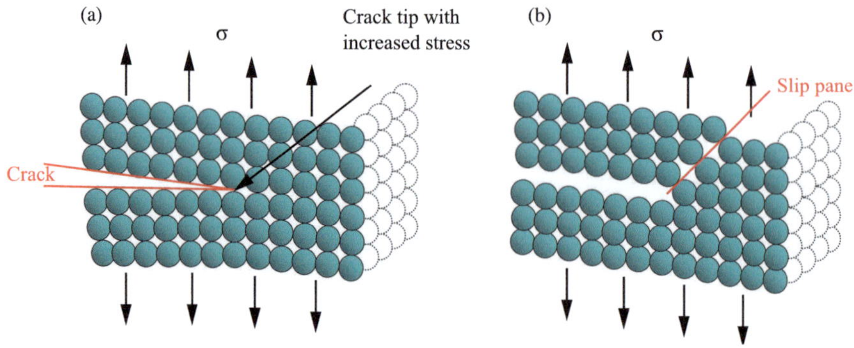

Fig. 4.5 Failure mechanisms for a uniaxial tensile stress state. **a** In brittle materials, stress concentrations at the crack tip (Mode I) lead to crack progression and failure. **b** In ductile materials, the lattice can slide along a sloping slip plane, relieving the stress at the crack tip point while undergoing relatively large and permanent deformations [19]

criteria for polymers used in glass and façade engineering are currently still subject of investigation [17, 18].

4.2 Glass Failure Criteria

4.2.1 Failure Mechanisms and Fracture Behaviour of Glass

Glass has an amorphous molecular structure in its glassy (solidified) state. As a result glass is isotropic and brittle. The latter is attributed to absence of long range order, slip planes and dislocations in its molecular structure, which are responsible for plastic deformations in other materials, see Fig. 4.5, such as construction grade steel. The formation of slip planes is not the absolute strength of the chemical bonds, but on whether parts of the lattice structure can slip relative to one another, while remaining in equilibrium with the external applied forces and changing the material's geometry while conserving its mass, see Fig. 4.5b. The amorphous structure of glass means that no such slip planes can be formed. Therefore glass fails in a brittle and catastrophic manner when its fracture strength is exceeded.

This indicates that fracture occurs when the major principal stress σ_1 exceeds the value of either the critical tensile fracture stress σ_f or the critical compressive fracture stress σ_c. Since the strength of glass is governed by the stress concentrations at the flaw tips, failure in tension occurs at much lower stress levels, in absolute terms, than failure in compression. Therefore, the failure criterion is most often written as:

$$\sigma_1 = \sigma_f. \tag{4.6}$$

The critical tensile stress (or tensile strength) of the glass depends on the surface quality (flaw density, flaw depth) and, as described in Sect. 5.3, flaws can grow slowly (subcritically) when subjected to stress; therefore the tensile strength is a function of stress history (load duration). Consequently, for engineering practice, the strength of the glass is not a material constant, but it is significantly influenced by sub-microscopic, microscopic and macroscopic surface flaws which can nucleate and grow during the service life of the glass component. In this context, a useful material constant is fracture toughness (also known as critical stress intensity factor), K_{Ic}, which is a measure of the maximum stress intensity at a given flaw tip that a material can withstand.

Upon fracture, the fragmentation characteristics (i.e. the size and distribution of the fragments) depends on the strain energy released during the fracture process. This energy is composed of the residual stresses in the glass due to thermal or chemical pre-treatments as well as the stress induced by the applied loads. The size of the fragments is inversely and non-linearly related with the strain energy released in the fracture process, i.e. a higher strain energy released results in smaller fragments. This explains why fracture of thermally toughened glass with high levels of surface pre-stress produces very small fragments of approximately 100 mm^2, whereas in (annealed) float glass with negligible levels of surface pre-stress (approximately –5 to –10 MPa [20, 21]) the fragments are very large. The fragmentation is an important characteristic of the glass in practice because the density of the fragments is often used for classifying glass in so-called fragmentation tests. For example, thermally toughened safety glass for building applications is classified as such if it meets the minimum fragment count and maximum fragment size according to EN 12150-1 [36], see Sect. 2.3.4.

4.2.2 Fundamentals of Linear-Elastic Fracture Mechanics

Linear-elastic fracture mechanics is useful for quantifying the stress concentrations at a crack tip in glass and the resistance of glass to crack growth. A (micro-)crack in a real world glass component can be subject to complex stress field that can be explained by superimposing three different types of crack opening modes, see Fig. 4.6. All modes cause a pre-existing crack in the x-z plane to extend in that plane: *Mode I* indicates a separation in the y-direction (perpendicular to the crack surfaces) and is therefore also known as a crack opening; *Mode II*, involves in-plane shear in the $x - y$ plane, sometimes referred to as sliding mode; *Mode III* consists of out-of-plane shear in the y-z plane and is also referred to as tearing mode. Modes *II* and *III* play a subordinate role in glass. For most applications, only Mode *I* is considered in glass engineering design.

A crack is a discrete discontinuity in a solid continuum, without a complete separation of the solid. The new surfaces created by the crack are referred to as *crack surfaces*, *crack flanks* or *crack edges* and are associated with some energy required for generating these surfaces. In brittle materials, such as glass, they are generally

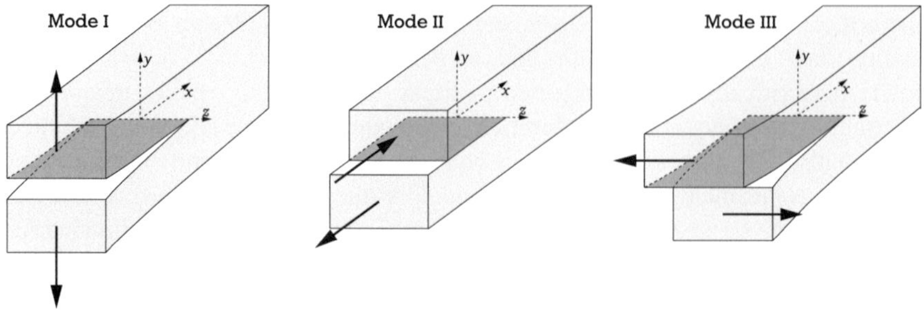

Fig. 4.6 Crack opening modes according to [22]: *Mode I* symmetrical separation orthogonal to the cracked surface; *Mode II* separation in the crack-line as a result of longitudinal in-plane shearing; *Mode III* separation as a result of out-of-plane shearing

stress-free and meet at the crack tip. Linear elastic fracture mechanics is based on the mechanical behaviour of a single crack in a linear-elastic, homogeneous and isotropic solid [23], so that in its application, the entire cracked body is considered as linear elastic. Linear elastic fracture mechanics is therefore ideally suited for representing the linear-elastic, brittle behaviour of glass. In fact, glass was originally used as a basic material in physical experiments to demonstrate and validate the fundamentals of linear fracture mechanics. Glass was chosen because of the reproducibility of surface defects and its relatively low cost. Its use dates back to the fundamental principles of Inglis [24] and Griffith [25]. Inglis who became interested in the fracture of metal plates in ship hulls and noticed that the rivet holes were deformed into an elliptical shape. This led to quantification of the stress concentrations of elliptical defects on uniaxially loaded elastic plates. For a plate with an internal elliptical crack of width $2c$, subjected to a uniformly distributed tensile stress σ_a, see Fig. 4.7a, the stress σ_r at the tip of the ellipse is given by:

$$\sigma_r/\sigma_a = 2\sqrt{c/r}\,, \tag{4.7}$$

where the ratio σ_r/σ_a represents the stress concentration and r is the distance from the tip. Using Eq. (4.7), it is possible to calculate the fracture stress $\sigma_a = \sigma_f$ when the theoretical tensile strength $\sigma_{f,theo}$ of the material is reached. This is given by:

$$\sigma_f = \frac{\sigma_{f,theo} \cdot r^{\frac{1}{2}}}{2 \cdot c^{\frac{1}{2}}}. \tag{4.8}$$

Inglis therefore provided a means for predicting the strength of the material containing a crack. Griffith developed a fracture criterion based on the principles of energy conservation (energy balance). Griffith's experiments demonstrated that the product of the failure stress, σ_f and the square root of (half) the crack length, c, was constant. From an energy conservation approach he arrived at:

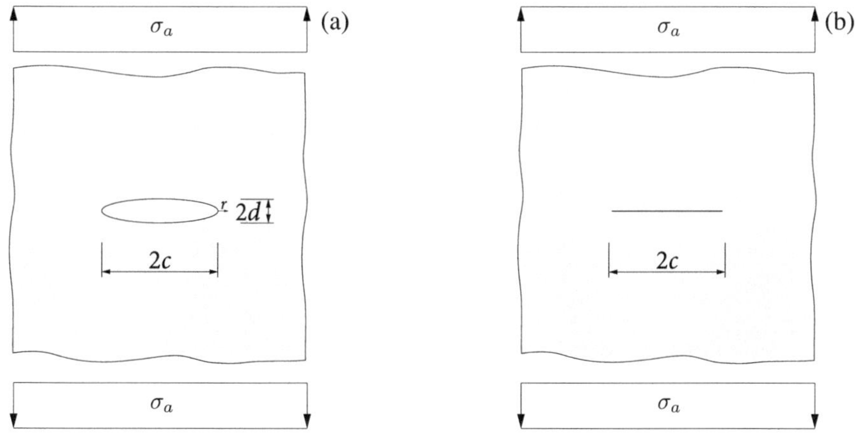

Fig. 4.7 **a** Elliptical flaw and **b** GRIFFITH flaw, each in an infinitely extended, uniformly loaded plate

$$\sigma_f^2 c = \frac{2E\gamma}{\pi}, \tag{4.9}$$

where γ is the energy required to create new surfaces. From this the energy release rate, G, can be defined as:

$$G = 2\gamma = \frac{\pi \sigma_f^2 c}{E}. \tag{4.10}$$

Although Griffiths' theory on brittle materials (glass) was very well proven, it was not until the 1950s that his work was further built upon through the work of Irwin [26] in the development of the stress intensity factor concept. This fracture concept follows the hypothesis that the state at the crack tip is influenced by the *stress intensity* (K-factors). The stress intensity factor is a measure of the stress state at the crack tip caused by a load on a body. The magnitude of K depends on sample geometry, the size and location of the crack, and the magnitude and the distribution of loads on the material.

The stress intensity factor approach is useful for describing a failure criterion for brittle materials. By neglecting sub-critical crack growth effects, see Sect. 5.3, separation occurs (formation of crack surfaces due to unstable crack progression or fracture), as soon as the respective stress intensity exceeds the critical stress intensity (also known as *fracture toughness*) for that specific material, depending on the crack opening mode:

$$K_{\mathrm{I}} = K_{\mathrm{Ic}} \quad , \quad K_{\mathrm{II}} = K_{\mathrm{IIc}} \quad , \quad K_{\mathrm{III}} = K_{\mathrm{IIIc}} \tag{4.11}$$

where K_{I}, K_{II} and K_{III} are the stress intensity factors and K_{Ic}, K_{IIc} and K_{IIIc} are the material- specific fracture toughness for modes I, II and III, respectively. The failure criteria according to Eq. (4.11) applies to a *single-mode-stress*, where the crack is

Table 4.1 Fractured toughness (Mode I) for soda-lime silica glass

Author	Year	Fracture toughness
[-]	[-]	K_{Ic} [$MPa\sqrt{m}$]
Wiederhorn [32]	1967	0.82
Gehrke, Ullner, and Hähnert [33]	1987	0.78
Menčik[a] [34]	1992	0.72–0.82
Ullner [35]	1993	0.76
Lawn [27]	1993	0.75

[a]Summary of different sources of literature

subjected to only one of the three opening modes. Stress intensity factors arising from the same crack opening method can be added due to the superposition principle applied in the linear elastic range, see e.g. [27]. Since modes II and III (sliding and tearing) only play a subordinate role in the fracture of glass, only a Mode I load is typically taken into account. The fracture toughness of glass is therefore usually referred to as a K_{Ic}. Table 4.1 summaries values for the fracture toughness of soda-lime silica glass, as experimentally identified by different authors. For comparison: the fracture toughness of steel is in excess of 100 $MPa\sqrt{m}$. This clearly demonstrates that the fracture toughness of glass is low, which is typical of brittle materials. There is also a direct relation between the energy release rate, G, and the critical stress intensity factor, which (for plane strain), can be written as:

$$G_{Ic} = \frac{K_{Ic}^2}{E}\left(1 - \nu^2\right) \tag{4.12}$$

The stress intensity, under inert environmental conditions, of a linear crack in a body (i.e. within the volume of the continuum) of length $2c$ subjected to a tension stress, σ_a, perpendicular to the crack edge (Mode I) or shear stresses, $\tau_{xy,a}$ (Mode II) and, $\tau_{yz,a}$ (Mode III), in a semi- infinite flat glass pane known as a Griffith Crack, see Fig. 4.7b, is defined as:

$$K_I = \sigma_a\sqrt{\pi \cdot c} \quad , \quad K_{II} = \tau_{xy,a}\sqrt{\pi \cdot c} \quad , \quad K_{III} = \tau_{yz,a}\sqrt{\pi \cdot c} \tag{4.13}$$

This general case may be extended to account for other crack geometries, load geometries and various load types, by introducing a dimensionless *geometry factor* Y into Eq. (4.13). In inert conditions and for a constant stress applied to a body, σ_a, see Fig. 4.8a, this can be represented as follows:

$$K_I = \sigma_a \cdot Y\ \sqrt{\pi \cdot c}. \tag{4.14}$$

A prerequisite for this approach is that the crack is significantly smaller than the solid that surrounds it. From a variety of relevant literature, such as [28] or [29], the

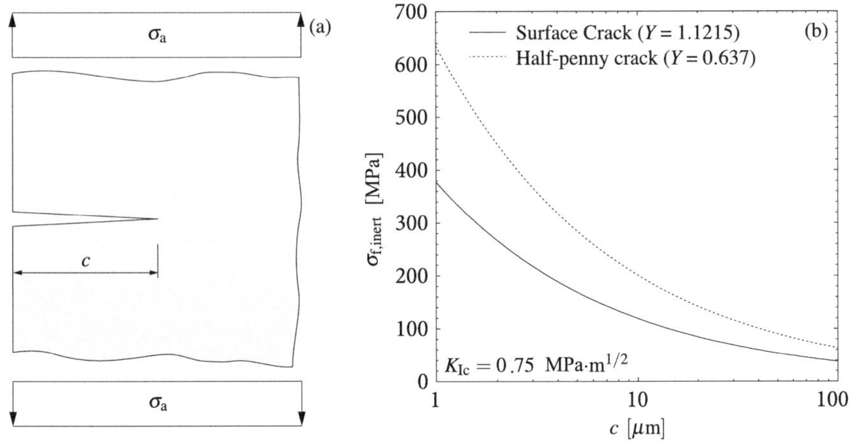

Fig. 4.8 **a** Surface crack in a semi-infinite body: geometry for a Mode I stress; **b** Fracture tensile stress $\sigma_{\mathrm{f,inert}}$ for a *long surface crack* and a *semi-elliptical surface crack* dependent on crack depth c in soda-lime silica glass

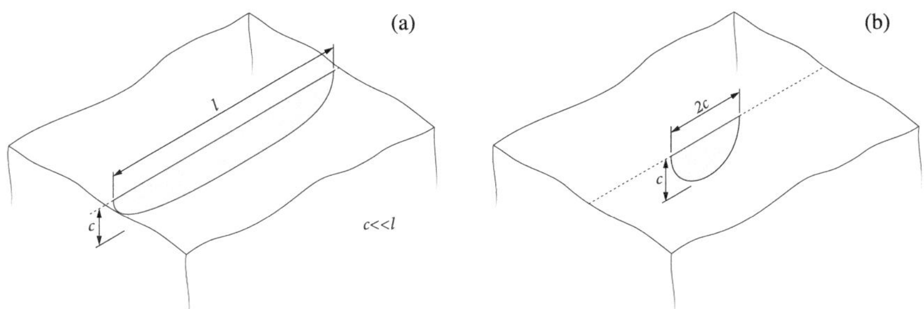

Fig. 4.9 Geometries of surface defects on the half-infinite body: **a** *long surface crack* and **b** *half-penny crack*

stress intensity factors can be obtained for different crack geometries and loading types.

In practice, to describe surface defects of glass in construction, such as, for example, scratches, the *long surface crack in semi-infinite body*, see Fig. 4.9a, under uniform, constant load σ_{a} with $Y = 1.1215$ in Eq. (4.14) can be used. The geometry factor for a half-penny crack (also known as thumbnail crack) is $Y = 2/\pi \approx 0.637$ at its bottom and $Y = 0.713$ on the surface, see Fig. 4.9b [27].

Under the assumption of inert conditions, and by neglecting sub-critical crack growth, the magnitude of the critical crack depth, $c_{\mathrm{f,inert}}$, that would lead to fracture under tensile loading can be calculated using Eqs. (4.11) and (4.14):

$$c_{\mathrm{f,inert}} = \left(\frac{K_{\mathrm{Ic}}}{Y \cdot \sigma_{\mathrm{a}} \sqrt{\pi}} \right)^2 . \tag{4.15}$$

The critical crack depth for glass components with a long surface crack for a given fracture toughness $K_{\mathrm{Ic}} = 0.75\ \mathrm{MPa}\sqrt{\mathrm{m}}$ and assuming an external tensile stress of $\sigma_{\mathrm{a}} = 45$ MPa leads to $c_{\mathrm{f,inert}} = 70^{-6}$ m. The inert strength $\sigma_{\mathrm{f,inert}}$ in correspondence with Eq. (4.15), see Fig. 4.8b, is:

$$\sigma_{\mathrm{f,inert}} = \frac{K_{\mathrm{Ic}}}{Y\sqrt{\pi \cdot c}}. \tag{4.16}$$

More elaborate formulations to derive the fracture strength by taking sub-critical crack growth into account are available in [30, 31].

The fracture strength of glass may therefore be determined by applying linear elastic fracture mechanics, but this requires precise knowledge of the critical flaw depth in the glass component. However, in practice it is unlikely that the critical flaw in a glass component is identified a-priori and that its precise depth and shape are known. This method should therefore be used with caution. On the other hand this approach can be particularly useful in forensic engineering applications (see Sect. 5.4) when the critical flaw is easier to identify and its depth can be measured accurately. Regardless of the application, fracture mechanics forms the basis for understanding the fracture of brittle materials such as glass.

References

1. F.R. Schwarzl, *Polymermechanik* (Springer, Berlin, 1990)
2. J. Simo, T. Hughes, *Computational Inelasticity* (Springer, Berlin, 1998)
3. T. Belytschko, W.K. Liu, B. Moran, *Nonlinear Finite Elements for Continua and Structures* (Wiley, 2000), XVI, 650 S (unknown)
4. O.C. Zienkiewicz, R.L. Taylor, *The Finite Element Method*, vol. 2, ed. by O.C. Zienkiewicz, R.L. Taylor. Solid Mechanics (Butterworth-Heinemann, 2002), XIV, 459 S (unknown)
5. N.S. Ottosen, M. Ristinmaa, *The Mechanics of Constitutive Modeling* (Elsevier, 2005)
6. H. Brinson, C. Brinson, *Polymer Engineering Science and Viscoelasticity* (Springer, New York, 2008)
7. J. Rösler, H. Harders, M. Bäker, *Mechanisches Verhalten der Werkstoffe*, 4th edn. (Springer Vieweg, 2012)
8. G.A. Holzapfel, *Nonlinear Solid Mechanics—A Continuum Approach for Engineering* (Wiley, New York, 2000)
9. H. Altenbach, *Kontinuumsmechanik - Einführung in die materialunabhängigen und materialabhängigen Gleichungen*, 2nd edn. (Springer, Berlin, 2012)
10. M. Kaliske, H. Rothert, On the finite element implementation of rubber-like materials at finite strains. J. Eng. Comput. **14**(2), 216–232 (1997)
11. A. Ali, M. Hosseini, B. Sahari, A review of constitutive models for rubber-like materials. Am. J. Eng. Appl. Sci. **3**(1), 232–239 (2010)
12. S. Kolling, P.A. Du Bois, D.J. Benson, W.W. Feng, A tabulated formulation of hyperelasticity with rate effects and damage. Comput. Mech. **40**(5), 885–899 (2007)
13. Ansys Inc., *Ansys Release 14.5* (Southpointe, 2012)
14. W. Grellmann, S. Seidler, *Kunststoffprüfung*, 2nd edn. (Carl Hanser Verlag, München, 2011)
15. J.D. Ferry, *Viscoelastic Properties of Polymers*, 3rd edn. (Wiley, New York, 1980)
16. N. Woicke, M. Keuerleber, B. Hegemann, P. Eyerer, Three-dimensional thermorheological behavior of isotactic polypropylene across glass transition temperature. J. Appl. Polym. Sci. **94**(3), 877–880 (2004)

17. M. Drass, *Constitutive Modelling and Failure Prediction for Silicone Adhesives in Façade Design*. Ph.D. Thesis (TU Darmstadt, 2020)
18. P.L. Rosendahl, *From Bulk to Structural Failure: Fracture of Hyperelastic Materials*. Ph.D. Thesis (TU Darmstadt, 2020)
19. M.E. Eberhardt, Warum Marmor bricht und Eisen nicht, in *Spektrum der Wissenschaft* (2000), pp. 44–50
20. N. Pourmoghaddam, M.A. Kraus, J. Schneider, G. Siebert, Relationship between strain energy and fracture pattern morphology of thermally tempered glass for the prediction of the 2D macro-scale fragmentation of glass. Glas. Struct. Eng. **4**(2), 257–275 (2019)
21. M. Zaccaria, M. Overend, Nondestructive safety evaluation of thermally tempered glass. J. Mater. Civ. Eng. **32**(4), 04020043 (2020)
22. D. Gross, T. Seelig, *Bruchmechanik - Mit einer Einführung in die Mikromechanik*, 5th edn. (Springer, Berlin, 2011)
23. F. Kerkhof, *Bruchvorgänge in Gläsern*. Deutsche Glastechnische Gesellschaft (1970)
24. C.E. Inglis, Stresses in a plate due to the presence of cracks and sharp corners, in *Transactions of the Royal Institution of Naval Architects*, vol. 55, issue 1 (1913), pp. 219–241
25. A.A. Griffith, The phenomena of rupture and flow in solids. Philos. Trans. R. Soc. Lond. **A221**(1), 163–198 (1921)
26. G.R. Irwin, D.C. Washington, Analysis of stresses and strains near the end of a crack traversing a plate. J. Appl. Mech. **24**, 361–364 (1957)
27. B.R. Lawn, *Fracture of Brittle Solids* (Cambridge University Press, Cambridge, 1993)
28. H. Tada, P. Paris, G.R. Irwin, *The Stress Analysis of Cracks Handbook*, 2nd edn. (Paris Productions & Del Research Corp., St. Louis, 1985)
29. Y. Murakami, *Stress Intensity Factors Handbook (in 2 Volumes)* (Pergamon Press, Oxford, 1987)
30. M. Haldimann, Fracture strength of structural glass elements—analytical and numerical modelling, testing and design. Dissertation, École polytechnique fédérale de Lausanne (2006)
31. M. Overend, K. Zammit, A computer algorithm for determining the tensile strength of float glass. Eng. Struct. **45**, 68–77 (2012)
32. S.M. Wiederhorn, Influence of water vapor on crack propagation in soda-lime glass. J. Am. Ceram. Soc. **50**(8), 407–414 (1967)
33. E. Gehrke, C. Ullner, M. Hähnert, Correlation between multistage crack growth and time-dependent strength in commercial silicate glasses. Glastechnische Berichte **60**(8), 268–278 (1987)
34. J. Mencik, *Strength and Fracture of Glass and Ceramics* (Elsevier, Amsterdam, 1992)
35. C. Ullner, *Forschungsvorhaben-Nr. 43 D: Untersuchungen zum Festigkeitsverhalten und zur Rissalterung von Glas unter dem Einfluss korrosiver Umgebungsbedingungen* (1993)
36. EN 12150-1, Glass in building - Thermally toughened soda lime silicate safety glass - Part 1: Definition and description (2015)

Chapter 5
Strength of Annealed Glass

This Chapter discusses the strength of annealed glass, which is not an intrinsic material constant because it is highly dependent on the condition of the surfaces, in particular, as discussed in Sect. 5.2, *surface flaws and scratches may reduce the strength of glass dramatically. Furthermore, glass typically becomes weaker over time as the number and size of flaws and cracks grow as a consequence of load duration and humidity and other factors, as described in Sect.* 5.3. *The most relevant property for practical applications is the glass (tensile) bending strength, which is covered in Sect.* 5.4. *Whereas, glass is typically stronger in compression, as addressed in Sect.* 5.5.

5.1 General

The strength of annealed glass, i.e. glass without thermal or chemical pre-stressing, is governed by the stress-raising flaws at the micro- and nanoscale. Such flaws are typically present on the surface, which leads to stress concentrations under tensile/bending loading. The compressive strength of glass is considerably higher than the tensile strength and is rarely critical for engineering applications. Therefore, in practice, the term 'strength' is often synonymous with the tensile strength of the glass. The reduction in strength is due to physical defects in the glass (bulk flaws) and surface defects (surface flaws) as shown in Fig. 5.1. Bulk flaws are typically introduced during the production process whereas surface flaws are caused during production, handling and throughout the service-life of the glass. Stress concentrations occur in the vicinity of flaws and are a function of the flaw geometry. The flaw tip radii in glass can be atomically sharp; therefore high stress concentrations are generated at relatively low levels of nominal tensile stress. As with other brittle materials, fracture in glass propagates at very high speeds and is perceived as instantaneous. The fracture strength of annealed glass, therefore, depends on the size and shape (mor-

J. Belis et al., *Engineering Glass Structures*,
https://doi.org/10.1007/978-3-032-03233-1_5

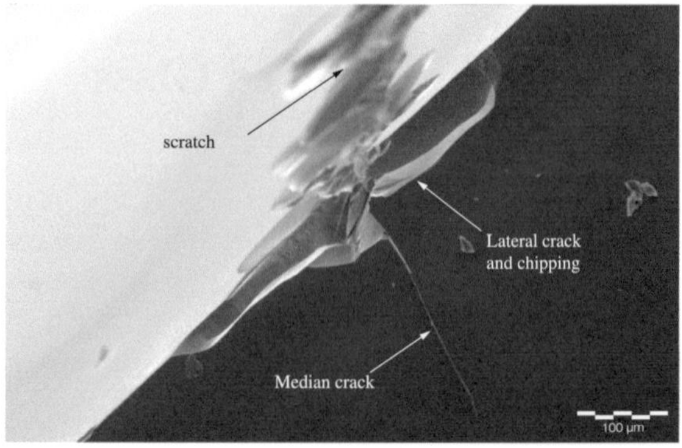

Fig. 5.1 Micrograph with scanning electron microscopic showing a micro-crack (deep crack) under a surface glass flaw (cross-section, the fractured area is shown in dark grey; and the original glass surface is shown in light grey)

phology) of the flaws. Flaws nucleate and accumulate on the surface of glass during its service life and therefore the strength reduces over time. In addition, if a tensile stress applied to a flaw in glass is insufficient to cause instantaneous (fast) fracture, the flaw could grow at a much lower speed until the size of the flaw becomes critical and fast fracture occurs. This slow crack growth, also known as sub-critical crack growth, occurs in the presence of tensile stress and moisture from the environment. This results in a significant reduction in tensile strength, particularly, when the glass is subjected to long-term loads [1].

The short-term strength of annealed glass with a surface exposed to damage accumulation in normal ambient conditions, is in the range of $\sigma_f \approx 30$–100 MPa, which is about 400 times smaller than the theoretical strength, see Fig. 5.2. The effect of the this sub-critical crack growth is taken into account in the codes and standards of many countries by means of a load duration modification factor, see Fig. 9.7.

The theoretical strength of the covalent molecular bonds within the glass is very high. The interatomic bond strength of the SiO network can theoretically be in the range of $\sigma_{f,theo} \approx 14.0$ GPa [3] to $\sigma_{f,theo} \approx 32.0$ GPa [2, 4]. From bending strength tests using liquid nitrogen (−196.2 °C) in pristine (almost no surface defects) glass fibres, Lower, Brow, and Kurkjian found the tensile strength experimentally to be $\sigma_f \approx 12.6$ GPa for pure silica glass and $\sigma_f \approx 10.0$–13.0 GPa for sodium silicate glass [5]. It was also observed that an increase in sodium content leads to an increase in glass strength. These values are in general agreement with the investigations of Proctor et al. [6], who found that the strength of fused silica glass at room temperature is $\sigma_f \approx 8.8$ GPa. The practical strength of glass is orders of magnitude lower than these values due to the presence of flaws.

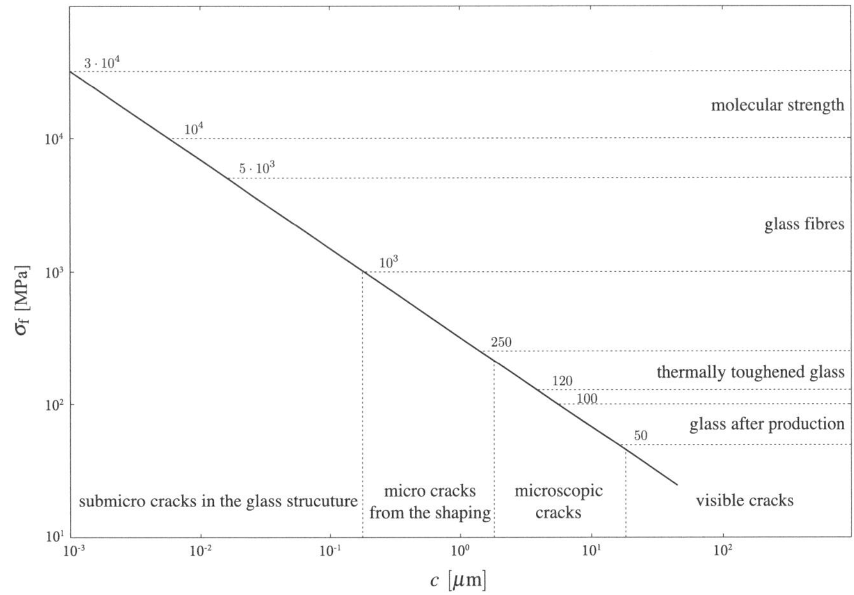

Fig. 5.2 Short-term strength of glass as a function of the effective crack depth c [2]

5.2 Defects/Flaws on Glass Surfaces

Contact between a hard body and a glass surface usually results in local damage. The latter is not limited to superficial scratches or indentations, but may include a more complex crack morphologies. Some idealised examples are shown in Fig. 5.3. In an indentation process, the indented body may exhibit either purely elastic or elastic–inelastic behaviour. The resulting crack morphology depends on the geometry and hardness of the indenter, as well as on the material properties of the receiving body, such as its elastic modulus and fracture toughness. The elastic stress distribution for two bodies in contact is described by the so-called *Hertzian contact stresses* [7, 8] and, according to [9], characterised by a conical crack *(Hertzian cone cracks)* below the contact area, see Fig. 5.3a. Such a crack runs circularly around the contact area and usually penetrates conically into the material's surface. This contact definition mainly refers to elastic contact. There is no permanent indentation on the glass surface after contact, which is typical for a so-called blunt contact. This kind of cracking can be observed, for example, in the local contact zone of ball drop tests to assess the response of laminated safety glass against hard impacts as described in EN356:2002 [10].

Elastic-inelastic material behaviour, which is also referred to as sharp contact, gives rise to a more complex crack morphology, see Fig. 5.4. Investigations into the scratch resistance of glass have shown that scratches and local indentations usually lead to this kind of visible damage [12, 13]. Due to the small contact area between damaging body and the receiving body and the relatively high load placed on the

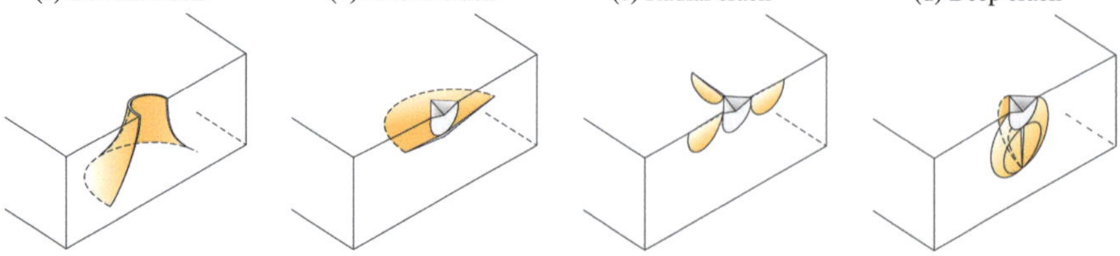

Fig. 5.3 Schematic representation of possible contact crack types on glass surfaces. An example of a hard indentation: **a** conical crack (*Hertzian cone crack*) due to elastic contact with a blunt penetrating body, not showing a permanently deformed contact zone, **b** lateral crack, **c** radial crack and **d** median crack [11]

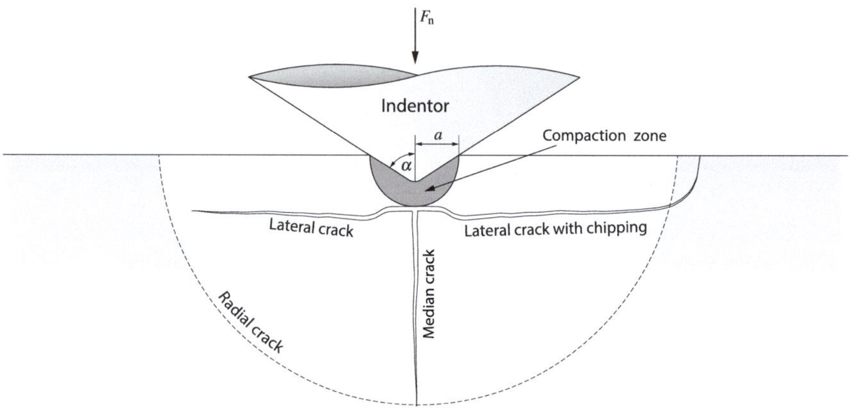

Fig. 5.4 Cross-sectional schematic representation of the crack morphology due to indentation: *median crack*, *radial crack* and *lateral cracks* close below the glass surface, finally running to the glass surface. As a result of the contact load, a compaction zone develops below the indenter [12]

former, a hydrostatic stress field with high shear stress develops on the glass surface and causes an irreversible deformation on the *compaction zone* (*scratch grooves* or *scratch marks*). Furthermore, according to this contact definition, *lateral cracks*, *radial cracks* and *median cracks* are induced [11, 14–16], see Fig. 5.3. While lateral cracks due to material chipping close to the surface significantly affect the visibility of a scratch, radial and median cracks are to a larger extent responsible for a reduction in the strength of the component [17]. The schematic representation in Fig. 5.3 depicts the formation of cracks in relation to the hardness of the surface and typical indentations caused by the indenter. In case of a faceted indenter, its geometry is in principle transferred to the indentation, see Fig. 5.1. A detailed examination of the local stress state below the indenter and the associated fracture mechanisms required for crack propagation can be found in [12]. Like other cracks in glass, the growth of lateral cracks is also time-dependent, due to the effects of subcritical crack growth, see Sect. 5.3. In engineering applications, due to the prevailing ambient conditions in combination with the residual stresses from the thermal pre-stressing process, thermally pre-stressed glass has a marginally higher optical scratch sensitivity compared to annealed float glass.

Other surface treatments such as acid etching and surface patterns can modfiy the surface flaw distribution and stress concentrations, thereby increasing enhancing

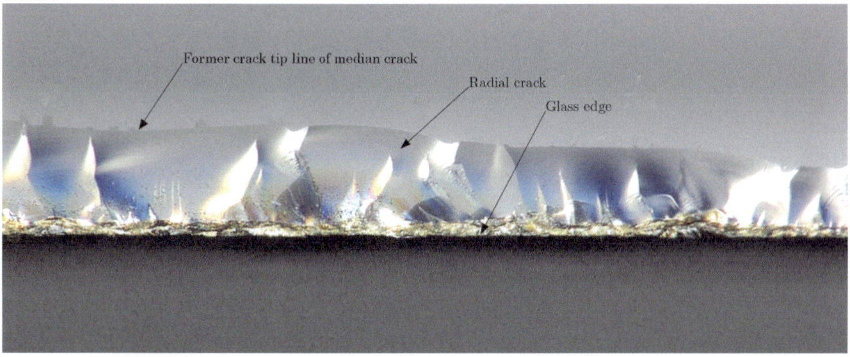

Fig. 5.5 Radial crack on an edge cut with a glass cutter (credits: J. Kleuderlein)

resistance to tensile cracking and localised contact (Hertzian) fracture [18]. Glass scoring (for the purpose of glass cutting) can be considered a special case of surface damage (scratching) occurring due to contact with a sharp object, see Sect. 2.3.1. As a result of the contact between the cutting wheel and the glass surface, lateral, radial and deep cracks occur. In the case of glass cutting, special liquids are used, which on the one hand lubricate the cutting wheel, and thus facilitate the rolling of the wheel over the glass, and on the other hand penetrate the crack, accelerate crack growth and avoid crack healing effects. If a small bending moment is applied, breaking of the glass along the scoring line occurs due to the strength-reducing effect of the median crack. During the cutting process, radial cracks are formed perpendicular to the freshly broken glass edge, and thus reduce the strength of the edge for stresses acting parallel to it, see Fig. 5.5. Although radial cracks can appear on the side of the cut edge, they can be removed by further edge processing (e.g., manual seaming or mechanical arrising, see Sect. 2.3.2). However, because of these additional mechanical processes, further cracks perpendicular to the glass edge might be created. Various authors [19–24] investigated the strength of the edge with respect to various edge qualities and different glass manufacturers. They demonstrated that the visual quality of the edge varied greatly, see Fig. 5.6, but these process-related differences do not always correlate to the edge strength. It was demonstrated that, especially when arrising the glass edges, the selected processing method (e.g., arrising by means of a cross-belt or grinding wheels) has significant influence on the edge strength.

5.3 Crack Growth and Static Fatigue

5.3.1 Subcritical Crack Growth

The fracture toughness of glass K_{Ic}, obtained from testing in inert environments is on its own insufficient to determine the glass resistance of any load history in

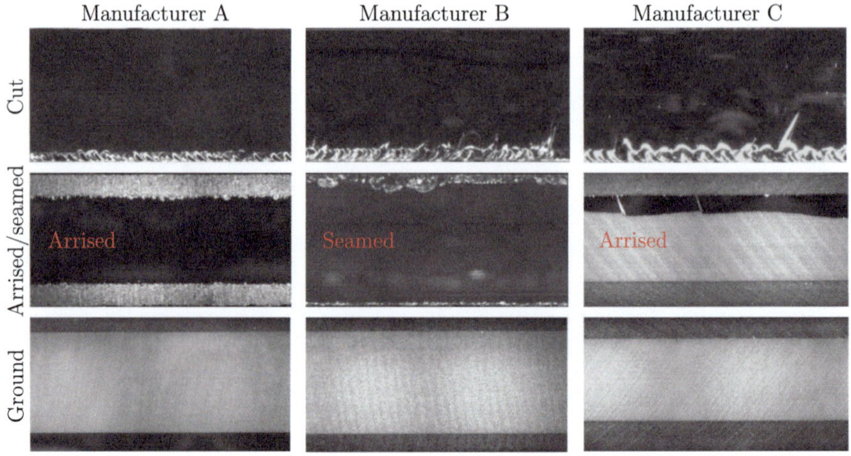

Fig. 5.6 8 mm thick glass edges from different glass manufacturers: Cut, arrised/seamed and ground, according to DIN 1249-11 [20, 104]

ambient conditions. In practice, processes at the crack tip at stress intensities lower than, K_{Ic}, also cause crack progression called *sub-critical crack* growth. Therefore, a time-delayed fracture may occur if glass is exposed to a load under ambient environmental conditions. The fracture strength increases with the strain rate as shown in the literature study by Meyland, Nielsen, and Kocer and also demonstrated by use of a modified split-Hopkinson setup in [25, 26]. Glass in engineering applications is subjected to a fluctuating Relative Humidity of up to 100% and can be exposed to direct moisture e.g., by rain), so that the effect of sub-critical crack growth has a significant impact on its strength, and thus also during strength measurements in ambient conditions.

The correlation between load duration and the fracture strength of glass was first examined by L. GRENET in 1899 as cited by [1, 27]. Fundamental experimental investigations into sub-critical crack growth effects in glass were conducted, amongst others, by Proctor, Whitney, and Johnson [6], Wiederhorn [28, 29], Gehrke, Ullner, and Hähnert (1987, 1990, 1993) [30–32]. While *glass corrosion* is based on chemical influences, see Sect. 1.3.3, the complex correlation between chemical and mechanical processes at the crack tip and its effect are referred to as *sub-critical crack growth*. Thus, the term 'sub-critical crack growth' refers to the process and the consequences of stress corrosion.[1]

The molecular description of stress corrosion in glass essentially refers back to Budd in [33] and Michalske and Freiman in [34, 35]. They describe the chemical reaction between pH-neutral water and other aqueous solutions on silicate glass. In the presence of these media, direct contact with the crack tip leads to a hydrolytic interaction between the atmospheric water molecules and the SiO-bonds in the zone

[1] Sub-critical crack growth is also referred to as *slow crack growth* or *static fatigue*.

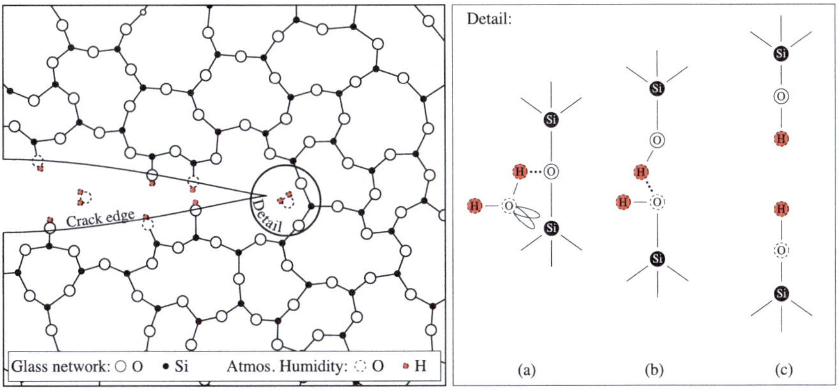

Fig. 5.7 Stress crack corrosion in glass: two-dimensional representation of the molecular reaction mechanism between water and silicate. Atmospheric water diffuses through the crack to the crack tip and leads to a separation of the glass network [35]

of the oxygen-bridge atom. The size of the reaction zone at the crack tip is, according to Wiederhorn [36], ~10 nm. The basic requirement for stress corrosion to take place is the expansion of the glass network by an adjacent mechanical tensile stress. Michalske and Freiman [35] explains that the corrosive process between water and a glass network takes place in three consecutive steps:

- **Step 1 (Adsorption)**
  Water molecules diffuse from the ambient air to the mechanically stressed glass. This is particularly critical at cracks tips that are subjected to opening stresses, because it leads to subcritical crack growth, see Fig. 5.7. Because of the dipolar nature of the H_2O molecule, one of the hydrogen atoms is directed towards the bridge ring oxygen atom and the oxygen atom towards the silicon atom of the glass network. Due to the resulting Van-der-Waals forces, the H_2O molecule is adsorbed at the SiO bonding point of the crack tip, Fig. 5.7a.
- **Step 2 (Reaction)**
  The resulting newly formed bonds are strengthened at the expense of the original SiO bonds. The adsorbed hydrogen atom of the H_2O molecule forms a strong bond with the bridge oxygen atom and the oxygen atom of the H_2O molecule forms an ionic bond with the silicon atom, see Fig. 5.7b. This causes a proton transfer from the hydrogen atom to the bridge oxygen and, simultaneously, an electron transfer of the oxygen atom of the H_2O molecule to the silicon atom. Since two hydrogen bonded silanol groups now exist, the original SiO bond no longer exists.

- **Step 3 (Separation)**
  This subsequently leads to the final separation of the previous hydrogen bridge bonds within the H_2O molecule, from the atomic bond of the glass network, see Fig. 5.7c. Two glass surfaces result from each silanol group surface. Since the hydrogen bridge binding the H_2O molecule has already been severely weakened in step 2, step 3 directly follows step 2.

Essentially, the reactive capability at the crack tip is determined by two factors. First, the penetrating medium must have the capacity to dispense electrons and protons. In addition, the glass network on the bridge bond must have a suitable polarity to enable the reaction. Invading molecules without ionisation and hydrogen behave inertly [37].

Due to the ring-like molecular structure of the glass (Fig. 5.7), a crack develops along the paths in the SiO bonds that have been corroded by the water molecules. The distance between the SiO bonds is 0.4–0.5 nm. The exact value depends on the number of tetrahedrons that form that particular ring-like molecular structure. Michalske and Bunker investigated the influence of the size of the molecules acting on the glass network [38]. It was that, with a molecular size of only 0.26 nm in diameter, water causes cracks to grow much faster than, for example, methanol (molecular diameter: 0.36 nm). It was also noted that the aniline molecule measuring 0.42 nm does not appear to have a measurable effect on crack growth. This explains the opening mechanism described above.

To account for the effect of load duration on the strength of glass, it is necessary to understand the underlying corrosive mechanism, characterised by the crack-growth velocity v and the stress-intensity factor K. Numerous studies were conducted to determine crack velocity in various chemical (non-inert) environments. The influence of water appears to be decisive for glass applications in engineering. Figure 5.8a shows, as an example for soda-lime silica glass, the influence of various relative humidity levels dependent on the stress intensity ($v(K)$-diagram). The position of these curves can vary depending on the chemical reaction capacity. The crack growth in ceramic materials and glass can be divided into four characteristic regions, see Fig. 5.8b.

- **Region 0: Stress corrosion limit or crack arrest**
  Especially in alkali-rich glass, below certain stress intensities, even with the presence of a reactive medium a *stress crack corrosion* limit or a *fatigue limit* can be observed. Depending on the composition of the glass, the crack growth rate this region 0 can be between $v_0 = 1 \times 10^{-8}$ m/s and $v_0 = 1 \times 10^{-12}$ m/s [39]. The stress intensity below which the stress crack corrosion limit is reached is denoted as K_{I0}. The stress corrosion limit in typical conditions in engineering applications lies within the range of $K_{I0} \approx 0.15$ MPa m$^{1/2}$ and 0.28 MPa m$^{1/2}$ [40].
- **Region I: Subcritical crack growth**
  Region I is mainly characterized by the molecular mechanisms of stress corrosion at the crack tip, see Fig. 5.7. The crack growth rate therefore depends on the stress

intensity and humidity level. The crack growth rate correlates strongly with the responsiveness at the crack tip, i.e., the interaction between the water molecules and the glass network. Although the fracture toughness K_{Ic} has not yet been reached, a slow crack propagation occurs. Crack propagation in region I is usually below $v = 1$ mm/s for soda-lime silica glass. In a logarithmic representation of the $v(K)$-diagram, see Fig. 5.8b, a linear correlation between crack propagation and stress intensity can be observed. Literature provides a variety of approaches to describe this correlation. For example, Wiederhorn [28] proposes a direct correlation to chemical responsiveness (kinetics) in region I, whereas Evans and Wiederhorn use an empirical power law to describe the rate of crack growth [41], see Eq. 5.1.

$$v^{\mathrm{I}} = \frac{dc}{dt} = A_{\mathrm{K}} \cdot K_{\mathrm{I}}^{N} \tag{5.1}$$

Here, A_{K}, refers to the crack propagation constant and N to a dimensionless crack growth parameter. According to Maugis [42], the crack growth rate can also be determined by inserting $A_{\mathrm{K}} = v_0 \cdot K_{\mathrm{Ic}}^{-N}$ into Eq. 5.1, which leads to the following:

$$v^{\mathrm{I}} = v_0 \left(\frac{K_{\mathrm{I}}}{K_{\mathrm{Ic}}} \right)^{N}, \tag{5.2}$$

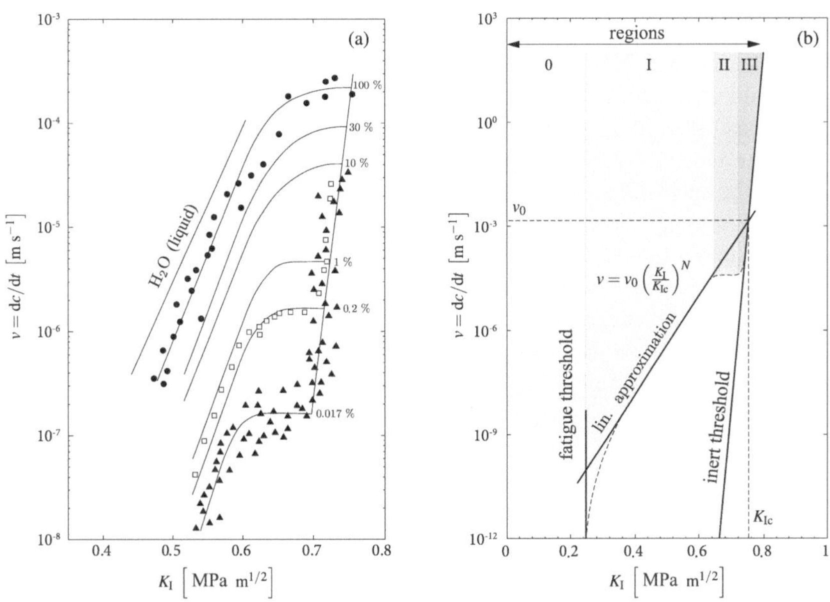

Fig. 5.8 **a** Influence of moisture on crack propagation, according to [28]; **b** characteristic $v(K)$ diagram for soda-lime silica glass for a relative humidity of 100% RH

where v_0, is referred to the initial crack growth rate. In a logarithmic representation, as shown in Fig. 5.8b, v_0 represents the ordinate value at K_{Ic}. The exponent, N, represents the slope of the linear approximation for region I. Typical crack growth parameters are listed in Table 5.1. This strong dependence of crack propagation on the environmental conditions is illustrated for relative humidity in Fig. 5.8a. The minimum crack velocity in region I is $v^{I}_{min} \approx 1 \times 10^{-10}$ m/s [43]b. For dry environments, the maximum subcritical crack growth velocity is $v^{I}_{RH=0.017\%} \approx 1 \times 10^{-7}$ m/s while the subcritical crack growth at maximum saturation of the ambient air at $v^{I}_{RH=100\%} \approx 1 \times 10^{-4}$ m/s. The crack growth parameter N, for soda-lime silica glass, has been investigated extensively. Values in the range of $N = 12$-100 have been reported and depend on the temperature and the chemical reaction capacity of the environment.

- **Region II: Critical crack growth**
  In region II, crack growth is still influenced by the environmental conditions, but is independent of the stress intensity. Since the crack growth is directly proportional to the relative humidity in the ambient air, it is described by a constant value of $v(K)$ as shown by the dashed line in Fig. 5.8b. Crack growth in this region changes from a subcritical crack growth to a critical crack growth. Lawn [37] assumes that the responsive medium through various mechanisms has only limited access to the crack tip. Only a very small plateau exists. For most ceramic materials, including glass, Region II occurs at high crack speeds. Crack propagation is thus mainly influenced by Region I [41].
- **Region III: Unstable crack growth**
  All experimental investigations at different humidities, at different temperatures and under inert conditions point to a limiting crack growth $v_0 \approx 1\,\mathrm{mm\,s^{-1}}$. The crack growth within region III is typically between $v^{III} = 1\,\mathrm{mm\,s^{-1}}$ and $v^{III} = 1e3\,\mathrm{mm\,s^{-1}}$. Crack growth in region III is unstable. As the crack accelerates, the surrounding reactive medium can no longer keep pace with the crack tip. Consequently, crack growth at values around K_{Ic} becomes independent of the ambient conditions. At larger stress intensities, the crack growth should theoretically reach the Rayleigh wave speed. However, the characteristic maximum speed for soda-lime silica glass is at $v^{III}_{max} \approx 1500\,\mathrm{mm\,s^{-1}}$, i.e. really only 40–50% of the theoretical value [44]. Instead of a further increase in crack growth, the excess energy is dissipated through an increase in roughness of the crack surface including micro-cracking and, ultimately, crack branching, see e.g. Sect. 10.3.1.

5.3.2 *Static and Dynamic Fatigue*

5.3.2.1 Basics

Sub-critical crack growth effects are essential for performing lifetime predictions of annealed glass under mechanical loading. Region I is of great importance for

Table 5.1 Subcritical crack growth parameters for prevailing ambient conditions of glass in engineering.

Literature source	Environmental	Crack growth parameters		
	Medium	N	A_K	v_0
[-]	[-]	[-]	$\left[\frac{m}{s\,(N\,mm^{-2}\,m^{\frac{1}{2}})^N}\right]$	$\left[mm\,s^{-1}\right]$
[28][b]	H_2O	17.4	0.6	3.8
	100.0% RH	20.8	1.4	3.6
	30.0% RH	22.6	1.1	1.7
	10.0% RH	21.4	0.29	0.6
	0.017% RH	27.2	0.22	$9{\cdot}10^{-2}$
	Vacuum	93.3	$6{\cdot}10^{7}$	0.13
[45][b]	H_2O	15.6	0.8	9.2
[46, 47]	H_2O	16	5.0	50.1[a]
	50.0% RH	18.1	0.45	2.5[a]
	Vacuum	70.0	250	$4.5{\cdot}10^{-4}$[a]
[48][b]	H_2O	19.2	4.0	16.0
[49][b]	H_2O	17.4	1.5	10.0
	Vacuum	98.0	$1{\cdot}10^{7}$	$1{\cdot}10^{-2}$
[50][b]	H_20	14.7	–	–
[30][b]	H_2O	15.5	0.28	3.3
	50.0% RH	16.8	0.12	0.95
[51][b]	H_2O	13.0	$4.7{\cdot}10^{-2}$	1.1
	50% RH	14.3	$1.0{\cdot}10^{-2}$	0.16
[31][b]	H_2O	18.4	0.42	2.14
[32][b]	H_2O	18.4	3.4	17.1
	50.0% RH	19.7	0.82	2.8
[52]	50.0% RH	17–21	–	–
[53]	H_2O	16.0	0.6[a]	6.0
	0.0–100.0% RH	16.0	0.6[a]	6.0

[a]Calculated in relation to $A_K = v_0 \cdot K_{Ic}^{-N}$, assuming a fracture toughness of $K_{Ic} = 0.75\,\text{MPam}^{1/2}$
[b]derived from linear regression according to Fig. 5.8a

lifetime predictions, since a significant part of crack growth takes place in this region. Equally relevant is the static fatigue limit (crack arrest, see Fig. 5.8). The crack growth in regions II and III can become relevant for stresses from dynamic loading (e.g., impact or blast overpressure). Evans and Wiederhorn [41], Kerkhof, Richter, and Stahn [46] and, later, also Munz and Fett [54] developed an analytical model from the empirical power law equation (Eq. 5.1) that accounts for the lifetime of glass components. Replacing Eq. 4.14 with Eq. 5.2 leads to:

$$\frac{dc}{dt} = v = v_0 \left(\frac{\sigma \cdot Y\,\sqrt{\pi \cdot c}}{K_{Ic}} \right)^N . \tag{5.3}$$

Using this differential equation, crack growth behaviour can be predicted by assuming that Eq. 4.14 applies to the entire range $K_{I0} \leq K_I \leq K_{Ic}$ and N is constant in both regions I and II [46, 54]. The characteristic plateau in region II is thus neglected, which is expressed by the linear extrapolation in Fig. 5.8b. Assuming that the critical crack resulting from subcritical crack growth is much greater than the initial crack depth ($c_f \gg c_i$), it follows that

$$\int_0^{t_f} \sigma^N \mathrm{d}t = \frac{2\, K_{Ic}^N}{(N-2)\, c_i^{\frac{(N-2)}{2}} \cdot v_0\, \left(Y\sqrt{\pi}\right)^N}. \tag{5.4}$$

Based on this equation, the lifetime of a glass component subjected to mechanical loading can be determined by neglecting the fatigue threshold K_{I0}, given by:

t_f	Life time expectancy	[s],
σ	Stress	[MPa],
K_{Ic}	Fracture toughness	[MPam$^{1/2}$],
N	Crack growth parameter	[−],
c_i	Initial crack depth	[m],
v_0	Initial crack growth	[m s^{-1}],
Y	Geometric factor	[−].

In general, the life time of a component can be determined for any stress history (Eq. 5.4) through knowledge of its fracture stress and load duration or of its fracture stress and stress rate, see Fig. 5.9. The conversion of a stress history $\sigma(t)$ to a constant stress is achieved through the introduction of an effective time t_{eff}. It defines the constant stress value $\sigma_{\max}$ applied over a period of time that would achieve the damage or fracture probability identical to the actual stress history $\sigma(t)$ over that time period, as follows:

$$\int_0^{t} \sigma^N \mathrm{d}t = \sigma_{\max}^N \cdot t_{\text{eff}}. \tag{5.5}$$

The crack growth parameter N has a crucial role in this relationship. However, the experimental identification and application of the latter may be flawed if:

- there is a pronounced crack arrest below region I during the experiments and this is not taken into account;
- in case of particularly fast loading, region II is included in evaluation of the measured data;
- crack healing effects in stress-free periods allow for an increase in strength (see Sect. 5.3.3).

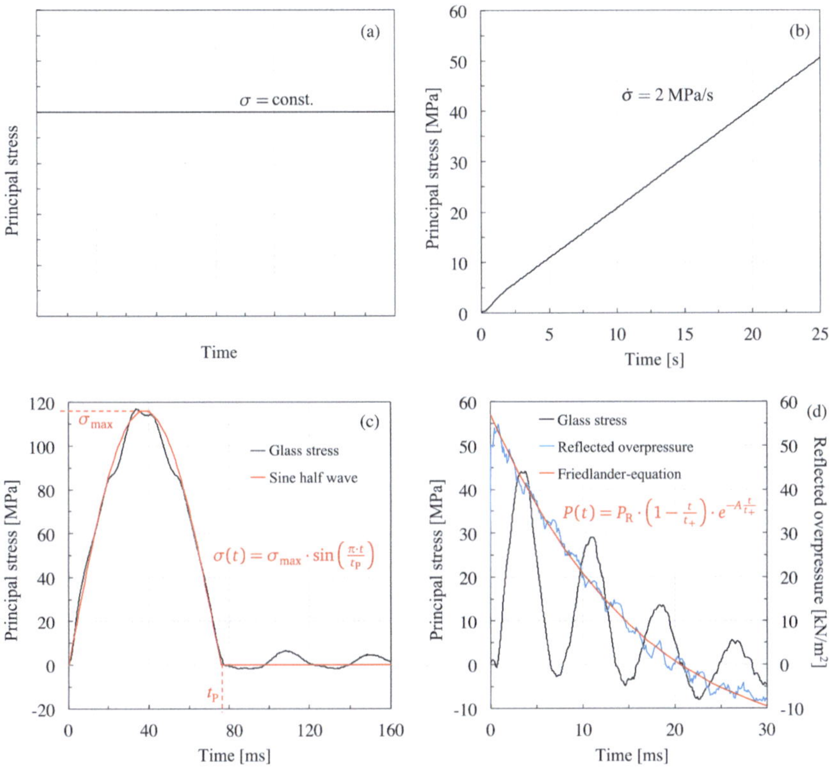

Fig. 5.9 Examples of typical stress-histories: **a** constant load, **b** standard test in a coaxial double-ring bending test with constant stress rate, **c** pendulum impact test and **d** reflected overpressure from a blast wave

5.3.2.2 Static Fatigue Due to Constant Loading

For a *constant load* (e.g., dead load, see Figs. 5.9a and 5.13a) with $\sigma(t) = \sigma_{f,s}$ the lifetime $t_{f,s}$ can be calculated from Section Eq. 5.4:

$$t_{f,s} = \frac{2\ K_{Ic}^N}{(N-2)\ c_i^{\frac{(N-2)}{2}} \cdot v_0\ \left(Y\sqrt{\pi}\right)^N \cdot \sigma_{f,s}^N}, \tag{5.6}$$

or by rearranging, the static fatigue strength $\sigma_{f,s}$ can be determined from:

$$\sigma_{f,s} = \left[\frac{2\ K_{Ic}^N}{(N-2)\ c_i^{\frac{(N-2)}{2}} \cdot v_0\ \left(Y\sqrt{\pi}\right)^N \cdot t_{f,s}}\right]^{\frac{1}{N}}. \tag{5.7}$$

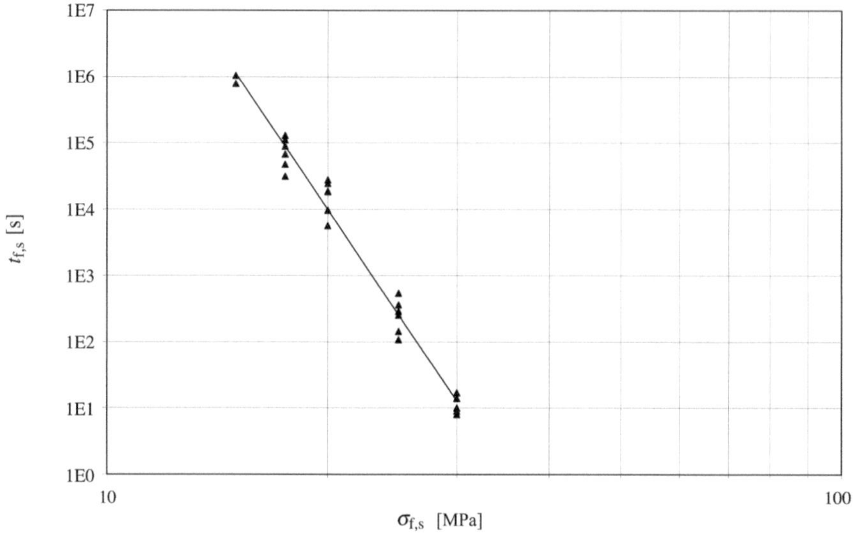

Fig. 5.10 Lifetime $t_{f,s}$ of annealed float glass with defined surface damage under constant ambient conditions and constant loading. Axes in logarithmic scale [55]

For two glass components under a constant stress $\sigma_{f,s,1}$ and $\sigma_{f,s,2}$ with equal surface damage, the relationship between lifetime and fracture stress can be obtained from:

$$\frac{t_{f,s,1}}{t_{f,s,2}} = \left(\frac{\sigma_{f,s,2}}{\sigma_{f,s,1}}\right)^N . \tag{5.8}$$

If there are different surface damages (initial crack depth c_i), then the relationship becomes:

$$\frac{t_{f,s,1}}{t_{f,s,2}} = \left(\frac{\sigma_{f,s,2}}{\sigma_{f,s,1}}\right)^N \cdot \left(\frac{c_{i,2}}{c_{i,1}}\right)^{\frac{(N-2)}{2}} . \tag{5.9}$$

Equation 5.8 demonstrates that for a typical crack growth parameter of $N = 16$ (soda-lime-silica glass) an increase in the stress by 10% results in a significant reduction of lifetime to 20% of the original value. Figure 5.10 shows results from load tests with constant stress levels by Fink [55], which were performed on annealed float glass panels in a four-point bending test (see Sect. 5.4.2.2). The graph shows the dependence of lifetime under different stress levels in a logarithmic representation. Thus, the load duration has a decisive influence on the glass strength in engineering applications, even for durations, which are generally referred to be of *short* or *medium* duration (e.g., wind loads, short-term traffic loads, see Sect. 9.3).

5.3.2.3 Quasi-Static and Dynamic Fatigue

For a *constant stress rate* $\dot{\sigma}$, which is commonly used in glass strength testing, the applied tensile stress increases constantly (see Figs. 5.9b and 5.13b) and is given by

$$\dot{\sigma} = \frac{d\sigma}{dt} = \text{constant}, \tag{5.10}$$

or

$$\sigma = \dot{\sigma} \cdot t. \tag{5.11}$$

The life time $t_{f,d}$ is determined by combining Eq. 5.10 with Eq. 5.4 to obtain

$$t_{f,d} = \left[\frac{2\, K_{Ic}^{N}\, (N+1)}{(N-2)\, c_i^{\frac{(N-2)}{2}} \cdot v_0\, \left(Y\sqrt{\pi}\right)^N \dot{\sigma}^N} \right]^{\frac{1}{(N+1)}}. \tag{5.12}$$

The quasi-static strength $\sigma_{f,d}$, is obtained by combining Eqs. 5.4 and 5.11 to obtain

$$\sigma_{f,d} = \left[\frac{2\, (N+1)\, K_{Ic}^{N}\, \dot{\sigma}}{(N-2)\, c_i^{\frac{(N-2)}{2}} \cdot v_0\, \left(Y\sqrt{\pi}\right)^N} \right]^{\frac{1}{(N+1)}}. \tag{5.13}$$

By transforming Eq. 5.13, the initial crack depth c_i can be obtained from

$$c_i = \left[\frac{2\, (N+1)\, K_{Ic}^{N}\, \dot{\sigma}}{(N-2)\, v_0\, \left(Y\sqrt{\pi}\right)^N \sigma_{f,d}^{(N+1)}} \right]^{\frac{2}{(N-2)}}. \tag{5.14}$$

Finally, for a constant stress rate and two glass components with a comparable surface damage, the following equation is valid:

$$\frac{\sigma_{f,d,1}}{\sigma_{f,d,2}} = \left(\frac{\dot{\sigma}_1}{\dot{\sigma}_2}\right)^{1/(N+1)}. \tag{5.15}$$

However, a doubling of the stress rate with a typical crack growth parameter of $N = 16$ only increases the strength by about 4%. To determine the crack growth parameter N experimentally, it is therefore necessary to increase the stress rate considerably.

5.3.2.4 Fatigue Due to Cyclic Loading

The fatigue of soda-lime silica glass under cyclic and periodically recurrent mechanical stress results from an accumulation of the crack growth of each individual oscil-

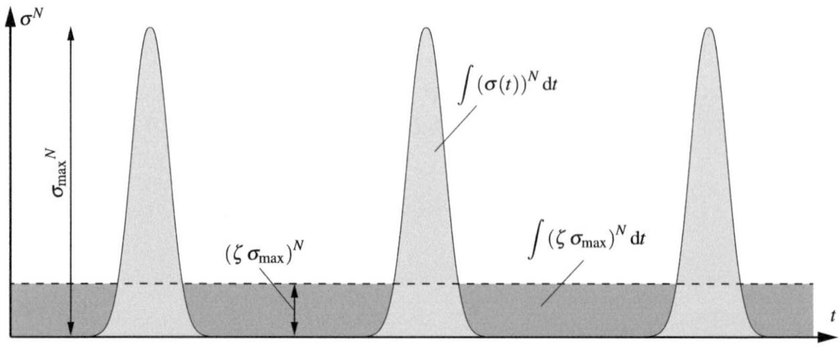

Fig. 5.11 Integration of an effective stress function for determining the stress coefficient ζ [40]

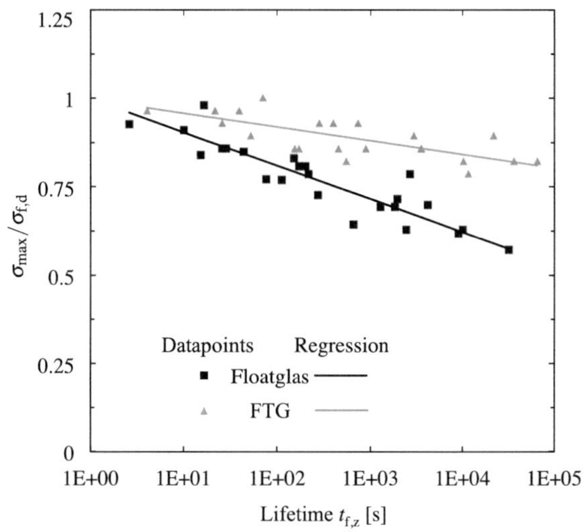

Fig. 5.12 Experimental studies of cyclic fatigue of glass: plot of ratio of peak stress σ_{max} to the fracture strength $\sigma_{f,d}$ against lifetime in cyclic tests $t_{f,z}$ [40]

lation. The lifetime for cyclic stress, just like the lifetime under static stress, can be predicted through integration of the stress over the stress history, which was shown by Overend and Zammit [56] and Hilcken [40]:

$$t_{f,z} = \frac{2\ K_{Ic}^N}{(N-2)\ c_i^{\frac{(N-2)}{2}} \cdot v_0\ \left(Y\sqrt{\pi}\right)^N\ (\zeta\sigma_{max})^N}. \tag{5.16}$$

Here, σ_{max} refers to the maximum value of the cyclic tensile stress, ζ is a stress coefficient that describes the relationship between the integral of the stress function of the cyclic stress and the static stress, see Fig. 5.11. For a sinusoidal, purely tensile stress history, for example, $N = 16$ and $\zeta = 0,9$.

$$\int \sigma(t)^N \mathrm{d}t = \int (\zeta\sigma_{max})^N\ \mathrm{d}t \tag{5.17}$$

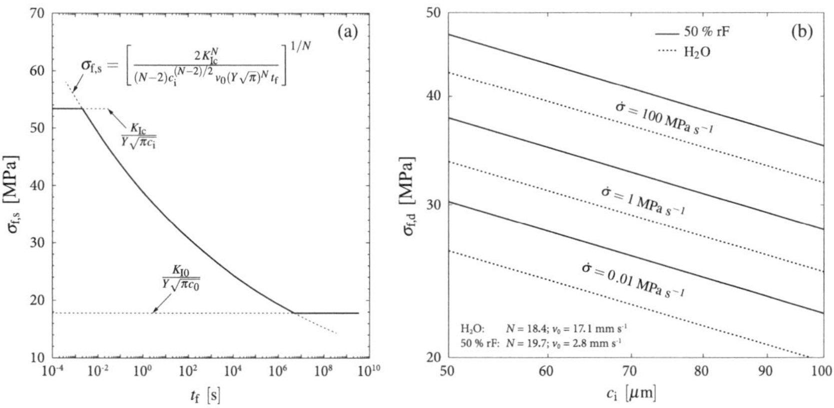

Fig. 5.13 **a** Relationship between static fatigue strength $\sigma_{f,s}$ and life time $t_{f,s}$ according to Eq. 5.7 [56]; **b** relationship between fracture strength $\sigma_{f,d}$ and initial crack depth c_i according to Eq. 5.14 for constant stress-rates $\dot{\sigma}$ [12]

According to experimental studies by Hilcken [40], cyclic fatigue can be calculated using the quasi-static tensile strength $\sigma_{f,d}$, see Fig. 5.12. However, when calculating the lifetime under cyclic stress with Eq. 5.16, a slightly smaller crack growth parameter is to be used than under static stress. Some of the earlier (relatively small number of) experimental investigations into the behaviour of glass under cyclic stress were carried out by Gurney and Pearson and Lü and [57, 58].

5.3.2.5 Fatigue Threshold and Inert Strength Limit

Physically, the range of values relating to fatigue strength is limited by an upper limit and a lower threshold: for a very small lifetime t_f (high crack speeds, the fatigue strength must converge with into a critical upper limit $\sigma_{f,\text{inert}}$ (Eq. 4.16). Also, no subcritical crack growth occurs below the fatigue threshold $\sigma_{f,I0} = K_{I0} / \left(Y \, (\pi \cdot c)^{\frac{1}{2}} \right)$ for very slow crack growth and small stress intensities ($K_I \leq K_{I0}$) must be met. The correlation between static fatigue strength and lifetime $\sigma_{f,s}$ is shown as an example in Eq. 5.7 and Fig. 5.13 for a typical surface crack in glass at a humidity of 50% RH. It is apparent from Eq. 5.7 that the fatigue strength would, incorrectly, tend towards infinity for very short stress durations ($\sigma_{f,I0} \rightarrow \infty$), and would, equally erroneously, approach zero below the fatigue limit. The upper and lower bounds must therefore be observed, as shown in Fig. 5.13, to avoid these errors.

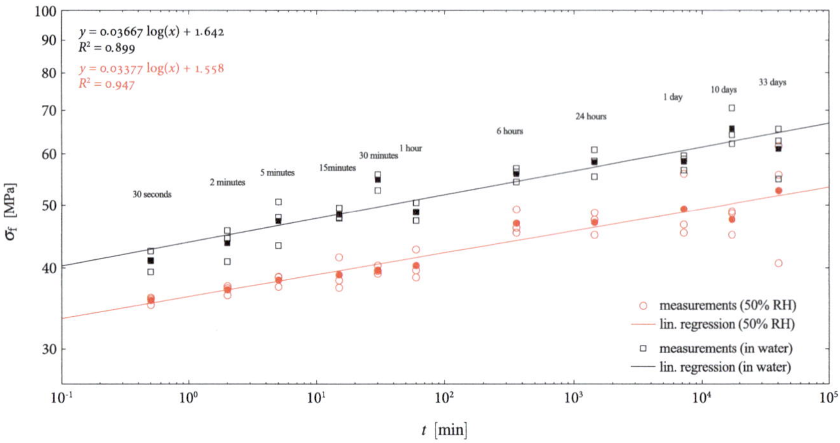

Fig. 5.14 Increase of bending strength depending on storage time [12]

5.3.3 *Crack Healing Effects*

In addition to stress corrosion (Sect. 5.3), the glass network reacts with H_2O in the absence of stress ($K \leq K_{I0}$) can also lead to *crack healing*. Levengood was the first to show from experiments that an increase in strength could also be observed by increasing the storage time, [59]. Further experiments with similar outcomes were conducted by Wiederhorn [60]. Ullner quantified the increase in inert strength of pre-damaged glass by a factor of 1.3–1.54 when immersed in water [32]. In this process, the strength approaches its limit over the course of several days. Figure 5.14 shows the results of strength tests of identically pre-damaged glasses. With increasing storage time (water immersion, at 50% RH), similar to [32], the strength increases observed after 24 h are 32% (50% RH) and 42% (water immersion), respectively. After a storage period of 33 d, a further increase in the strength of 50% was observed. However, the physical background of crack healing effects has not yet been sufficiently studied and understood for engineering applications. Therefore, they are currently not taken into account for the design of glass components.

Michalske systematically examined crack growth behaviour close to the stress corrosion limit (aka crack growth threshold or crack arrest), see Region 0 in Fig. 5.8b, [61]. In crack growth tests identical to those of [28], a stress intensity of $K_I = 0.45 \text{MPam}^{1/2}$ was first selected for cracks in soda-lime silica glass (Region I in Fig. 5.8b), to trigger subcritical crack growth. The stress intensity was then reduced to $K_I = 0.225\,\text{MPa m}^{1/2}$, below the stress corrosion limit. After a storage period of 16 hours and no further crack growth, the stress intensity was increased back to the original value. It was discovered that the sub-critical crack growth continued, but with a time lag. The higher the stress intensity chosen for the re-loading, the shorter the time lag became before sub-critical crack growth occurred again. The measured time lag was within the range of a few minutes to an hour; Michalske could not record a time lag at an immediate re-loading.

There are two interpretations of the underlying mechanisms of crack heal: one assumes *rounding of the crack tip (blunting)*, and the other is a chemical interpretation based on an *ion leaching process* in the vicinity of the crack tip and along the crack flanks.

The first interpretation postulated by Ito and Tomozawa and Bando, Ito, is based on the rounding of the crack tip according to Charles and Ito [62, 66], from a resolution and local deposit of the glass network in the area of the crack tip, see Fig. 5.15a, Bando's investigations have, as yet, turned out to be unique [63]. Examination of stress-free, water-saturated test specimens using high-resolution Environmental Scanning Electron Microscope[2] TEM revealed crack tip rounding, see Fig. 5.15b, which can be explained by the work of Ito [62]. Michalske also justifies the time lag before sub-critical crack growth by the fact that a crack tip rounding has taken place that must first be compensated for by a "sharpening the crack tip" [61]. Lawn et al. [67] doubted Bando's results [63]. Lawn et al. calculated, for the initial tip rounding of 1.5 nm, a corresponding fracture toughness of $K_{Ic} = 2.7 \pm 0.2\,\text{MPam}^{1/2}$, which is about three to four times greater than the fracture toughness specific to the material.

The second interpretation emerges from the initial works of Lawn, Jakus, and Gonzalez [68] and later works of Wiederhorn, Dretzke, and Rödel [64]. Lawn, Jakus, and Gonzalez could not find any evidence for crack tip rounding and therefore suspected that the experiments conducted by Bando, Ito, and Tomozawa [63] contained a systematic error. Lawn, Jakus, and Gonzalez noticed that no crack tip rounding occurs at stress intensities below the crack corrosion limit [68] and that a strength increase is rather caused by the absence of weakening ions (Na^+, K^+) in the vicinity of the crack tip. In this case, crack growth seemed to occur below the stress corrosion threshold. However, test specimens with residual stresses as a result of the indentation test did not show crack growth below the stress corrosion limit (K_{I0}) after annealing. In addition, neither measurable increase in strength nor crack healing effects occurred. Wiederhorn, Dretzke, and Rödel finally demonstrated that the absence of weakening ions and the corresponding local change in material properties is due to leaching and leads to a change in the crack path direction and an apparent crack healing, see Fig. 5.15c and d [64].

Investigations using the double-torsion (DT) test and the double-Cantilever (DC) test, according to Stavrinidis and Wiederhorn, [60, 69], reveal that these methods, in addition to causing an increase in strength and storage at room temperature (50 %RH to 70 %RH), as well as heating the specimens to about 120 °C, can reset the behaviour of specimens to the original degree of damage. In these experiments, specimens with a defined crack were provided; then, the (original) crack growth velocities of the cracks under load were measured, the specimens with cracks were stored for different durations ("healed") and then they were loaded again. In the case of a new

[2] The transmission electron microscope is a special kind of scanning electron microscope, which operates through irradiation of the specimen by the electrons. However, the specimen must be thin enough.

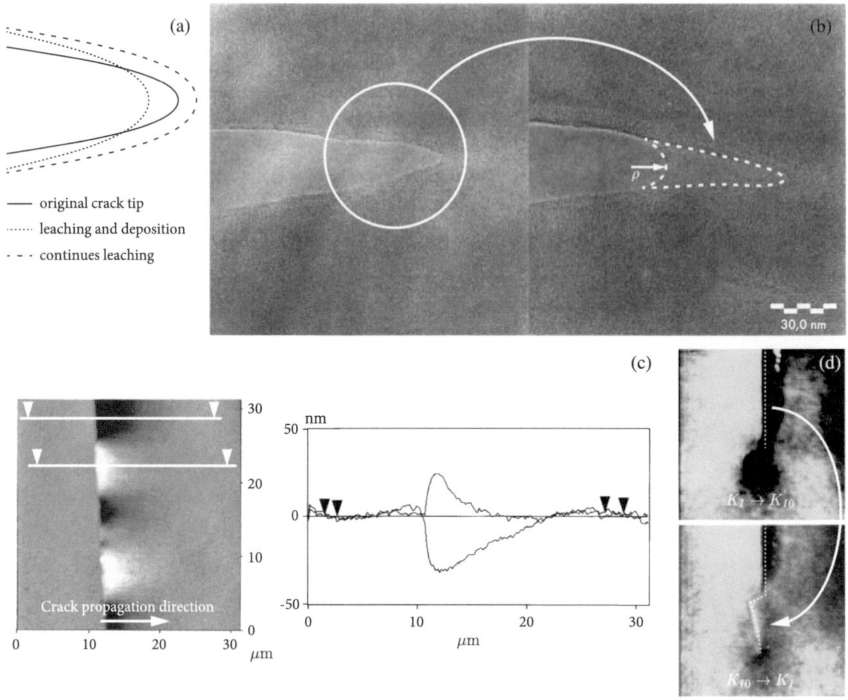

Fig. 5.15 Observed crack healing effects in soda-lime glass: **a** Crack tip rounding (blunting): Phenomenological representation according to [62]; **b** Crack tip rounding: (left) before and (right) after, stress-free storage in water shown through a transmission electron microscope [63]. **c** Crack path directional change caused by a local change in material properties: (left) Topography of a crack front, which, after storage below the stress corrosion limit, is increased again to a stress intensity above the stress corrosion limit and (right) its profile, as shown in the topography ([64]). **d** Crack path directional change caused by a local change in material properties: Crack tip observed with an Atomic Force Microscope (AFM) [65]: Stress intensity was reduced below the crack corrosion limit, (top) the crack growth was interrupted for about one hour and then (bottom) reloading continued over its original stress intensity. In (**c**) and (**d**), despite a pure mode I loading, initial deviations from the original crack path direction are observed

load, a lower fracture energy W_{break}, see Fig. 5.16, with increasing storage time or at temperature treatment, almost reached the level of the samples with the original degree of damage. No difference in crack growth behaviour was found in these tests, even in the case of multiple loading and unloading. It is believed that the temperature treatment removed water that had previously reacted with the fracture surface, thereby enabling more effective crack healing [70]. In theoretical models, such as the original one by Griffith [71], total crack healing is possible under ideal conditions.

Gehrke [39] justified the crack healing effect through the formation of an extended alkaline layer caused by leaching, as expressed by Eq. 1.6, along the crack flanks and at the crack tip. These leached areas have a higher fracture toughness than that of the glass network. According to their experimental findings, the time lag for continued

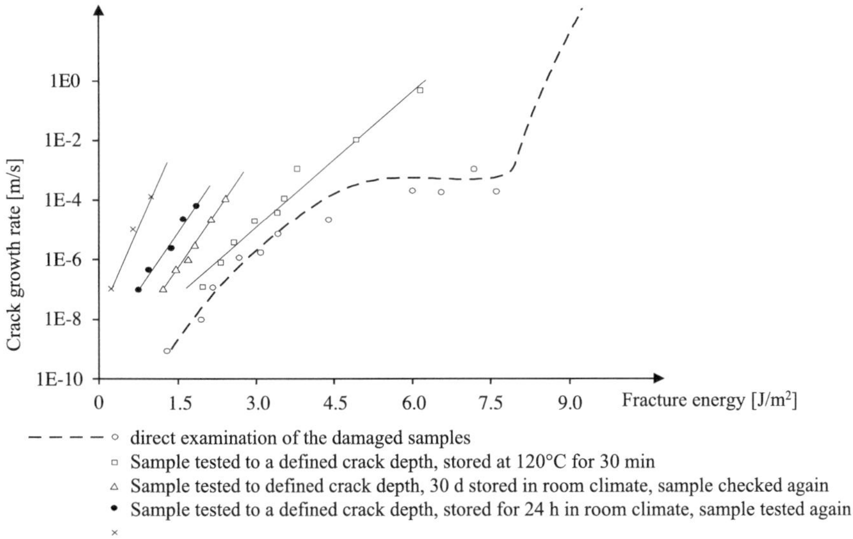

Fig. 5.16 Crack growth rate of differently healed specimens of soda-lime silicate glass, depending on the fracture energy at room temperature [69]

sub-critical crack growth next to stress intensity and the ageing period is also due to the leaching behaviour of the glass. Thus, a proportion of alkaline content serves as an important influencing parameter on crack healing. Henaux, Wiederhorn, Dretzke, and Rödel and Guin investigated sub-critical crack growth after storage under stress corrosion limits using high-resolution AFM recordings [64, 65, 72, 73]. Although there was a pure mode I stress in their experiments, they were able to show clearly that the crack propagation at the crack tip is first subjected to a change in direction, which leads to an apparent strength increase. Subsequently, the crack propagates along the original crack path direction, see Fig. 5.15c–d, by which the gain in strength is lost again. The origin of the crack progression is due to the high stress intensity at the crack tip, but no rounding of the crack tip in the experiments.

5.4 Bending Strength of Glass

5.4.1 Standards and Regulations for the Bending Strength of Glass

The strength of glass is significantly influenced by the morphology and orientation of flaws present at the surface as well as in the bulk material. Significant surface defects can form and accumulate on exposed surfaces during the service-life of the glass, but some are already induced during glass production and through further processing and handling. On the other hand, bulk flaws such as detrimental local inconsistencies in material composition, inclusions and air bubbles are typically only induced during the production of glass. All these flaws have a negative effect on the strength of glass, although in practice the strength of glass is typically governed by surface flaws, because they are more substantial and are generally found at locations of high stress.

Furthermore, the bending strength of glass is influenced by the tensile stress distribution on the surface of the glass. A large area of glass subjected to a tensile stress has a larger likelihood of containing a large flaw than a smaller area subjected to the same stress (see Sect. 5.4.4). The chemical processes at the tip of the flaw and the associated stress corrosion also cause slow crack growth, thereby also influencing the strength of the glass, particularly for longer duration loads. For these reasons, the characteristic strength of annealed float glass is not a material constant, but is a function of many case-specific factors (see Sect. 9.3), namely: the surface and bulk material quality; the stress intensity and duration (aka stress history); the stress distribution; the chemical effects caused by the surrounding environment and the ambient temperature (see Sect. 5.3).

Most of the glass used in engineering applications is soda-lime silica glass, which is mainly produced by the float process (see Sect. 2.2.1). Since the design life of glass components in practice can exceed 50 years, it is therefore very difficult to estimate the severity of strength-reducing surface flaws. Strength values derived directly from the strength testing of glass immediately after production do not seem representative in this respect. A relatively small amount of systematic testing of aged or weathered was performed in the development of national codes of practice such as those by Blank, Mellmann and Maultzsch [47, 74] that form the basis for design and construction rules for glass in the building industry in Germany (DIN 18008 [112]) and those by Beason and Morgan [75] and Dalgliesh and Taylor [76] for the design strength guidelines in the US (ATSM E-1330) and Canada (CAN- CGSB 12.20), respectively.

The strength of (aged) glass is commonly determined by means of four-point bending tests or coaxial double-ring bending tests (see Sect. 5.4.2). Some attempts have been made to quantify the strength of glass in its damaged state e.g. testing of naturally weathered glass [77, 78] and testing artificially aged glass using carborundum or sand abrasion [47, 79]. Results from some of these tests have been incorporated into design rules, but it is not clear whether the ageing conditions used in the tests are truly representative of the broad range of conditions to which glass could be exposed during its service life. In particular whether the tested glass specimens actually had a suitable flaw size and flaw distribution that are representative of all real-world conditions.

On the other hand, characteristic bending strengths for thermally pre-stressed glasses (HSG and TTG) are typically assessed using four-point bending tests (see Sect. 5.4.2.2). Standardised values of characteristic bending strengths f_k (5%-fractile) for annealed and pre-stressed glass are listed in Table 5.2, according to the respective product standards.

The bending strength of chemically pre-stressed glass is even more sensitive to the surface condition than annealed or thermally pre-stressed glass. This is due to

Table 5.2 Characteristic bending strength f_k of glass used in construction

Glass	Standard	Characteristic bending strength[a] f_k
[-]	[-]	[MPa]
Annealed glass (ANG)	EN 572-1 [115]	45
Heat strengthened glass (HSG)	EN 1863-1 [114]	70
Enamelled HSG	EN 1863-1 [114]	45
Thermally toughened glass (TTG)	EN 12150-1 [119]	120
Enameled TTG	EN 12150-1 [119]	75
Chemically toughened glass (CTG)	EN 12337-1 [106]	150

[a]5%-fractive value at 95% confidence coefficient

the limited Depth of Layer (DoL) of the zones subjected to compressive pre-stress (usually about 40 µm). Consequently, if the surface damage is deeper than the DoL, the bending strength is in fact marginally lower than that of annealed glass. For other glass products, such as ornamental glass according to EN 572-5 [117] or polished glass according to EN 572-3 [116], the specified strength for EN 572-1 [115] of $f_\mathrm{k} = 45$ MPa is not applicable. For these glass products it is commonplace to assume a reduced characteristic bending strength, e.g. $f_\mathrm{k} = 25$ MPa [80].

The cutting process induces large flaws along the edges of glass and this can have a significant effect on the strength, particularly if the edges are not treated after cutting. Glass strength guidelines often account for an edge strength reduction, e.g. DIN 18008-1 specifies a reduction of 20% while PrEN 16612 [118] recommends a reduction of 20% for as-cut, arrissed or ground edges, 10% for seamed edges and no reduction for polished edges of annealed float glass. Whereas the edge strength of other glass types (patterned glass, polished wired glass and wired patterned glass) is approximately 20% weaker than the surface strength. The research of Vandebroek, Müller-Braun et al. [21, 22, 24, 81] provides further insights on the edge strength of glass and the crack formation originating from the edge processing. Furthermore, Jotz, Schneider, and Rädlein [82, 83] developed a new method to cut, test and evaluate the edge strength of ultra-thin glass that is used in electronic consumer goods.

It is more challenging to establish the strength of curved glass than that of flat glass. Here too it is important to distinguish between the bending strength of the glass surface and that at the edge [84]. Depending on the geometry of the curved glass, it may be challenging to apply the double-ring bending test (see Sect. 5.4.2.3). Müller- Braun and Schneider [85] have shown that the ring-on-ring test yields accurate results in curved glass with large radii. Alternatively, the four-point bending test (see Sect. 5.4.2.2) can be used to determine the bending strength for cylindrical bent glass. In principle, the bending strength of curved glass strongly depends on the quality of the manufacturing process. Curved annealed glass produced by sag-bending, building should achieve a characteristic bending strength at the surface of $f_\mathrm{k} = 40$ MPa (5 %- fractile at 95 % confidence level). A strength reduction of 30 % is commonplace at the edges of curved glass.

Table 5.3 Comparison of characteristic strengths f_k (partial safety concept) and permissible bending stresses σ_{perm} (global safety concept) for glasses used in construction in Europe and North America and the related design codes

Type of glass	Europe	Germany	Austria	Switzerland	North America
	CEN-TS 19100	DIN 18008	ÖNORM B 3716-1	SIA 2057	ASTM E1300[a]
	f_k	f_k	f_k	f_k	σ_{perm}
	MPa	MPa	MPa	MPa	MPa
Ann. float glass	45	45	45	45	23.3
HSG	70	70	70	70	46.6
TTG	120	120	120	120	93.1

[a]for a load duration of 3 s (gust wind load) and a failure probability of $P_f = 8/1\,000$

For pre-stressed curved glass, Bucak et al. [86] determined $f_k = 55$ MPa for heat strengthened curved glass and $f_k = 105$ MPa for thermally toughened curved glass, each for a 5%-fractile value at 95% confidence level. It should be noted, however, that the resulting bending strength of curved thermally toughened glass strongly depends on the quality of the manufacturing process and thus requires special attention. Technical approvals for manufacturers were granted for curved TTG in Germany, for example, since 2016 [87] (Table 5.3).

As the engineering glass often needs to follow local guidelines and specific national guidelines in addition to the design methodology in the Eurocode, for example in Germany DIN 18008-1 [112], in Austria ÖNORM B 3716-1 [111] and more recently in Switzerland, see also [88]. In this regard, the review of relevant standards and regulations in Europe by Feldmann and Kasper [89] from 2014 is a very useful overview and forms the basis of the new Eurocode [90–92].

5.4.2 *Testing Methods*

5.4.2.1 General

The experimental determination of the characteristic bending strength f_k for glass is typically carried out according to EN 1288 by means of four-point bending tests (see Sect. 5.4.2.2) or by coaxial double- ring tests (see Sect. 5.4.2.3). In order to normalise the effects from the ambient environment, the tests need to be carried out under controlled climatic conditions. For example, in EN 1288, the ambient temperature for a bending test should be $T = 23(5)$ °C and the relative humidity should be between 40 and 70%. The test process should induce a constant stress rate of 2.0 ± 0.4 MPa s^{-1}. It is important to account for any geometric nonlinear response that may occur. In such instances, the applied force is no longer directly proportional

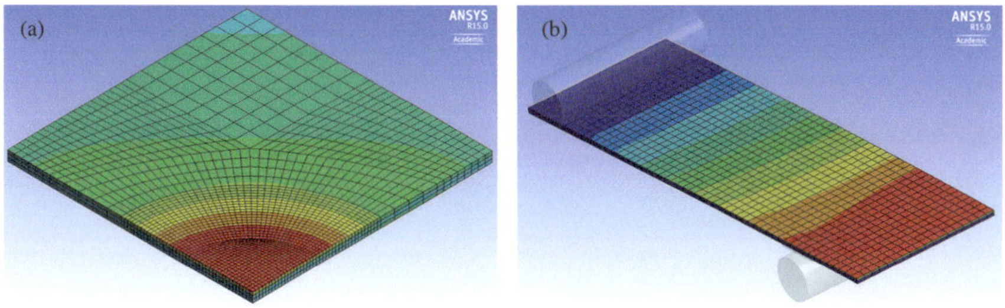

Fig. 5.17 FEM modelling for determining the principal tensile stresses on the glass surface: **a** coaxial double-ring bending test and **b** four-point bending test. The tension sides of a quarter model using double symmetry conditions are shown. The areas with the highest and lowest principal tensile stresses are indicated in red and blue, respectively

to the stress induced in the glass surface. This typically occurs at large deflections and associated higher loads. As a rule of thumb, non-linear effects become significant when the deflection exceeds the thickness of the plate.

The actual thickness of glass can also deviate from its stated "nominal" thickness within a tolerance defined by manufacturing standards. In fact, most glass is typically at the lower end of the tolerance limit of the nominal thickness. The reason for this is that the relatively large tolerance ranges date back to old production processes for flat glass that have yet to be updated to reflect the improved precision capabilities of modern process technologies. Regardless of the higher precision, it is still necessary to measure the thickness of the glass at multiple locations before calculating the fracture stress of a test specimen. It is also advisable to locate the fracture origin after testing, by applying a transparent self-adhesive polymer film to the top (compression) surface of the specimen. This will hold the fragments together after fracture with negligible effects on the fracture strength. In this context it is important to bear in mind that glass does not necessarily break at the location of the maximum stress due to different surface conditions and pre-treatment.

Direct contact between glass and steel during the bending test is possible, but this requires the loading jigs to be perfectly aligned with the specimen. Such alignment can be achieved by introducing an articulated joint above the loading jigs so that the loading edge makes perfect contact with the glass specimen. In addition, in four-point bending tests it is recommended to use relatively soft metal (e.g. aluminium) strips between the loading piece and the glass specimen to reduce stress concentrations from the contact.

The fracture stress can be calculated by means of analytical relationships, but it is often useful to add further insights from numerical (FEA) analyses that provide a more complete view of the effects of the test setup, see Fig. 5.17 and provides a useful correlation between the load applied and the related principal stresses. For more complex or unusual setups it is useful to apply strain gauges to the glass surface to validate the numerical models, particularly when a new test jig or new test configuration is used.

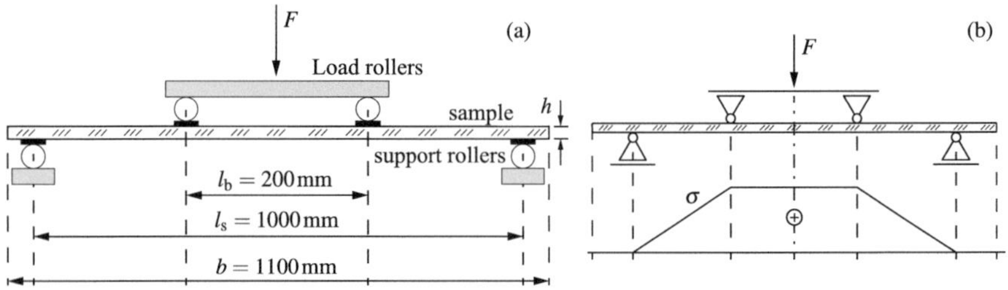

Fig. 5.18 **a** Test configuration for a four-point bending test, in accordance with EN 1288-3 [109] and **b** the corresponding stress distribution according to beam theory

5.4.2.2 Four-Point Bending Test

In the case of according to EN 1288-3 [109], the test specimen is mounted onto two line-support rollers and loaded by two line-load rollers, each 100 mm to the left and right of mid-span. This produces a constant bending moment between the load rollers (aka the load span), and thus a state of almost uniform tension stress on the bottom surface of the glass specimen between the load rollers, see Fig. 5.18. This will invariably induce a very similar tensile stress level along the edges of the glass specimen in this region. According to simple beam theory the stress at the bottom surface of the load span is constant and uniaxial, but this is no longer valid for wide beams and plates because of the lateral transverse strain (contraction) arising from the Poisson's ratio effect. The tensile stress level is not uniform, but is largest at the edges which experiences the smallest transverse contraction and smallest at the centre of the plate which is subjected to the smallest transverse contraction, thus the plate adopts a convex shape in transverse direction (see Fig. 5.17b).

The glass specimen dimensions for carrying out the test according to EN 1288-3 [109] are $(a \times b)$ 360 ± 5mm $\times$ 1100 ± 5mm. The total span (i.e. between the supports) is $l_s = 1\,000$ mm, and between the span between the load rollers is $l_b = 200$ mm.

The large deflections of specimens of thin glass or curved glass must be taken into account for testing. Experimental studies have also shown that, in order to determine the correct fracture stress, the friction in the rollers or between the bearing rollers and the specimen must be considered carefully.

In addition to the four-point bending test, there is also the three-point bending method, in which the test specimen is loaded with only one line-load roller at mid-span. Due to the variation in bending moment along the span, the stress at failure will depend on the location of the fracture origin, which will typically vary considerably between specimens. Furthermore, only a very small area below the load roller is stressed to the maximum stress, so the resulting strength values are difficult to compare to each other and to the results obtained from the four-point bending test. Consequently, four point bending tests are generally preferable to three point bending tests. Both there and four point bending tests induce high tensile stresses in the edge of the glass and unless the edges are not processed very carefully (e.g. polished)

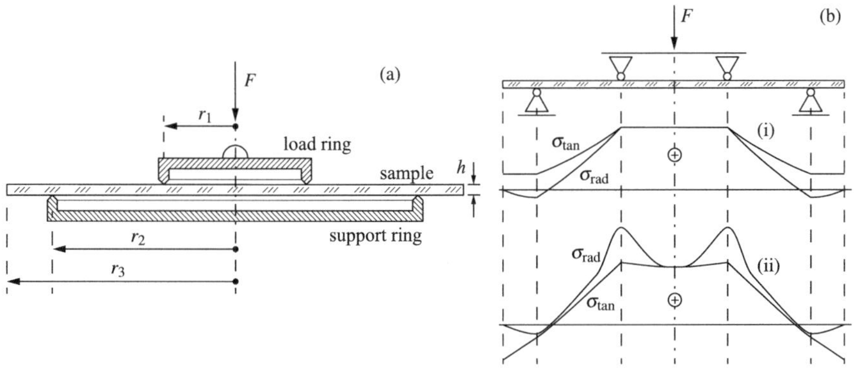

Fig. 5.19 **a** Test arrangement in the double-ring bending test in accordance with EN 1288-2 [108] and EN 1288-5 [110] and **b** the stress distribution for the radial and tangential stresses on the bottom surface of the specimen, assuming small deflections (linear plate theory, i) and large deflections (geometrically non-linear plate theory, ii)

the fracture strength in both these test set-ups is very often governed by edge defects (i.e. the fracture origin is often located on the edges), therefore what is being tested is the edge strength of the glass, rather than its surface strength.

5.4.2.3 Coaxial Double Ring Bending Test

The coaxial double-ring is an axisymmetric four-point bending test, which induces a high and almost constant stress level in the circular area confined by the load ring (smaller of the two rings) and a linearly decreasing stress in the region between the load ring and the support ring. The edges of the glass specimen are thus subjected to negligible stress levels (Fig. 2.23). The test is therefore ideal for determining the tensile strength of the glass surface i.e. independent of the influence of the glass edges.

EN 1288 provides two alternative test setup sizes for double-ring bending tests, see Table 5.4. The procedure according to EN 1288-2 [108] requires very large test specimens that lead to large deflections, and the pressure in the ring chambers must be actively controlled, thereby making this test difficult to perform and relatively unpopular. Conversely, the smaller setup in EN 1288-5 [110] is often considered to be too small to be truly representative of typical architectural and engineering applications. Moreover, the glass specimens are so small that they cannot be easily thermally treated in conventional tempering furnaces. Therefore, researchers have often selected glass specimens and ring sizes that deviate from the two sizes provided in EN 1288, see e.g. [52, 93].

For the coaxial double ring bending test, the maximum bending stresses on the glass surface σ_{rad} and σ_{tan} inside the load ring according to EN 1288-1 [107] should be calculated using

Table 5.4 Geometries of the double-ring bending test setup, according to EN 1288-2 [108] and EN 1288-5 [110]

Reference name	Type/Name	Radius of load ring	Radius of support ring	Test surface[a]	Radius of specimen[b]
		r_1	r_2	A	r_3'
[-]	[-]	[mm]	[mm]	[mm^2]	[mm]
EN 1288-2	–	300 ± 1	400 ± 1	240 000	500 ± 2
EN 1288-5	R45	9	45	254	50 ± 1
	R30	6	30	113	33 ± 0.5

[a]Surface under quasi-uniform stress distribution ($\sigma_{rad} = \sigma_{tan} =$ constant)
[b]The inscribed circle radius r_3' of a square specimen can be used instead of the radius r_3 of a circular specimen

$$\sigma_{rad} = \sigma_{tan} = \frac{3(1+\nu)}{2\pi}\left[\ln\frac{r_2}{r_1} + \frac{1-\nu}{1+\nu}\frac{r_2^2 - r_1^2}{2r_3^2}\right]\frac{F_{max}}{h^2} \tag{5.18}$$

given by

r_1	Radius of load ring	[mm],
r_2	Radius of support ring	[mm],
r_3	Radius of circular specimen	[mm],
ν	Poisson's ratio of glass	[−],
F_{max}	Fracture load	[N],
h	Glass thickness	[mm].

Due to the axi-symmetric stress state, the tensile stresses on the glass surface in the area of the load ring are equi-biaxial i.e. $\sigma_1 = \sigma_2 = \sigma_{rad} = \sigma_{tan}$. Since Eq. 5.18 is based on an analytical solution from linear plate theory, the solution is strictly applicable only to rotationally symmetrical (circular) test specimens and for small deflections. Membrane stresses originating from geometrically non-linear behaviour and large deflections are neglected here. In the case of thin test specimens, the use of Eq. 5.18 therefore can lead to significant deviations from the real mechanical response (see Sect. 9.6). Furthermore, depending on the ratio of the rings to the thickness of the specimen, a local increase of σ_{rad} occurs in the vicinity of the load ring, see Fig. 5.19b. Also, for ease of fabrication, square plates are typically used as specimens instead of circular plates. This has a negligible effect on the test results. A FE-model, validated using strain gauges, is recommended for determining the bending stresses from tests, particularly when large deflections (> thickness of the glass plate) are expected.

5.4.3 *Statistical Analysis and Characteristic Bending Strength*

5.4.3.1 General

Technical guidelines and standards, such as the Eurocodes [113], require very low probabilities of structural failure. for structural components. Reliability class *RC 2* is usually used for such components. This is associated with consequence class *CC 2*, characterised by a moderate consequence on human life, and significant economic, social or environmental consequences (e.g. residential and office buildings, public buildings, see Sect. 9.2.1). The probability of failure acceptable within a period of one year is $P_{\mathrm{f,RC2}} = 1.3 \times 10^{-6}\,\mathrm{a}^{-1}$ and accumulates over 50 year to $P_{\mathrm{f,RC2}} = 7.2 \times 10^{-5}(50\,a)^{-1}$. This reliability is mainly determined by taking into account the probability density functions of the actions and the material resistance (see Sect. 9.2). For the tensile strength of pristine annealed float glass the coefficient of variation is typically up to $V = 0.3$, indicating that tests on a small number of specimens are not statistically representative. Further details on appropriate sample sizes for glass testing are available in [94]. Statistically, these uncertainties can be expressed in the form of confidence intervals within which the value sought is within a given probability. If this value is in the middle range of the bending strength distribution, for example, the mean value, this confidence interval can be rather narrow, even with only a few specimens, see Fig. 5.21. On the other hand, to determine a characteristic bending strength f_{k}, (low fractile value) a large number n of randomly selected and nominally identical specimens $\{x_1, \ldots, x_{\mathrm{n}}\}$ needs to be tested according to a uniform procedure; this is necessary so that the lower tail of the distribution can be populated with sufficient data. Therefore, a comprehensive statistical evaluation requires a database of strength values obtained from carefully executed tests such as those described in Sect. 5.4.2. According to EN 1990 [113], the characteristic bending strength corresponds to the 5% fractile of the bending strength using a confidence interval of 95%. Since the method for statistical evaluation in EN 1990, Appendix D [113], is based on a confidence interval of 75%, this is not applicable for the determination of a characteristic bending strength for glass, therefore an evaluation based on a statistical distribution function (see Sect. 5.4.3) is required. In scientific literature, the two-parameter Weibull distribution is typically used as a distribution function. Behr et al. [95], Schneider [52] and Schula [12] investigated several statistical distribution functions for the evaluation of bending strengths. *Log-normal* and *Weibull distribution* probability theories proved to be especially suitable in considering the strength values of glass. According to EN 1990 [113] both distribution functions can be applied. Datsiou and Overend [94] identified the Weibull parameters and estimators that provide the best fit for glass strength. Another approach is provided by Ballarini, Pisano, and Royer-Carfagni [97] in which a left-truncated Weibull distribution is proposed to account for the effects of manufacturing quality control of the lower fractile values for the glass strength.

5.4.3.2 Normal Distribution

The *normal distribution*, also known as *standard normal distribution* or *Gaussian normal distribution* is the probability distribution most commonly used in engineering design. The probability density f_{N}, see Fig. 5.20a, of a normal distribution is defined as

$$f_{\mathrm{N}}(x) = \frac{1}{\sigma\sqrt{2\pi}} \int_{-\infty}^{x} e^{-\frac{1}{2}\left[\frac{x-\mu}{\sigma}\right]^2} \mathrm{d}x. \tag{5.19}$$

This can be expressed as a cumulative distribution function F_{N} Eq. 5.20, by integration of Eq. 5.19:

$$F_{\mathrm{N}} = \frac{1}{2}[1 + e^{\left(\frac{x-\mu}{\sigma\sqrt{2}}\right)}]. \tag{5.20}$$

The normal distribution is characterised by the mean value μ and the standard deviation σ. The mean value defines expected value and the standard deviation represents the scatter. The points of inflection for probability density can be found at point $x = \mu \pm \sigma$, and 68.27% of the total area lies within this interval, where 90% can be found at interval $x = \mu \pm 1,65 \cdot \sigma$, 95% at interval $x = \mu \pm 1,96 \cdot \sigma$ and 99% at interval $x = \mu \pm 2,58 \cdot \sigma$ of the total area.

5.4.3.3 Log-Normal Distribution

Properties that cannot be smaller than a certain value can be described very well using the asymmetric, skewed, log-normal distribution. For example, the experimentally determined bending strength $(\{x_1, \ldots, x_n\})$ for normally distributed values is expressed in logarithmic form. A continuous random variable $X (> 0)$ is thus normally logarithmically distributed if $\ln(x)$ is distributed normally [98]. The probability density f_{LN}, see Fig. 5.20a of a log-normal distribution is

$$f_{\mathrm{LN}}(x) = \begin{cases} \frac{1}{\sigma_{\mathrm{LN}} x\sqrt{2\pi}} \exp^{-\frac{(\ln x - \mu_{\mathrm{LN}})^2}{2\sigma_{\mathrm{LN}}^2}} & \text{for } x > 0, \\ 0 & \text{for } x \leq 0. \end{cases} \tag{5.21}$$

Here, μ_{LN} and σ_{LN}^2 are the characteristic parameters for describing the log-normal distribution. By integrating Eq. 5.21, the cumulative distribution function is:

$$F_{\mathrm{LN}}(x) = \begin{cases} \frac{1}{\sigma_{\mathrm{LN}}\sqrt{2\pi}} \int\limits_0^x t^{-1} \exp^{-\frac{(\ln t - \mu_{\mathrm{LN}})^2}{2\sigma_{\mathrm{LN}}^2}} \mathrm{d}t & \text{for } x > 0, \\ 0 & \text{for } x \leq 0. \end{cases} \tag{5.22}$$

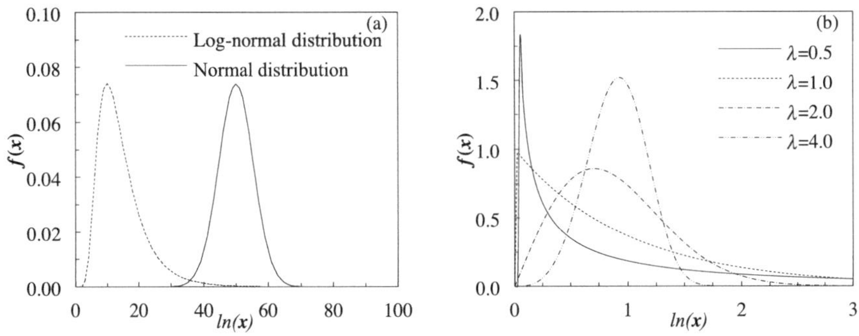

Fig. 5.20 Comparison of **a** the probability densities of normal and log-normal distribution and **b** two-parameter Weibull distribution, taking into account various form parameters

5.4.3.4 Weibull Distribution

The Weibull distribution, named after Waloddi Weibull [99, 100], is used to describe lifetimes, failure probabilities and material strength(weakest-link model). It is often used for brittle materials such as ceramics and glass, especially in the field of materials science. A continuous random variable $X (> 0)$ is referred to as a two parameter Weibull-distributed if the probability density f_{WB}, see Fig. 5.20b, is given by

$$f_{\mathrm{WB}}(x) = \begin{cases} \left(\frac{\lambda}{\beta}\right)\left(\frac{x}{\beta}\right)^{\lambda-1} \exp\left[-\left(\frac{x}{\beta}\right)^{\lambda}\right] & \text{for } x > 0, \\ 0 & \text{for } x \leq 0. \end{cases} \tag{5.23}$$

Here, β is the *scale parameter* and λ the *shape parameters*. The scale parameter β expresses the characteristic value of the random variable X with a probability of $P_{\mathrm{f}} = 0{,}6321$. The shape parameter λ is a measure of the scatter. Integrating Eq. 5.23, the cumulative distribution function F_{WB} of the Weibull distribution can be expressed as follows:

$$F_{\mathrm{WB}}(x) = \begin{cases} 1 - \exp\left[-\left(\frac{x}{\beta}\right)^{\lambda}\right] & \text{for } x > 0, \\ 0 & \text{for } x \leq 0. \end{cases} \tag{5.24}$$

The Weibull distribution is characterised by its special adaptability and is able to describe either rising, constant or falling failure rates. Varying λ, the Weibull distribution can express similarity to a variety of distribution functions (see Fig. 5.20b). For a constant failure rate of $\lambda = 1$, the Weibull distribution is expressed in the form of an exponential function. For a linearly rising failure rate $\lambda = 2$, it is expressed in the form of a Rayleigh-distribution, and for a value of $\lambda = 3, 6$, it results in a distribution with low skew similar to the normal distribution.

5.4.3.5 Determination of Characteristic Values

The cumulative distribution function is particularly useful for engineering design, because the characteristic values for any given probability of failure P_f can be obtained directly from it. In order to plot the distribution function, the test values $(x_1 \leq x_2 \leq \cdots \leq x_n)$ are sorted in rank and the cumulative probability of failure $P_{f,i}$ is assigned to each x value using an estimator. According to Makkonen [101], only the estimator [99] is a suitable consideration for an extreme value and is given by

$$P_{f,i} = \frac{i}{n+1}. \tag{5.25}$$

where i is the rank number of the sorted test values x_i and n the sample size. A representation in the probability plot is a good way to assess the goodness-of-fit of a distribution function to the test values (see Figs. 5.22 and 5.23). On the ordinate axis, the cumulative probability of failure, according to Eq. 5.25 and/or the linearisation of the distribution function, is applied equidistantly. The scaling of the axis is determined by the respective distribution function. The corresponding values of the sample are applied on the abscissa, which is logarithmically scaled for the log-normal and Weibull distribution. If the data can be fitted by a linear regression, the selected distribution function may be considered suitable. R^2 (coefficient of determination) is used to express the deviation between the sample values x_i and the regression line. R^2 is defined as

$$R = \frac{\sum_{i=1}^{n}(x_i - \bar{x})(y_i - \bar{y})}{\sqrt{\sum_{i=1}^{n}(x_i - \bar{x})^2 \sum_{i=1}^{n}(y_i - \bar{y})^2}}. \tag{5.26}$$

A linearised version of the log-normal distribution can be plotted by adjusting each of the test data, for example fracture stresses $\sigma_{f,i}$, as follows:

$$\begin{aligned} x_i &= \ln\left(\sigma_{f,i}\right), \\ y_i &= F^{-1}. \end{aligned} \tag{5.27}$$

In this case, F^{-1} is the inverse function of the cumulative distribution function of the Gaussian normal distribution, using the mean value μ and the standard deviation $\sigma = 1$ for the associated probabilities $P_{f,i}$. The parameters of the log-normal distribution $(\mu_{LN}, \sigma_{LN}^2)$ do not match the expected value and variance of the corresponding random variables. Since they cannot be determined by a linear regression in the probability plot [98], an estimation of the parameters or the estimated values $\hat{\mu}$ and $\hat{\sigma}$ is

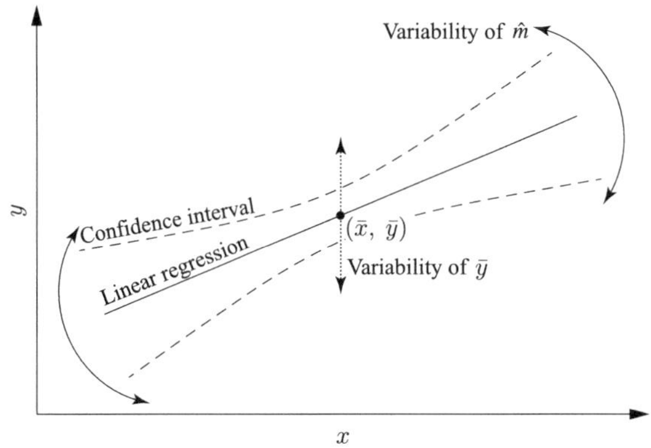

Fig. 5.21 Schematic representation of the confidence interval for linear regression [98]

$$\hat{\mu} = \frac{1}{n}\sum_{i=1}^{n} \ln(x_i),$$
$$\hat{\sigma}^2 = \frac{1}{n}\sum_{i=1}^{n} \left(\ln(x_i) - \hat{\mu}\right)^2. \tag{5.28}$$

The data points x_i and y_i can be approximated using the least squares method.

A two parameter Weibull distribution can be plotted in a probability plot as follows [98]:

$$x_i = \ln\left(\sigma_{\mathrm{f},i}\right),$$
$$y_i = \ln\left(\ln\left(\frac{1}{1 - P_{\mathrm{f},i}}\right)\right). \tag{5.29}$$

The cumulative distribution function of the Weibull distribution by Eq. 5.24 is given as a linearised representation:

$$y = -\lambda \cdot \ln(\beta) + \lambda \cdot x. \tag{5.30}$$

Equation 5.30 can be approximated using the least squares method. An estimate of the x-axis section $\hat{a}$ and slope $\hat{m}$ of the linear equation $y = a + mx$ leads, with $\hat{\lambda} = \hat{m}$ and$\hat{\beta} = \exp(-\hat{a}/\hat{m})$, directly to the estimates of the two parameters λ and β of the Weibull distribution.

The determination of a statistically significant 5% fractile value of a characteristic value, e.g. the bending strength f_{k}, is possible using the corresponding confidence interval. The confidence interval and the prediction interval are measures of the degree of certainty that the test data provide an accurate basis for a prediction and the probabilities of failure of interest (e.g. the 5% fractile values required for engineering design). The confidence interval provides an average predicted value based on the

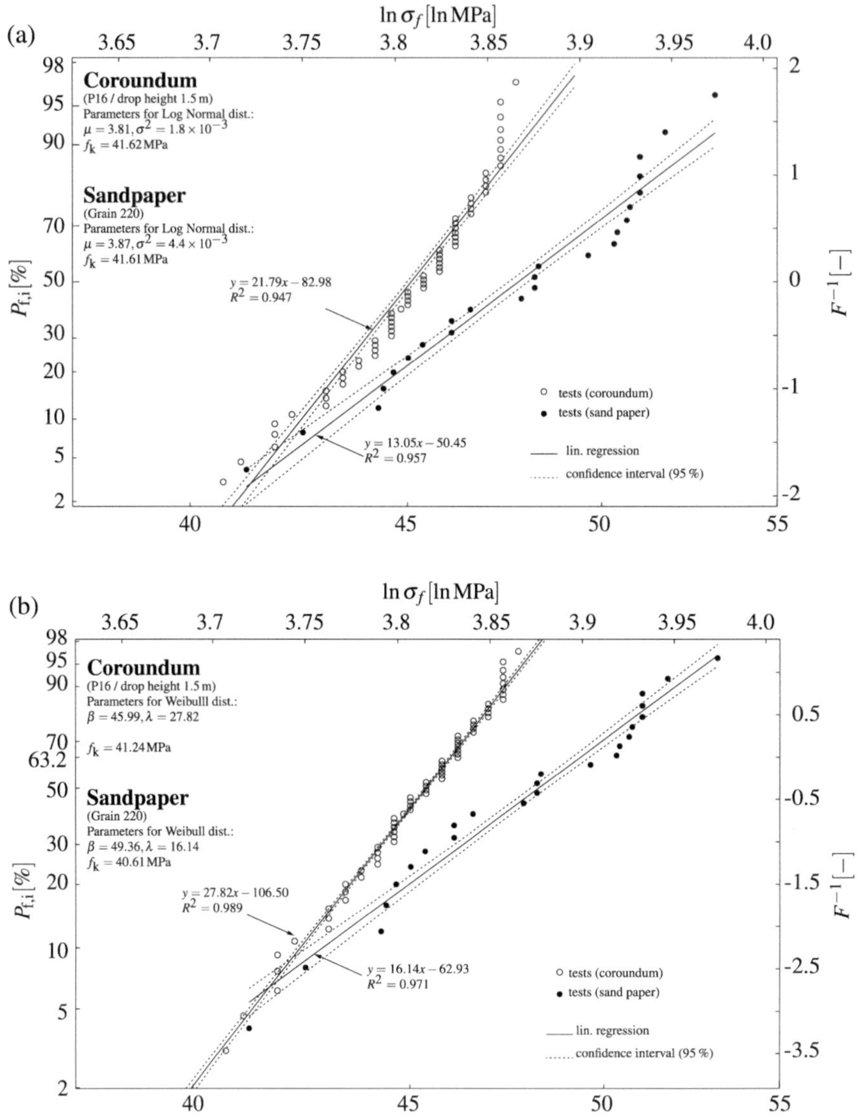

Fig. 5.22 Plot of bending strengths of pre-damaged glass (damaged with corundum P16 (drop distance: 1.5 m) and damaged with sandpaper (grit: 220), single values from Blank [47]): The plots are based on Sect. 5.4.3.5 as a linear regression of **a** a log-normal and **b** a two parameter Weibull distribution; in addition, the confidence intervals for a 95% confidence level is shown

calculated regression level, while the prediction interval provides a predicted value based on the individual measured values. According to EN 1990 [113], the confidence interval is used to determine the characteristic bending strength of materials. Each given regression line is defined by the mean value $\bar{y}$, and a vertical parallel shift or a change in the slope of the regression rotated about its centre $(\bar{x}, \bar{y})$ (Fig. 5.21). The confidence interval C thus provides an average predicted value, and thereby captures the variability of the regression line in the probability plot. It is calculated by

$$C = t_{\mathrm{l;\,n-2}} \cdot s_e \sqrt{\frac{1}{n} + \frac{(x-\bar{x})^2}{\sum\limits_{i=1}^{n}(x_i-\bar{x})^2}}. \tag{5.31}$$

In this case, $t_{\mathrm{l;\,n-2}}$ is the percentile of the student's (t-)distribution (the index refers to the confidence level: $l = 1 - \alpha$ for a one-sided and $l = 1 - \alpha/2$ for a two-sided demarcation), s_{e} the standard error of the estimated value as a result of the approximation to a regression line, n the random sample size, $\bar{x}$, the arithmetic mean of the observed sample values and x the independent variable at the point $X = x$. To determine a 5% fractile value is, for a one-side demarcation $t_{1-\alpha;\,\mathrm{n}-2}$, a confidence level of 95%, or for a two-sided demarcation $t_{1-\alpha/2;\,\mathrm{n}-2}$ a confidence level of 90%. The standard error is calculated by

$$s_{\mathrm{e}} = \sqrt{\frac{\sum\limits_{i=1}^{n}\left(|y_i| - |a + b \cdot x_i|^2\right)}{n-2}}. \tag{5.32}$$

In the probability plot, the upper and lower limits of $\breve{y}_{\mathrm{C}}$ result in the confidence interval.

$$\breve{y}_{\mathrm{C}} = a + m \cdot x \pm C. \tag{5.33}$$

Figure 5.22 provides an example of the above method for the statistical evaluation of bending strength. The example consists of two sets of artificially aged glass specimens generated by: (i) dropping corundum (P16 drop distance: 1.600 mm) onto the glass surface; (ii) pre-damaging it with sandpaper (grit 220) [47]. The aim of this example is to compare the results of the probability plot in general and particularly the characteristic bending strength (f_{k}) of the two sets of data, in particular at the confidence interval of 95% using a log-normal distribution (Fig. 5.22a) and a Weibull distribution (Fig. 5.22b). It is apparent that, based on R^2, the data is well captured by both distribution functions. In contrast to the pre-damage with sandpaper, the pre-damage with corundum shows a smaller dispersion in the identified bending strengths. However, the distribution function should always be chosen carefully. In particular it is essential to obtain an adequate overall correlation coefficient for the entire distribution and achieve a good correlation at lower fractiles (i.e. the tail of the distribution). For this, recently Berlinger, Kolling, and Schneider [102] developed a new method to analyse the goodness-of-fit using an Anderson–Darling test.

Figure 5.23 presents the adaptation to a log-normal distribution of the bending strengths of 48 year old and weathered glass (measurement data from Fink [55]) and of commercially available float glass (tin and air side; measurement data from Mellmann and Maultzsch [74]), taking into account a one-sided confidence level of

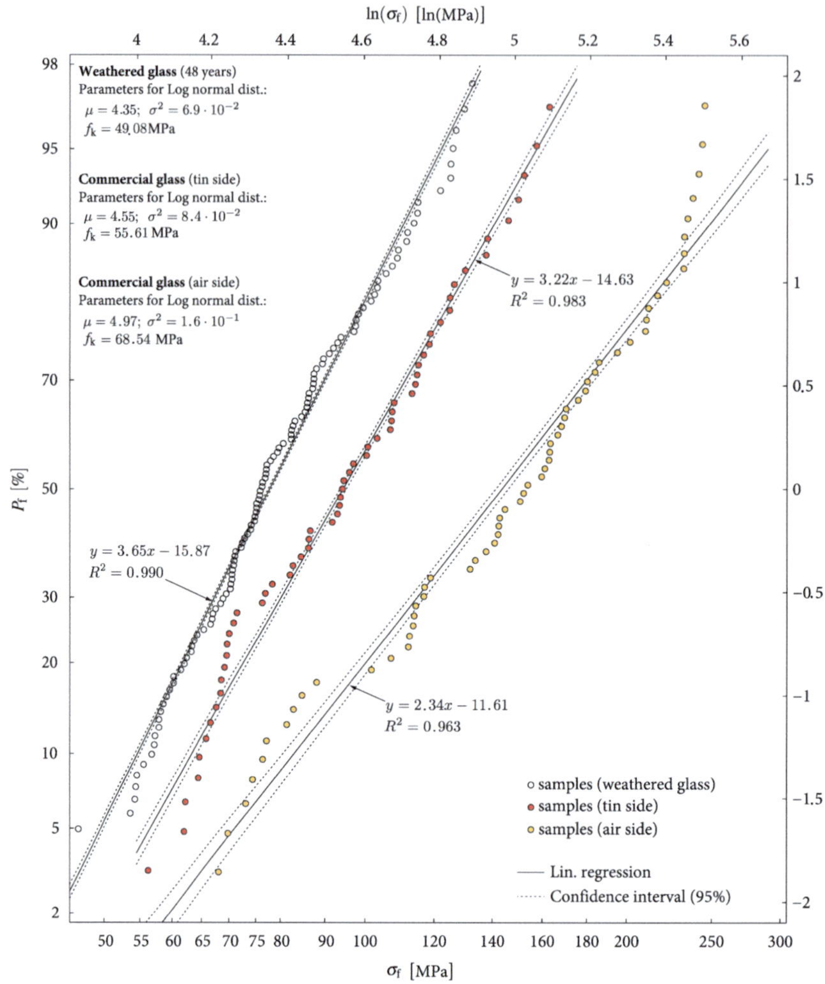

Fig. 5.23 Plot of the bending strengths of weathered glass (from [55]) and of commercial pristine glass (from [74]) with differentiation of the tin bath side and the air side by means of a linear regression to a log-normal distribution with representation of the confidence intervals for a one-sided confidence level of 95%

95%. The bending strength of the tin side, which is significantly lower in comparison to the air side, can be explained by a higher defect density due to the contact of the glass surface with the conveyor rollers immediately after the tin bath. However, this difference is usually neglected in engineering design, as the surface condition of the air side gradually approached that of the tin side as a result of the further processing and use.

For the statistical analysis, it should be noted that the low scatter caused by uniform pre-damage is not taken into consideration, typically, for the determination of partial safety factors according to the semi-probabilistic concept (see Sect. 9.2.1). Furthermore, for all ranges in Figs. 5.22 and 5.23, it is evident that the distribution functions do not provide a good fit for high and low strength values. More recent works further improved the knowledge by using a truncated Weibull-distribution [97] and by using more advanced methods for estimating the parameters of the Weibull distribution [94] or by using a bi-modal Weibull distribution [79]. Further research is needed to clarify the underlying failure mechanisms.

5.4.4 *Influence of the Stressed Area on the Bending Strength*

The probability of encountering a significant surface flaw is larger on a large glass surface area than on a small one. This dependence on the stressed area is a well-known phenomenon in brittle materials and can be derived directly from the Weibull distribution as follows:

The probability P that a specimen with a test surface of A_0 will not fail, is:

$$P(\sigma_{\mathrm{f,A_0}}) = 1 - F_{\mathrm{WB}}(\sigma_{\mathrm{f,A_0}}) = \exp\left[-\left(\frac{\sigma_{\mathrm{f,A_0}}}{\beta_{\mathrm{A_0}}}\right)^{\lambda}\right]. \tag{5.34}$$

A larger test surface $A_1 > A_0$ will thus also not fail if none of the sub-surfaces A_0 fail. If the probability of fracture P is constant, the dependence of the test surfaces A_0 and A_1 can be described as

$$P(\sigma_{\mathrm{f,A_1}}) = 1 - F_{\mathrm{WB}}(\sigma_{\mathrm{f,A_1}}) = \exp\left[-\frac{A_1}{A_0}\left(\frac{\sigma_{\mathrm{f,A_0}}}{\beta_{\mathrm{A_0}}}\right)^{\lambda}\right], \tag{5.35}$$

so that the distribution function is

$$F_{\mathrm{WB}}(\sigma_{\mathrm{f,A_1}}) = 1 - \exp\left[-\frac{A_1}{A_0}\left(\frac{\sigma_{\mathrm{f,A_0}}}{\beta_{\mathrm{A_0}}}\right)^{\lambda}\right]. \tag{5.36}$$

If it is possible to statistically record the distribution of the experimentally measured test strength $\sigma_{\mathrm{f},i}(A_0)$ by means of the Weibull distribution, as defined in Eq. 5.24, then the equivalent test strength (i.e. with an identical probability of failure P) can be converted to a different test surface A_1:

$$\frac{\sigma_{\mathrm{f,A_1}}}{\sigma_{\mathrm{f,A_0}}} = \left(\frac{A_0}{A_1}\right)^{1/\lambda}. \tag{5.37}$$

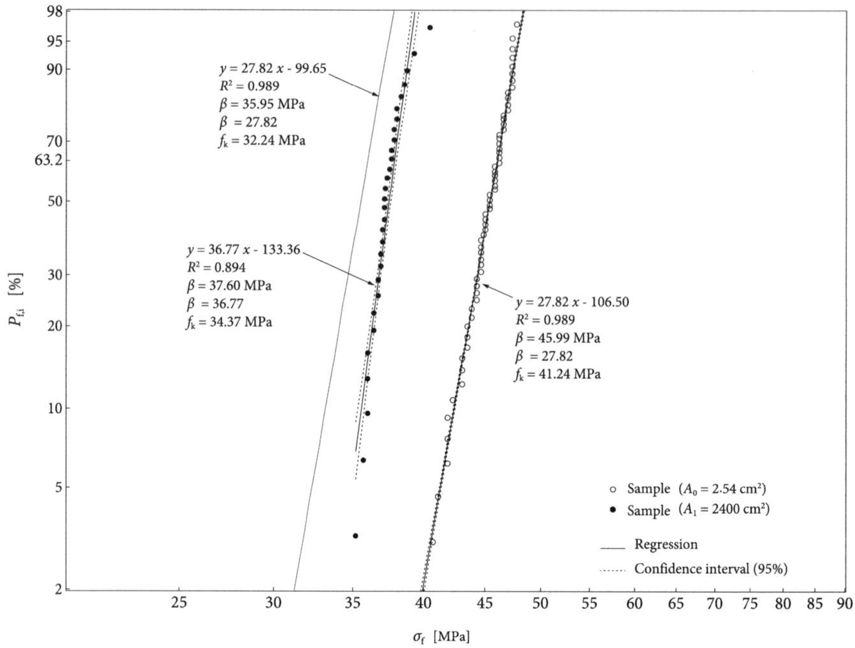

Fig. 5.24 Example of Weibull distributions of the bending strength of pre-damaged annealed float glass (treated with corundum; test data from [47]): The tests were carried out in different coaxial double ring bending test setups (test area $A_0 = 2.54\,\text{cm}^2$ and $A_1 = 2400\,\text{cm}^2$); the red line indicates the horizontal shift due to surface conversion according to Eq. 5.37 of the bending strength of surface A_0 in surface area A_1

This is shown in the example in Fig. 5.24. For this purpose, the summarised test strengths of the specimens that were pre-damaged with corundum, as in the probability network of the Weibull distribution (see Sect. 5.4.3.4) [47] were statistically evaluated. The tests were carried out in two different test setups using the coaxial double ring bending test (test surface $A_0 = 2.54\,\text{cm}^2$, according to EN 1288-5 [110], and $A_1 = 2400\,\text{cm}^2$, according to EN 1288-2 [108]). It can be seen that the distribution of the test strengths $\sigma_{\text{f, A}_0}$ over the measured surface A_0 according to Eq. 5.37 can be shifted horizontally and the bending strength of the larger stress area $A_1 = 2400\,\text{cm}^2$ can be approximated. However, it should be noticed that this method assumes similar pre-damage, which is often not the case in practice.

An extension to this probability-based approach is the *method of effective stresses*, in which a glass pane A with a non-constant stress distribution is theoretically divided into evenly stressed sub-areas A_i. The distribution function of the probability of fracture of the glass pane made up of the individual sub-areas A_i is given through the multiplication of the individual fracture probabilities. This fracture probability can be transferred over a surface integral into an evenly distributed effective stress across the entire glass pane [75].

$$\sigma_{\text{eff}} = \left(\frac{1}{A} \int_A (k(x, y) \cdot \sigma_1(x, y))^{\lambda} \, dx \, dy \right)^{\frac{1}{\lambda}} \tag{5.38}$$

The effective stress is represented by an "average" stress, which, is derived from the principal tensile stresses acting over the entire surface, and results in the same fracture probability as the stress distribution given in Eq. 5.38. Since such stresses are only effective on surface defects perpendicular to the cracks that assume a uniform distribution to the orientation of the surface defects, this can only be determined by the introduction of a bi-axial modification function $k(x, y)$. This modification factor function is dependent on a prior given stress state. A bi-axial stress state with $\sigma_1 = \sigma_2$ results in a modification factor of $k = 1$. Under other conditions, k is dependent on the form parameters of the Weibull distribution and takes on values between 0 and 1 (cf. [75]). The areas that are in compression are excluded from the evaluation of the integral in Eq. 5.38.

The design of the glass component can then be carried out on the basis of the effective stress. The accuracy of this observation depends on the size of the sub-areas. Thus, the procedure is particularly suitable for numerical considerations (e.g. FE calculations) and subsequent post-processing with appropriately programmed algorithms according to Eq. 5.38. A detailed explanation of this approach and an associated example is provided in Overend and Zammit [56]. This approach assumes that the surface defects are randomly distributed over the glass surface. This is not always the case (e.g. in the area of the glass edge or near holes a much larger number of defects is present which depend on the processing). In such cases it is possible to use a modified approach described in Watson, Nielsen and Overend [103] or to conservatively use the maximum principal stress values for the design.

5.5 Compressive Strength of Glass

The compressive strength of glass has not as yet been the focus of interest, because the load bearing resistance of glass used in construction is often governed by the bending (tension) resistance. However, knowledge of the compressive strength of glass can be of interest for localised areas in glass components which are subjected to high compressive forces such as glass supports. Glass components subjected to high compressive stresses are typically made of pre-stressed glass. So far, systematic assessment of the load capacity of glass in compression exist. Although DIN 1249-10 [105][3] provides approximate values for the compressive strength of float glass (roughly 700–900 MPa), tests by Fink [55] show that these values do not apply to annealed glass (ANG) (see Fig. 5.25). At present, no scientifically reliable data is available for the compressive strength of thermally toughened glass.

[3] DIN 1249-10 [105] has now been replaced by EN 572-1 [115]; the mention of compressive strength has subsequently been omitted.

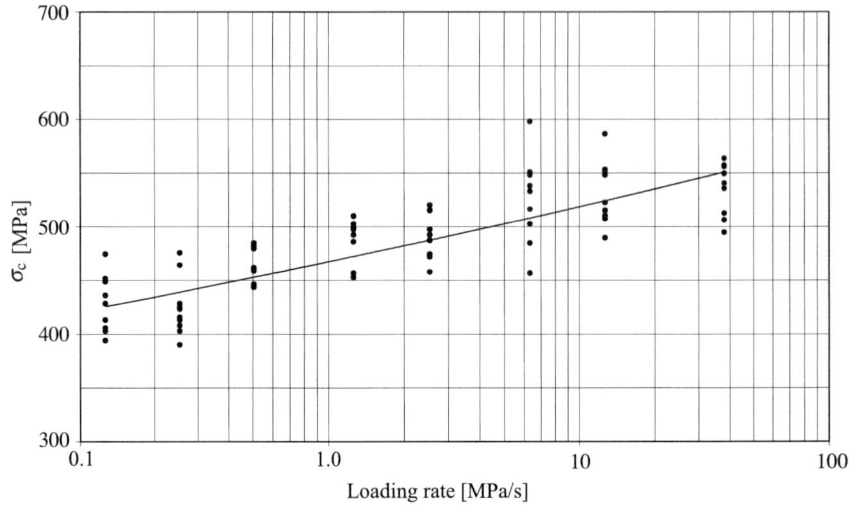

Fig. 5.25 Compressive strength σ_c of float glass depending on the stress rate [55]

Fig. 5.26 Cracks developing inside glass cylinders after constant compressive load

According to test specifications in DIN 1249-10 [105], the compressive strength of glass is to be determined by at least five cylindrical specimens of 10 mm diameter and height with a load rate of 3–4 kN/s. Also in the case of glass under compression, it is evident that the strength is a function of the load (or stress) rate. Depending on the load velocity (or stress rate), the test strength shows a similar characteristic, as found in glass bending tests. The strength under permanent load is typically also clearly lower than the strength found from short time loading. In the case of a compressive load, lateral deformation occurs in the amorphous body due to Poisson's ratio effect. The molecular bonds are therefore broken by lateral strain, similar to what is observed when testing concrete cubes or cylinders in compression (splitting tension). This explains why the compressive strength is sensitive to sub-critical crack growth (Fig. 5.26). Therefore the strength of the glass specimens in compression is also a function of the size of the specimens, the stiffness of the load introduction and the friction between the load introduction and the glass specimen. If a load duration of roughly 50 years is assumed, the strength can be expected to decrease by a factor of about 3 compared to the short-term test strength [55].

References

1. M.J. Meyland, J.H. Nielsen, C. Kocer, Tensile behaviour of soda-lime-silica glass and the significance of load duration–a literature review. J. Build. Eng. **44**, 102966 (2021)
2. A. Petzold, H. Marusch, B. Schramm, *Der Baustoff Glas: Grundlagen, Eigenschaften, Erzeugnisse, Glasbauelemente, Anwendungen* (Hofmann, Berlin, 1990)
3. E.K. Pavelchek, R.H. Doremus, Fracture strength of soda-lime glass after etching. J. Mater. Sci. **9**(11), 1803–1808 (1974)
4. F.W. Preston, The structure of abraded glass surfaces. Trans. Opt. Soc. **23**(3), 141–164 (1922)
5. N.P. Lower, R.K. Brow, C.R. Kurkjian, Inert failure strain studies of sodium silicate glass fibers. J. Non-Cryst. Solids **349**, 168–172 (2004)
6. C. Gurney, S. Pearson, Fatigue of mineral glass under static and cyclic loading. Proc. R. Soc. Lond. **192**(1031), 537–544 (1948)
7. W. Weibull, A statistical distribution function of wide applicability. J. Appl. Mech. **18**(3), 293–297 (1951)
8. H. Hertz, *Miscellaneous Papers* (Macmillan and Co., Ltd., New York, 1896)
9. B.R. Lawn, S.M. Wiederhorn, H.H. Johnson, Strength degradation of brittle surfaces: blunt indenters. J. Am. Ceram. Soc. **58**(9–10), 428–432 (1975)
10. R.J. Charles, W.B. Hillig, The kinetics of glass failure by stress corrosion (Charleroi, Belgien, 1962)
11. R.F. Cook, G.M. Pharr, Direct observation and analysis of indentation cracking in glasses and ceramics. J. Am. Ceram. Soc. **73**(4), 787–817 (1990)
12. S.M. Wiederhorn, Influence of water vapor on crack propagation in soda-lime glass. J. Am. Ceram. Soc. **50**(8), 407–414 (1967)
13. F. Schneider, Lösungen für energieeffiziente multifunktionale Verglasungen. Stahlbau **81**, 291–304 (2012)
14. B.R. Lawn, E.R. Fuller, S.M. Wiederhorn, Strength degradation of brittle surfaces: sharp indenters. J. Am. Ceram. Soc. **59**(5–6), 193–197 (1976)
15. B.R. Lawn, A.G. Evans, D.B. Marshall, Elastic/plastic indentation damage in ceramics: the median/radial crack system. J. Am. Ceram. Soc. **63**(9–10), 574–581 (1980)
16. D.B. Marshall, B.R. Lawn, A.G. Evans, Elastic/plastic indentation damage in ceramics: the lateral crack system. J. Am. Ceram. Soc. **65**(11), 561–566 (1982)
17. R.F. Cook, D.H. Roach, The effect of lateral crack growth on the strength of contact flaws in brittle materials. J. Mater. Res. **1**(4), 589–600 (1986)
18. K.C. Datsiou, M. Overend, Surface tensile strength and Hertzian fracture resistance of patterned acid-etched glass. J. Arch. Eng. **29**(3), 04023015 (2023). https://doi.org/10.1061/JAEIED.AEENG-1464
19. M. Vandebroek, J. Belis, C. Louter, G. Van Tendeloo, Experimental validation of edge strength model for glass with polished and cut edge finishing. Eng. Fract. Mech. **96**, 480–489 (2012)
20. B.R. Lawn, S.M. Wiederhorn, H.H. Johnson, Strength degradation of brittle surfaces: blunt indenters. J. Am. Ceram. Soc. **58**(9–10), 428–432 (1975)
21. M. Vandebroek, C. Louter, R. Caspeele, F. Ensslen, J. Belis, Size effect model for the edge strength of glass with cut and ground edge finishing. Eng. Struct. **79**, 96–105 (2014)
22. S. Müller-Braun, M. Seel, M. König, P. Hof, J. Schneider, M. Oechsner, Cut edge of annealed float glass: crack system and possibilities to increase the edge strength by adjusting the cutting process. Glas. Struct. Eng. **5**(1), 3–25 (2019)

23. S.M. Wiederhorn, Mechanisms of subcritical crack growth in glass, in *Fracture Mechanics of Ceramics—Crack Growth and Microstructure*, vol. 4 (Plenum Press, New York, 1978)
24. B.R. Lawn, A.G. Evans, D.B. Marshall, Elastic/plastic indentation damage in ceramics: the median/radial crack system. J. Am. Ceram. Soc. **63**(9–10), 574–581 (1980)
25. M.J. Meyland, J.H. Nielsen, *High strain rate characterisation of soda-lime-silica glass and the effect of residual stresses* (Glas. Struct, Eng, 2022)
26. C.J. Simmons, S.W. Freiman, Effect of corrosion processes on subcritical crack growth in glass. J. Am. Ceram. Soc. **64**(11), 683–686 (1981)
27. F.W. Preston, The structure of abraded glass surfaces. Trans. Opt. Soc. **23**(3), 141–164 (1922)
28. S.M. Wiederhorn, Influence of water vapor on crack propagation in soda-lime glass. J. Am. Ceram. Soc. **50**(8), 407–414 (1967)
29. S.M. Wiederhorn, Moisture assisted crack growth in ceramics. Int. J. Fract. Mech. **4**(2), 171–177 (1968)
30. E. Gehrke, C. Ullner, M. Hähnert, Correlation between multistage crack growth and time-dependent strength in commercial silicate glasses. Glastech. Ber. **60**(8), 268–278 (1987)
31. E. Gehrke, C. Ullner, M. Hähnert, Effect of corrosive media on crack growth of model glasses and commercial silicate glasses. Glastech. Ber. **63**(9), 255–265 (1990)
32. B. Stavrinidis, D.G. Holloway, Crack healing in glass. Phys. Chem. Glas. **24**(1), 19–25 (1983)
33. S.M. Budd, The mechanisms of chemical reaction between silicate glass and attacking agents–Part 1: electrophilic and nucleophilic mechanims of attack. Phys. Chem. Glas. **2**(4), 111–114 (1961)
34. T.A. Michalske, S.W. Freiman, A molecular interpretation of stress corrosion in silica. Nature **295**(5849), 511–512 (1982)
35. T.A. Michalske, S.W. Freiman, A molecular mechanism for stress corrosion in vitreous silica. J. Am. Ceram. Soc. **66**(4), 284–288 (1983)
36. S.M. Wiederhorn, *Mechanisms of subcritical crack growth in glass, in Fracture Mechanics of Ceramics-Crack Growth and Microstructure*, vol. 4 (Plenum Press, New York, 1978)
37. B.R. Lawn, *Fracture of Brittle Solids* (Cambridge University Press, Cambridge, 1993)
38. T.A. Michalske, B.C. Bunker, Wie Glas bricht. Spektrum der Wissenschaft **2**, 114–121 (1985)
39. E. Gehrke, C. Ullner, M. Hähnert, Fatigue limit and crack arrest in alkali-containing silicate glasses. J. Mater. Sci. **26**(20), 5445–5455 (1991)
40. H.S. Norville, J.E. Minor, Strength of weathered window glass. Am. Ceram. Soc. Bull. **64**(11), 1467–1470 (1985)
41. A.G. Evans, S.M. Wiederhorn, Proof testing of ceramic materials–an analytical basis for failure prediction. Int. J. Fract. **10**(3), 379–392 (1974)
42. D. Maugis, Subcritical crack growth, surface energy, fracture toughness, stick-slip and embrittlement. J. Mater. Sci. **20**(9), 3041–3073 (1985)
43. S.M. Wiederhorn, L.H. Bolz, Stress corrosion and static fatigue of glass. J. Am. Ceram. Soc. **53**(10), 543–548 (1970)
44. J.H. Nielsen, J.F. Olesen, H. Stang, The fracture process of tempered soda-lime-silica glass. Exp. Mech. **49**(6), 855–870 (2009)
45. S.W. Freiman, Effect of alcohols on crack propagation in glass. J. Am. Ceram. Soc. **57**(8), 350–353 (1974)
46. F. Kerkhof, H. Richter, D. Stahn, Festigkeit von Glas - Zur Abhängigkeit von Belastungsdauer und -verlauf. Glastech. Ber. **54**(8), 265–277 (1981)
47. W.A. Dalgliesh, D.A. Taylor, The strength and testing of window glass. Can. J. Civ. Eng. **17**(5), 752–762 (1990)
48. C.J. Simmons, S.W. Freiman, Effect of corrosion processes on subcritical crack growth in glass. J. Am. Ceram. Soc. **64**(11), 683–686 (1981)
49. S.M. Wiederhorn, S.W. Freiman, E.R. Fuller Jr., C.J. Simmons, Effects of water and other dielectrics on crack growth English. J. Mater. Sci. **17**(12), 3460–3478 (1982)
50. B.R. Lawn, D.B. Marshall, T.P. Dabbs, *Fatigue strength of glass: a controlled flaw study, in Strength of Inorganic Glass* (Plenum Press, New York, 1985), pp.249–260

51. R. Behr, M. Karson, J. Minor, Reliability analysis of window glass failure pressure data. Struct. Saf. **11**(1), 43–58 (1991)
52. E. Gehrke, C. Ullner, M. Hähnert, Fatigue limit and crack arrest in alkali-containing silicate glasses. J. Mater. Sci. **26**(20), 5445–5455 (1991)
53. K. Blank, *Dickenbemessung von vierseitig gelagerten rechteckigen Glasscheiben unter gleichf ö rmiger Fl ä chenlast* (Gelsenkirchen, 1993)
54. D. Munz, T. Fett, *Ceramics: Mechanical Properties, Failure Behaviour, Materials Selection* (Springer, Berlin, 1999)
55. C. Ullner, *Forschungsvorhaben-Nr. 43 D: Untersuchungen zum Festigkeitsverhalten und zur Rißszlig; alterung von Glas unter dem Einflußszlig; korrosiver Umgebungsbedingungen* (1993)
56. M. Overend, K. Zammit, A computer algorithm for determining the tensile strength of float glass. Eng. Struct. **45**, 68–77 (2012)
57. C. Gurney, S. Pearson, Fatigue of mineral glass under static and cyclic loading. Proc. R. Soc. Lond. **192**(1031), 537–544 (1948)
58. B.-T. Lü, Fatigue strength prediction of soda-lime glass. Theor. Appl. Fract. Mech. **27**(2), 107–114 (1997)
59. W.C. Levengood, Effect of origin flaw characteristics on glass strength. J. Appl. Phys. **29**(5), 820–826 (1958)
60. S.M. Wiederhorn, P.R. Townsend, Crack healing in glass. J. Am. Ceram. Soc. **53**(9), 486–489 (1970)
61. T.A. Michalske, *The stress corrosion limit: its measurement and implications, in Fracture Mechanics of Ceramics-Surface Flaws, Statistics, and Microcracking*, vol. 5 (Plenum Press, New York, 1977)
62. S. Ito, M. Tomozawa, Crack blunting of high-silica glass. J. Am. Ceram. Soc. **65**(8), 368–371 (1982)
63. Y. Bando, S. Ito, M. Tomozawa, Direct observation of crack tip geometry of SiO_2 glass by high-resolution electron microscopy. J. Am. Ceram. Soc. **67**(3), C36–C37 (1984)
64. S.M. Wiederhorn, A. Dretzke, J. Rödel, Crack growth in soda-lime-silicate glass near the static fatigue limit. J. Am. Ceram. Soc. **85**(9), 2287–2292 (2002)
65. S. Hénaux, F. Creuzet, Kinetic fracture of glass at the nanometer scale. J. Mater. Sci. Lett. **16**(12), 1008–1011 (1997)
66. R.J. Charles, W.B. Hillig, *The kinetics of glass failure by stress corrosion* (Charleroi, Belgien, 1962)
67. B.R. Lawn, D.B. Marshall, A.H. Heuer, Comment on Direct observation of crack-tip geometry of SiO_2 glass by high-resolution electron microscopy. J. Am. Ceram. Soc. **67**(11), c253–c253 (1984)
68. B.R. Lawn, K. Jakus, A.C. Gonzalez, Sharp versus blunt crack hypotheses in the strength of glass: a critical study using indentation flaws. J. Am. Ceram. Soc. **68**(1), 25–34 (1985)
69. B. Stavrinidis, D.G. Holloway, Crack healing in glass. Phys. Chem. Glas. **24**(1), 19–25 (1983)
70. M.P. Brungs, Fracture and failure of glass. Mater. Forum **19**, 227–239 (1995)
71. A.A. Griffith, The phenomena of rupture and flow in solids. Philos. Trans. R. Soc. Lond. **A221**(1), 163–198 (1921)
72. S.M. Wiederhorn, A. Dretzke, J. Rödel, Near the static fatigue limit in glass. Int. J. Fract. **121**(1), 1–7 (2003)
73. J.-P. Guin, S.M. Wiederhorn, T. Fett, Crack-tip structure in soda-lime-silicate glass. J. Am. Ceram. Soc. **88**(3), 652–659 (2005)
74. G. Mellmann, M. Maultzsch, *Forschungsbericht 161 - Untersuchung zur Ermittlung der Biegefestigkeit von Flachglas f ü r bauliche Anlagen, Technical report* (Bundesanstalt für Materialforschung und -prüfung (BAM), Berlin, 1989)
75. W.L. Beason, J.R. Morgan, Glass failure prediction model. J. Struct. Eng. **110**(2), 197–212 (1984)
76. W.A. Dalgliesh, D.A. Taylor, The strength and testing of window glass. Can. J. Civ. Eng. **17**(5), 752–762 (1990)

77. H.S. Norville, J.E. Minor, Strength of weathered window glass. Am. Ceram. Soc. Bull. **64**(11), 1467–1470 (1985)
78. K.C. Datsiou, M. Overend, The strength of aged glass. Glas. Struct. Eng. **2**(2), 105–120 (2017)
79. K.C. Datsiou, M. Overend, Artificial ageing of glass with sand abrasion. Constr. Build. Mater. **142**, 536–551 (2017)
80. EN 1990, Eurocode: basis of structural design (2010)
81. F. Ensslen, J. Schneider, S. Schula, Produktion, Eigenschaften und Tragverhalten von thermisch gebogenen Floatgläsern für das Bauwesen—Erstprüfung und werkseigene Produktionskontrolle im Rahmen des Zulassungsverfahrens. Stahlbau-Spezial (Glasbau) **79**(1), 46–51 (2010)
82. EN 1863-1, Glass in building—heat strengthened soda lime silicate glass—Part 1: definition and description (2012)
83. EN 572-1, Glass in building—basic soda-lime silicate glass products—Part 1: definitions and general physical and mechanical properties (Beuth, 2012)
84. F. Ensslen, J. Schneider, S. Schula, Produktion, Eigenschaften und Tragverhalten von thermisch gebogenen Floatgläsern für das Bauwesen-Erstprüfung und werkseigene Produktionskontrolle im Rahmen des Zulassungsverfahrens. Stahlbau-Spezial (Glasbau) **79**(1), 46–51 (2010)
85. S. Müller-Braun, J. Schneider, Biaxially curved glass with large radii-determination of strength using the coaxial double ring test. Glas. Struct. Eng. **2**(2), 121–131 (2017)
86. Ö. Bucak, M. Feldmann, R. Kasper, M. Bues, M. Illguth, Das Bauprodukt warm gebogenes Glas - Prüfverfahren. Festigkeiten und Qualitätssicherung. Stahlbau-Spezial (Glasbau) **78**(1), 23–28 (2009)
87. L. Sachs, J. Hedderich, *Angewandte Statistik*, 14th edn. (Springer, Berlin, 2012)
88. F. Schneider, Lösungen für energieeffiziente multifunktionale Verglasungen. Stahlbau **81**, 291–304 (2012)
89. M. Vandebroek, J. Belis, C. Louter, G. Van Tendeloo, Experimental validation of edge strength model for glass with polished and cut edge finishing. Eng. Fract. Mech. **96**, 480–489 (2012)
90. prEN 16612, Glas im Bauwesen—Bestimmung des Belastungswiderstandes von Glasscheiben durch Berechnung und Pr ü fung (2013)
91. DIBt, Allgemeine bauaufsichtliche Zulassung Z-70.3-197: Verbund-Sicherheitsglas mit einer Verbundfolie der Produktfamilie EVASAFE mit Ansatz eines Schubverbunds (2014)
92. M. Feldmann, R. Kasper, Guidance for European structural design of glass components—support to the implementation, harmonization and further development of the Eurocodes (SaT-report) (2014)
93. M. Overend, G.A. Parke, D. Buhagiar, Predicting failure in glass-a general crack growth model. J. Struct. Eng. **133**(8), 1146–1155 (2007)
94. K.C. Datsiou, M. Overend, Weibull parameter estimation and goodness-of-fit for glass strength data. Struct. Saf. **73**, 29–41 (2018)
95. R. Behr, M. Karson, J. Minor, Reliability analysis of window glass failure pressure data. Struct. Saf. **11**(1), 43–58 (1991)
96. K.C. Datsiou, D. Hall, M. Overend, Repair of soda-lime-silica glass. Glas. Struct. Eng. **3**(2), 125–137 (2018)
97. J. Hilcken, Zyklische Ermüdung von thermisch entspanntem und thermisch vorgespannten Kalk-Natron-Silikatglas. Dissertation, Technische Universität Darmstadt (2015)
98. L. Sachs, J. Hedderich, *Angewandte Statistik*, 14th edn. (Springer, Berlin, 2012)
99. W. Weibull, A statistical theory of the strength of materials. Ingeni ö rsvetenskapsakademiens handlingar **151**, 1–45 (1939)
100. W. Weibull, A statistical distribution function of wide applicability. J. Appl. Mech. **18**(3), 293–297 (1951)
101. L. Makkonen, Problems in the extreme value analysis. Struct. Saf. **30**(5), 405–419 (2008)
102. M. Berlinger, S. Kolling, J. Schneider, A generalized Anderson-Darling test for the goodness-of-fit evaluation of the fracture strain distribution of acrylic glass. Glas. Struct. Eng. **6**(2), 195–208 (2021)

103. J. Watson, J.H. Nielsen, M. Overend, A critical flaw size approach for predicting the strength of bolted glass connections, Eng. Struct. **57**, 87–99 (2013). https://doi.org/10.1016/j.engstruct.2013.07.026
104. S. Müller-Braun, J. Schneider, Biaxially curved glass with large radii-determination of strength using the coaxial double ring test. Glas. Struct. Eng. **2**(2), 121–131 (2017)
105. K.C. Datsiou, D. Hall, M. Overend, Repair of soda-lime-silica glass. Glas. Struct. Eng. **3**(2), 125–137 (2018)
106. K.C. Datsiou, M. Overend, Weibull parameter estimation and goodness-of-fit for glass strength data. Struct. Saf. **73**, 29–41 (2018)
107. M. Jotz, J. Schneider, E. Rädlein, Introducing the "Cylinder Fit Test", a simplified edge strength measurement method for ultra thin glass, in *Engineered Transparency 2018. International Conference at glasstec* (Ernst & Sohn, Berlin, 2018), pp. 65–73
108. S. Müller-Braun, M. Seel, M. König, P. Hof, J. Schneider, M. Oechsner, Cut edge of annealed float glass: crack system and possibilities to increase the edge strength by adjusting the cutting process. Glas. Struct. Eng. **5**(1), 3–25 (2019)
109. P. Bukieda, K. Lohr, J. Meiberg, B. Weller, Study on the optical quality and strength of glass edges after the grinding and polishing process. Glas. Struct. Eng. 1–18 (2020)
110. T. Wüest, A. Luible, Glass design in Switzerland, in *Challenging Glass Conference Proceedings* (2020)
111. M. Berlinger, S. Kolling, J. Schneider, A generalized Anderson-Darling test for the goodness-of-fit evaluation of the fracture strain distribution of acrylic glass. Glas. Struct. Eng. **6**(2), 195–208 (2021)
112. CEN/TS-19100-1:2021 Design of glass structures—Part 1: basis of design and materials. Technical report (2021)
113. CEN/TS-19100-2:2021 Design of glass structures—Part 2: design of out-of-plane loaded glass components. Technical report (2021)
114. CEN/TS-19100-3:2021 Design of glass structures—Part 3: design of in-plane loaded glass components and their mechanical joints. Technical report (2021)
115. FprSIA 2057:2021-01—glass structures (2021)
116. M.J. Meyland, J.H. Nielsen, C. Kocer, Tensile behaviour of soda-lime-silica glass and the significance of load duration—a literature review. J. Build. Eng. **44**, 102966 (2021)
117. M. Jotz, Kantenfestigkeitsoptimierte (Weiter-) Entwicklung eines Verfahrens zum Trennen von ultradünnem Glas. Ph.D. thesis, Ilmenau (2022)
118. M.J. Meyland, J.H. Nielsen, High strain rate characterisation of soda-lime-silica glass and the effect of residual stresses. Glas. Struct. Eng. (2022)
119. S. Müller-Braun, *Risssystem und Festigkeit der geschnittenen Kante von Floatglas* (Springer Fachmedien Wiesbaden, 2022)
120. M.J. Meyland, R.N.W. Eriksen, J.H. Nielsen, A modified split-Hopkinson pressure bar setup enabling stereo digital image correlation measurements for flexural testing. Int. J. Impact Eng. **173** (2023)

Chapter 6
Pre-stressed Glass

Pre-stressed glass is commonly used in architectural applications because controlled changes in the residual stress state produce useful characteristics such as higher strengths and safer fracture patterns compared to annealed glass. Common glass pre-stressing techniques are either thermal pre-stressing or chemical pre-stressing. These two processes and their effects on the performance of the glass are discussed in Sects. 6.2 *and* 6.3, *respectively. Measuring the amount of residual stresses using photoelasticity is discussed in Sect.* 6.4.

6.1 General

The primary aims of pre-stressing glass is to increase its apparent tensile strength and induce a failure mode which reduces the risk of human injury. This is achieved by creating a residual stress state in which the glass surfaces are in compression and are in equilibrium with the tensile stresses in the core of the glass.

The most common techniques to pre-stress glass are thermal or chemical treatment. These two processes and products are described further in this chapter and referred to as thermally pre-stressed glass and chemically pre-stressed glass, respectively.

Figure 6.1 shows the superposition of residual and bending stresses in a thermally pre-stressed glass. The residual stress state, induced by the pre-stressing, increases the apparent strength considerably by putting the surface flaws in a state of compression which has to be overcome by external loads before the flaws start to grow and glass fracture can occur.

As discussed in previous chapters, critical flaws are generally located on the surface rather than in the core of the glass, consequently the tensile stress in the glass core has no overall strength-reducing effect.

J. Belis et al., *Engineering Glass Structures*,
https://doi.org/10.1007/978-3-032-03233-1_6

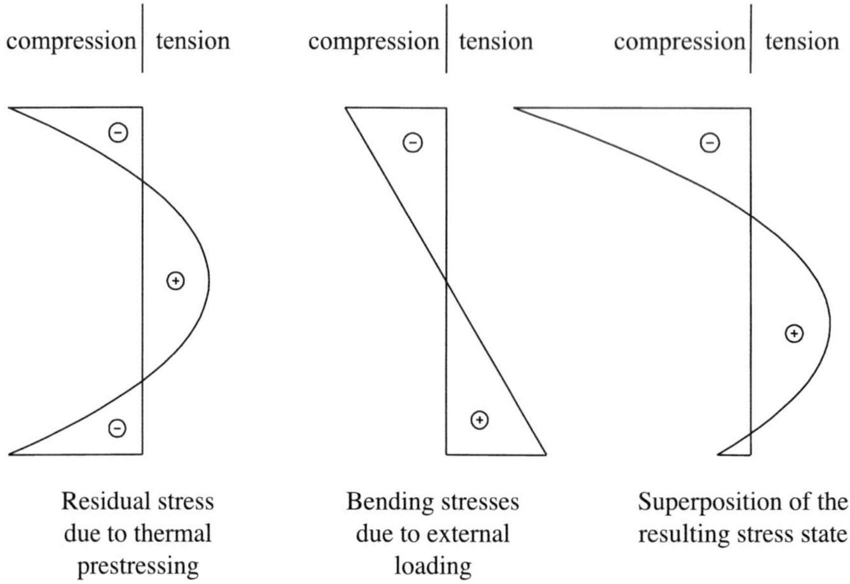

Fig. 6.1 Residual stress state of thermally tempered glass and overlapping bending stresses

In addition to the residual stress profile, the thermal pre-stressing process also induces a degree of crack healing in the glass, leading to a further increase in strength [1, 2]. After the thermal pre-stressing is complete, the glass is also more resistant to the formation of median lateral scratches compared to annealed glass [3]. Finally, a further advantage, apart from the increase in apparent strength, is that no subcritical crack growth, see e.g. Sect. 5.3.1, occurs while the surface flaws are in compression and thus the glass strength is not affected by load duration as long as the surface of the glass remains in compression.

6.2 Thermal Pre-stressing

During the process of thermal tempering, the glass is heated and then quenched (cooled quickly) to obtain pre-stressed glass within seconds to minutes, depending on the glass thickness. During the cooling, surface compression up to ≈ 200 MPa, balanced by internal tension, is created in the glass. This results in an increase in apparent strength of up to five times that of the base product, annealed float glass. In zones far from edges of teh glass, the residual stress distribution through the glass thickness is nearly parabolic and the depth of the compression zone is approximately 20% of the glass thickness on both sides. This depth is affected by a various factors [4].

6.2.1 *Process and Residual Stresses*

Thermal pre-stressing exploits the temperature dependency of the material properties. The glass is heated up to approximately 100 °C above the glass transition temperature, T_g, followed by quenching with air jets from both the top and bottom surface. When the glass temperature is above T_g, it is in the rubbery state with the characteristics of a viscoelastic material. The uneven cooling through the thickness during the quenching phase causes the surfaces to contract before the core which is warmer. This leads to transient tensile stresses at the surfaces, see Fig. 6.2. If the cooling is too fast and/or the edge quality of the pane is not sufficient, these transient stresses lead to fracture during the process. The temperature at the surfaces during the cooling process is lower than within the core and therefore the surfaces cool down below T_g, entering the glassy state, before the core. Due to the significantly lower stiffness in the rubbery state, the core follows the contraction of the surfaces without generating internal stresses. As the temperature in the core drops below the transition temperature, T_g, the stiffness increases and the core continues to contract, thereby generating a compressive stress at the surfaces and a state of tension in the core. If the glass is initially heated to a temperature below T_g it will remain elastic and any stress from the thermal treatment will be temporary and vanish when the temperature gradient return to zero. A sufficiently high heating temperature and an appropriate high cooling rate are essential for achieving the desired residual stresses. The high temperature ensures that the glass is fully in its rubbery state and reduces the risk of fracture due to transient tensile stresses. The drawback is that the glass becomes softer at the surfaces and maintaining the surface flatness of glass becomes more challenging. For example, the glass sags over the transport rollers in the tempering oven leading to what is known as roller-wave distortions.

The magnitude of the residual surface stress depends on the temperature and the temperature gradient. The temperature gradient is induced by the rapid cooling and the faster heat can be removed from the surface, the larger is the gradient. Obtaining a large temperature gradient is easier in thick glass than in thin glass and therefore the cooling rate required to produce a particular pre-stress level in thin glass is higher than for thick glass. Therefore, glass thicknesses such as 2 or 3 mm can be difficult to thermally pre-stress using traditional tempering equipment.

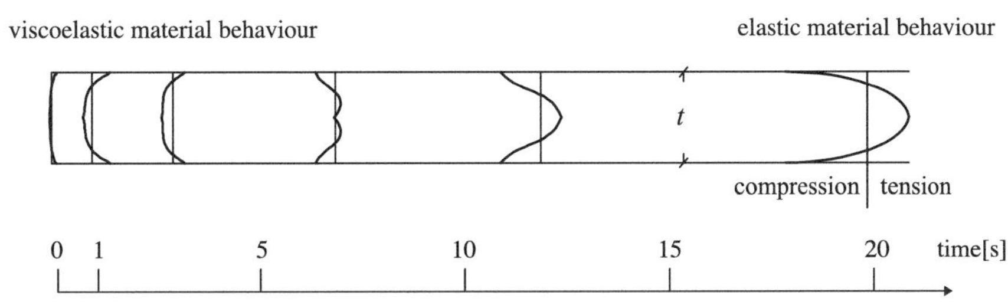

Fig. 6.2 Example of the evolution of stresses during the thermal pre-stressing process, according to [5]

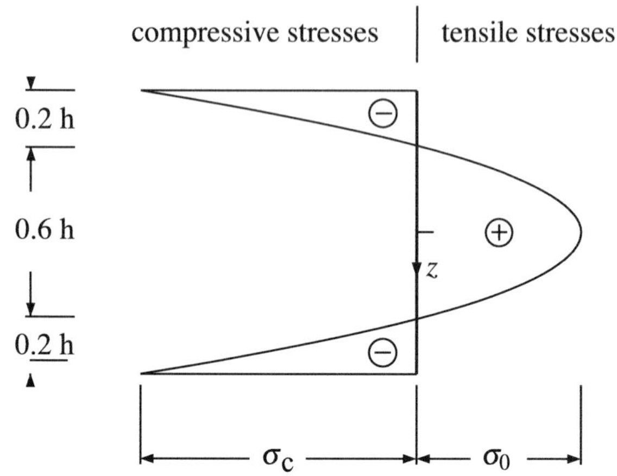

Fig. 6.3 Distribution of the residual stresses in thermally pre-stressed glass $\sigma_c = -2\sigma_0$

Away from edges, the final residual stress profile through the thickness of thermally pre-stressed glass process can be approximated by a second order parabola Fig. 6.3:

$$\sigma(z) = \sigma_0 \cdot \left(1 - 3 \cdot \zeta^2\right) \tag{6.1}$$

where

$$\sigma_0 = -\frac{\sigma_c}{2}\ \zeta = \frac{2 \cdot z}{h}. \tag{6.2}$$

There are two standard grades of thermally pre-stressed glass, the so-called thermally toughened (or fully tempered) glass (TTG) and heat-strengthened glass (HSG), as defined in product standards [6, 7]. Typical values for the surface compressive stress, σ_c, are in the range of 75–140 MPa for TTG and 40–60 MPa for HSG. European standards do not explicitly specify any requirements for the minimum surface compressive stress, but specify an overall glass bending strength. In contrast, the US standard ASTM C1048, prescribes the bending strength as well as the minimum residual surface stress: –69 MPa away from the edges of TTG and –67 MPa close to the edges of TTF; –52 MPa away from the edges of HSG and –24 MPa close to the edges of HSG. The residual stresses can be measured by means of optical methods, see e.g. Sect. 6.4.

Away from edges or holes in the glass, the residual stress state is often assumed to be equibiaxial i.e. planar and with equal stress in all planar directions. This assumption is valid when the distance from the edge is more than three to four times the thickness of the glass. Closer to the edges, the stress state becomes more complex because by definition the free edge means that the stress state is not equibiaxial. This also implies that corners of tempered glass will be stress free due to the presence of two adjacent edges, see Fig. 6.4. Edges and corners in thermally pre-stressed glass are therefore more sensitive to localised tensile stresses from external loading. Since

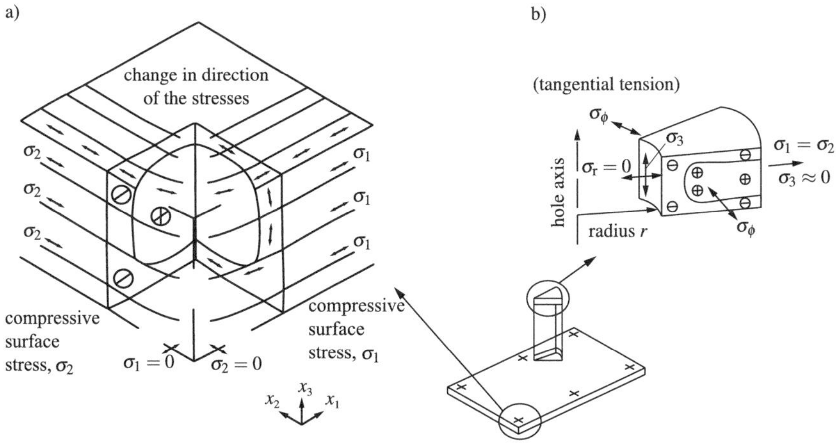

Fig. 6.4 Distribution of the residual stresses at a corner (**a**) and a hole (**b**) of thermally pre- stressed glass, according to [8]

a significant part of the apparent strength is due to the equibiaxial residual stresses, the apparent strength is low in the transverse direction.

Numerical simulations have shown that the tangential residual stresses around holes varies [9, 10]. This should be considered in the design of bolted connections, see Chap. 7.

When the glass is thermally pre-stressed a relatively large amount of (strain) energy is stored in the glass. If the magnitude of the residual stresses is sufficiently large, drilling, cutting or edge-treatment releases the energy locally which results in a "chain-reaction" where the propagation of a crack is fed by stored energy resulting in branching and fragmentation of the glass, see Sect. 6.2.8. Therefore, any (significant) work on the edges, holes, etc. must be done before the pre-stressing process. Small adjustments of the geometry along edges, etc. are possible after tempering, however, this will alter the residual stress state locally [11, 12].

6.2.2 *Bending Strength*

The bending strength of thermally pre-stressed glass can be determined by a four-point bending test according to EN 1288-3 [93], see Sect. 5.4.2.2. In particular, edge work has considerable influence on bending strength, because this often leads to flaws at the edges, which will typically experience the highest loading stress during a four-point bending test. Edge stresses from the thermal pre-stressing process therefore also have considerable influence on the bending strength. However, in statistical evaluations of the tests according to EN 1288-3, there is usually no differentiation made between *fracture at the edge and fracture at the surface*, so that the evaluation is in fact a mixture of the various effects. It is therefore sensible to use the coaxial double-ring test, see Sect. 5.4.2.3 for determining the surface bending strength of glass. It is also possible to non-destructively measure the residual stresses using optical methods [8, 13]; see e.g. Sect. 6.4.

The coaxial double-ring bending test is also very suitable for determining the strength of pre-stressed glass in the area of holes. The specimens are placed with the hole centred within the load ring. Maximum principal stresses appear in the area of the chamfer, where the maximum principal stresses also usually occur in point-fixings of glass plates. Experimental investigations with the coaxial double-ring test have shown that the strength of pre-stressed glass at the vicinity of the chamfer is typically well above the values defined in the product standard manuals [8, 13]. However, care must be taken for in-plane loading (e.g. bolted connections) as the pre-stress in the mid-plane of a hole is typically lower than at the chamfers [14, 15].

A modified four-point bending test based on EN 1288-3 is also being used in determining the edge strength of glass. Specimens are placed vertically and tested about their major axis, while being horizontally restrained against lateral torsional buckling [16–18].

The apparent strength of thermally pre-stressed glass resulting from today's production processes can be significantly higher than the values found in the product standards, see Sect. 2.3.4. Values for the surface strength can sometimes be more than 220MPa [3].

6.2.3 Process Technology and Pre-stressing Furnaces

The process of thermal pre-stressing of glass had already been invented in 1874 by François Royer de la Bastie and was later patented [19–21]. In this process, molten glass was quenched in warm oil. The technology available today is a significant improvement over that from the 1960s and allows for thermal pre-stressing of thin glass down to 1.8 mm in thermally toughened quality. Until the mid-1980s, pre-stressing systems were widespread, in which the glass was suspended vertically from tongs during the process. The optical quality of this glass was good, however, the tongs left marks in the glass. The cooling process was also severely impeded in the areas shielded by the tongs. This could lead to undesirable local tensile stresses on the surface. The horizontal pre-stressing systems that had been established since the 1970s have today almost completely replaced the vertical systems. In these horizontal systems (see Figs. 6.5 and 6.6), glass is transported on ceramic rollers into the furnace and on kevlar-coated rollers into the cooling (quenching) area.

The glass is first cleaned and then heated uniformly to a temperature between 620 °C and 700 °C, i.e. approximately 100 °C above T_g, and then quenched. The heating process in the furnace is monitored by infrared temperature sensors. The evenness of the temperature in the glass plates can be controlled by selective firing up of the furnace, because the peripheral areas are usually somewhat cooler. Heating is carried out in a radiation furnace (using electrical resistance heaters), a convection

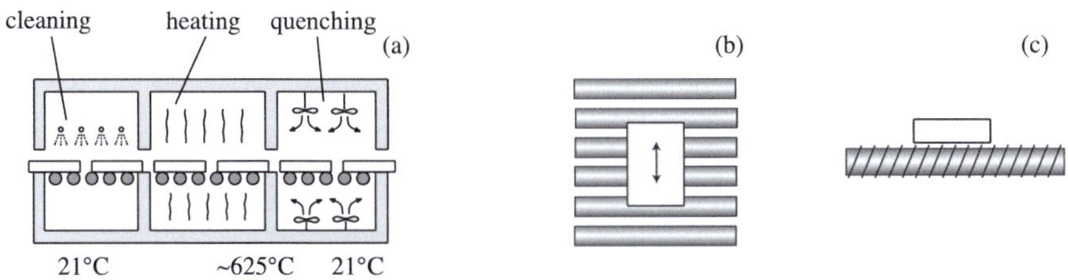

Fig. 6.5 **a** Horizontal thermal toughening process on rollers; Movement of glass as on rollers: **b** Top view and **c** cross-sectional view

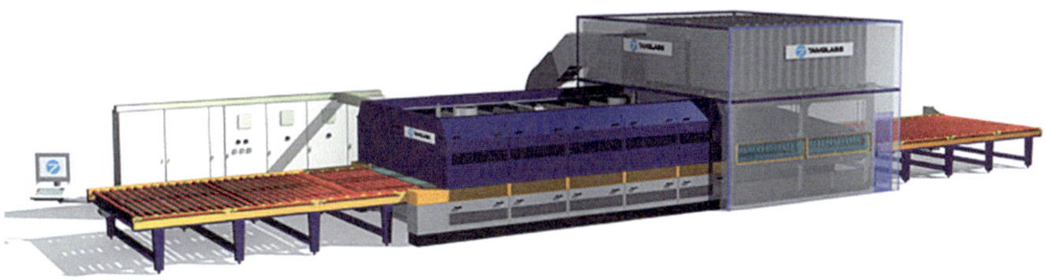

Fig. 6.6 Pre-stressing furnace for horizontal toughening of glass (Photo credits: Glaston)

furnace (heating with hot air through convection) or a combination of both techniques. The different types of furnaces available are as follows: [22]:

- Radiation furnace;
- Radiation furnace with external convection support;
- Full convection furnace;
- Hybrid furnace, completely convectional at the top and radiational below, with additional convectional support if necessary.

Convection furnaces are especially suitable for thermally pre-stressing coated glass (low-e glass), which has become more prevalent on the market. Heating-up times are approximately 30–60s per millimeter thickness of glass. Air is used to cool the glass. In the cooling area, the glass is quenched by compressed air from nozzles. An important aspect during thermal pre-stressing is a uniform temperature distribution, during heating but especially during cooling, which can be achieved by continuously moving the glass by a few centimeters back and forth in the quenching chamber.

Furthermore, special systems have been developed over the last decade to thermally pre-stress thin glass, i.e. glass with a thickness down to 1.8 mm, where the glass is heated and cooled on an air cushion rather than on rollers. Very high heat transfer coefficients of over $\bar{h} = 1\,000 \mathrm{W} m^{-2} \mathrm{K}^{-1}$ can be reached using this method. In the procedure developed by LiSEC GmbH, the plates are tilted to just below 15° and transported via the edges at the bottom, Fig. 6.7. Another method developed by Glaston Corporation tilts the glass to 1.5° and heats it on the transport rollers, then

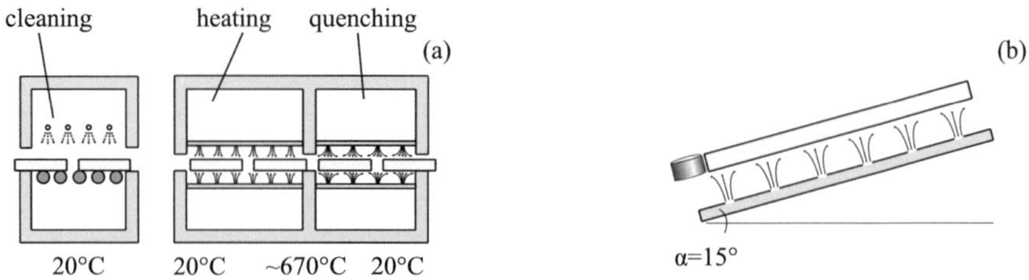

Fig. 6.7 LiSEC method: **a** process of thermal pre-stressing of glass on an air cushion; **b** position of glass during pre-stressing.

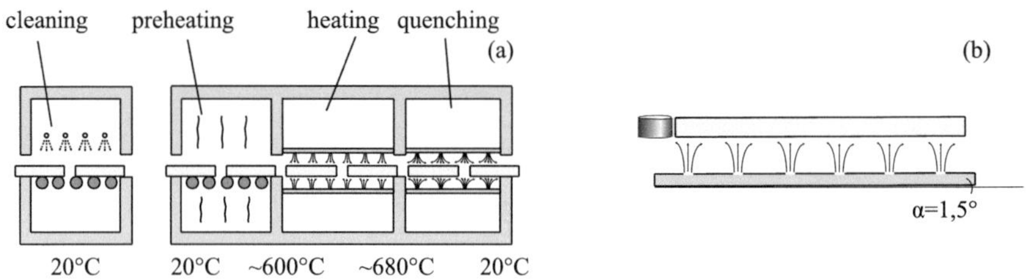

Fig. 6.8 Method according to Glaston: **a** process of thermal pre-stressing on air cushion; **b** position of glass during pre-stressing.

overheat it on air cushions and cools it down again on air cushions, in the cooling area Fig. 6.8.

Thermal pre-stressing processes can lead to a reduction in dimensional accuracy and optical quality as follows:

- Global bow;
- Roller waves (horizontal process) ;
- Tong marks due to hanging of the plates (vertical process);
- Local edge lift, edge kink or frame effect (horizontal process and air cushion process);
- Local bow (vertical process);
- Surface undulation (air cushion process);
- Optical anisotropies (all processes).

The above-listed defects can be caused by an uneven temperature distribution of the glass in the furnace, if, for example, through coatings or contact with the rollers, a zone of the glass is heated up faster (or slower) than the other. This is particularly problematic in radiation furnaces. The appearance of roller waves can generally be attributed to the temperature of the glass being too high when moving between the furnace and the cooling area, the distance of the transport rollers being too large or the speed being too low during movement of the glass on the rollers. The global bow is measured over the entire length of the glass and in the case of float glass should be limited to 3mm/m. Local bow is measured over relatively short distances

(300 mm) along the glass edges and for horizontally pre-stressed glass should not exceed 0.3/300mm [6, 7].

The local uneven thickness of edges and deformation along the sides of the glass can often be traced back to the following causes:

- The temperature of the glass before quenching is too high;
- The back and forth motion of the glass occurs at the same reversal point;
- The spacing between the rollers is too large;
- The heating of edges is non-uniform, especially in coated glass with edge deletion;
- The transfer from furnace to the cooling area is too slow.

Relatively large *roller waves* are permitted by European standards in heat-strengthened and thermally pre-stressed glass. These are therefore very visible in the glass when it is installed in the building which can be aesthetically disturbing when the glass is used in highly visible areas and particularly where the glass can reflect regular patterns e.g. in façade applications. Therefore for these aesthetically sensitive applications it is commonplace to produce samples of the glass and install them in visual mock-ups for approval of the client before producing the glass panels for that particular application. Usually, "half standard values" are agreed upon with the glass manufacturer.

6.2.4 Optical Anistropy and Iridescence

Optical anisotropy and iridescence is an optical defect that may occur in thermally toughened glass. This effect is due to an in-plane, non-uniform residual stress distribution caused by uneven cooling in the tempering process. These unwanted optical phenomena give rise to hue changes with the angle of observation and the angle of illumination appearing as white or rainbow-like spots (leopard spots), patches or patterns. They are very visible to the human eye and the effect can be magnified by polarised light or when coatings are present on the surface. Examples of optical anisotropies are shown in Figs. 6.9 and 6.10.

The heating and cooling are not homogeneous during the toughening process due to e.g. the position of the glass in the furnace and the arrangement of the rollers and air nozzles in the quench, which leads inhomogeneous residual stresses. Stress differences of about 1 MPa are sufficient to produce a visible effect, [23, 24]. Even if the requirements of the product standards for thermally toughened glass EN 12150-1 are met, optical anisotropy is usually not considered a fault, but rather a visual effect. Research on how to evaluate the optical anisotropy can be found in e.g. [23, 25].

One of the major challenges in developing a fast and reliable detection method is that the differences in surface stresses can only be evaluated with complex measurements techniques, see e.g. Sect. 6.4. However, the development of a methodology with criteria for assessing optical anisotropies, taking into account human perception, will be indispensable for the future [26, 27].

Fig. 6.9 Anisotropies on a glazed façade with thermally toughened glass (TTG)

Fig. 6.10 Anisotropies on curved TTG, more visible on the right because of use of a polarising filter

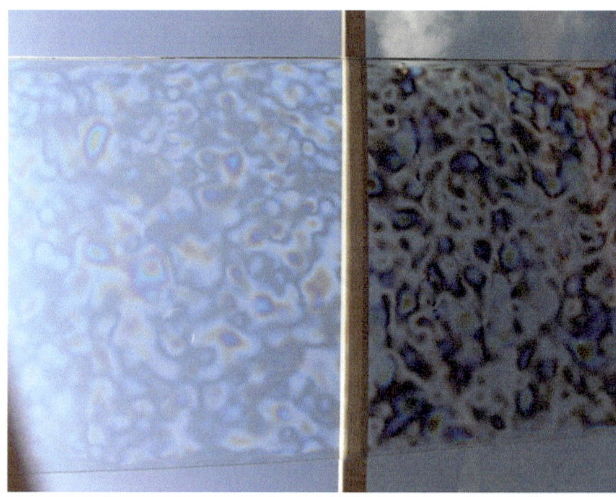

6.2.5 Mathematical Models for the Thermal Pre-stressing Process

6.2.5.1 Basics

The residual stresses in thermally pre-stressed glass are governed by the evolution of temperature during cooling. The material is soft at high temperatures and stiff at low temperatures which is an essential for achieving a pre-stress by cooling the surfaces before the core. The first mathematical models developed to predict the levels of pre-stress assumed the transition from soft to stiff to be instant (instant freeze models) [28, 29]. These were followed by more sophisticated models to include viscoelasticity [30] and structural relaxation [5].

The effects of stress relaxation due to viscoelasticity, as well as the effects from structural relaxation and the associated cooling speed of the glass on the residual stresses (see Sect. 1.2.3.4), should be taken into account to establish an accurate model. According to available literature [5], ≈ 24 % of thermal stresses are caused by structural relaxation. Basic models for this were developed as early as the 1970s [5, 31, 32]. This is where the principle of a time-temperature shift for basic thermo-rheological materials, see Sect. 4.1.5.2, is also applied to glass and extended to structural relaxation. Today numerical simulations typically adopt generalized Maxwell models (Prony series), see [33, 34]. Experimental data for stress and structural relaxation can be found in literature [35–38].

6.2.5.2 Temperature Calculations

In order to calculate the temperature during the quenching process, the heat exchange between glass and the surrounding medium must be considered. Heat exchange is generally a combination of convection, radiation and conduction within the glass and through contact with rollers, see Fig. 6.11. Since heat exchange is dominated by convection, much of the available literature simplifies all of the effects into the parameters for convection, see e.g. [34]. More accurate approaches, that consider heat radiation in glass can be found in [39, 40]. The convective heat transfer coefficient is required to determine the heat transfer by convection. This coefficient is a measure for the amount of heat removed (or added) by convection, including the effect of boundary layers at the surface of the glass. For a simplified temperature calculation, in addition to the heat transfer coefficient, the following must also be known: the temperature difference between the body and the cooling medium (e.g. the intake air); the starting temperature of the glass; the density of the glass; the specific heat capacity of the glass; and the thermal conductivity of the glass Table 6.1.

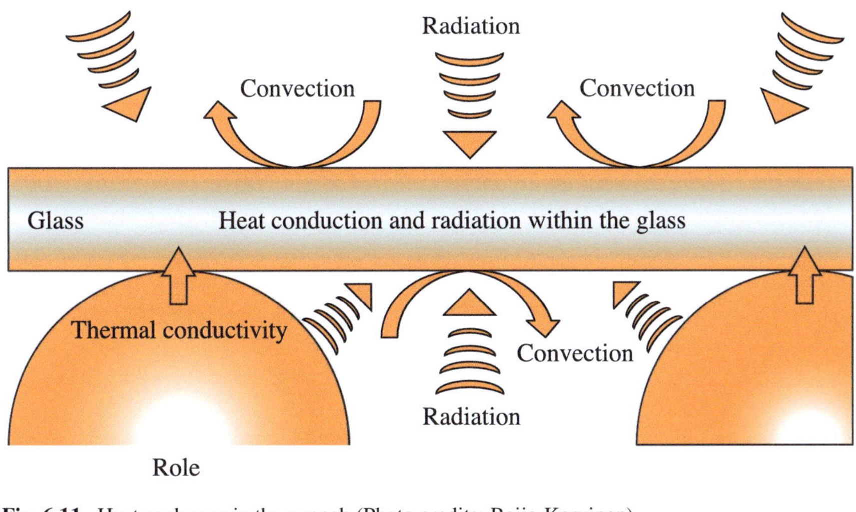

Fig. 6.11 Heat exchange in the quench (Photo credits: Reijo Karvinen)

Table 6.1 Material properties of glass for computational simulation of the residual stress state [37, 38, 45]

Properties	Value
Young's modulus of elasticity E	70 000 MPa
Poisson ratio ν	0.23
Thermal expansion coefficient ($T < T_g$) $\alpha_{T,g}$	9×10^{-6} K^{-1}
Thermal expansion coefficient ($T > T_g$) $\alpha_{T,l}$	32×10^{-6}K^{-1}
Thermal conductivity λ	$0.975 + 8.58 \times 10^{-4} \cdot T$ W (mK)$^{-1}$ (T in °C)
Heat capacity ($T < T_g$) $c_{p,g}$	$893 + 0.4 \cdot T$ J (kgK)$^{-1}$ (T in K)
Heat capacity ($T > T_g$) $c_{p,l}$	1.433J (kgK)$^{-1}$
Ratio E_A/R_m	66 000 K
Constant x	0.5
Density ρ	2 530 kg m^{-3}

In the case of air cooling, the heat transfer coefficient depends on the size and geometry of the air nozzles and the blowing pressure and is difficult to determine for typical cooling processes at temperatures found in the quench [41]. If a glass plate is freely cooled in still air, the convective heat transfer coefficient is approximately $\bar{h} = 12.5\,\text{Wm}^{-2}\text{K}^{-1}$ [42]. In the case of cooling by forced convection (i.e. blowing air), the value can be increased to $\bar{h} = 500\,\text{Wm}^{-2}\text{K}^{-1}$, while cooling with liquids is an order of magnitude higher [43]. Modern systems for thin glass can reach values up to $\bar{h} = 1\,000\text{W}m^{-2}\text{K}^{-1}$. For regular glass thicknesses of 4–19 mm of soda-lime silica glass, heat transfer coefficients between $\bar{h} = 200\,\text{Wm}^{-2}\text{K}^{-1}$ and $\bar{h} = 80\,\text{Wm}^{-2}\text{K}^{-1}$, respectively, are required to achieve thermally toughened quality. Research indicates that considerably higher pre-stressing can be obtained if the quench procedure is controlled and optimised [44].

6.2.5.3 Material Model for Glass During Quenching

The amount of residual stresses generated during the tempering process is governed by the temperature relationship described in the previous section. At temperatures well above the glass-transition temperature the viscosity of the glass is low and the glass is unable to sustain significant stresses over time. It is therefore often assumed that all stresses are relaxed at such high temperature.

Due to the high temperatures during the tempering process, the glass can be assumed to be isotropic linear viscoelastic. Due to the linear properties of the material, the principle of superposition is valid and the governing equations can be written in terms of a hereditary integral. Furthermore, the stress and strain state can be split into a hydrostatic and deviatoric part and due to the isotropic assumption, two material functions are sufficient to describe the material. More specifically, these are known as the bulk relaxation modulus and the shear relaxation modulus. A temperature dependency on the bulk- and shear modulus can be elegantly incorporated by shifting the time-scale. This is known as Time-Temperature Superposition (TTSP) and indicates that all relaxation curves found at different temperatures are identical by adjusting the time-scale. Experiments confirm that glass belongs to the group of materials where this is valid [35]. Such materials are often referred to as thermo-rheologically simple materials. The constitutive equations can therefore be written as:

$$\sigma_{ij}(t) = \delta_{ij} \int_0^t K(t-t') \frac{\partial\left(\varepsilon_{ii} - 3\varepsilon^{\text{th}}\right)}{\partial t'} \mathrm{d}t' + 2\int_0^t G\left(t-t'\right) \frac{\partial e_{ij}}{\partial t'} \mathrm{d}t'. \quad (6.3)$$

where:

σ_{ij}	Stress tensor;
δ_{ij}	Kroneckers delta;
K	Bulk relaxation function;
G	Shear relaxation function;
$\varepsilon_{ii} = \varepsilon_{\text{xx}} + \varepsilon_{\text{yy}} + \varepsilon_{\text{zz}}$	Trace of the strain tensor;
ε^{th}	Thermal strain;
$e_{ij} = \varepsilon_{ij} - \delta_{ij}\frac{1}{3}\varepsilon_{ii}$	Deviatoric strain tensor.

The bulk- and shear relaxation functions can be determined directly through experimental work and may, in principle, be a set of tabulated data. However, there are significant advantages by representing the relaxation by means of a series of exponential functions (Prony series). Firstly, the numerical computations become less demanding since integration of the complete strain history can be avoided [34] and secondly the material can be interpreted as a mechanical model consisting of springs and dashpots, a so-called rheological model. The particular model associated with the Prony series is named a generalised Maxwell model or a Maxwell-Wiechert model. The Prony series for the shear and bulk modulus, respectively, can be written as:

$$G(t) = G_0 \sum_{i=1}^{\text{N}} g_{1i} \cdot e^{-\frac{t}{\tau_{1i}}}, \quad (6.4)$$

$$K(t) = K_\infty + (K_0 - K_\infty) \sum_{i=1}^{\text{M}} k_{2i} \cdot e^{-\frac{t}{\tau_{2i}}}. \quad (6.5)$$

where index 0 and ∞ refers to the instantaneous properties and the long-term properties, respectively. The parameters g_{1i}, k_{2i}, τ_{1i} and τ_{2i} are constants to be determined by experiments.

By assuming that glass is a thermo-rheologically simple material, relaxation times can be determined at different temperatures using the shift function ϕ.

In glass, the *Arrhenius equation*, see Sect. 4.1.5.2 is often used as the shift function:

$$\phi(t) = \frac{\tau_{\mathrm{ref}}}{\tau} = e^{\frac{W_{\mathrm{A},\sigma}}{R_{\mathrm{m}}}\left(\frac{1}{T_{\mathrm{ref}}}-\frac{1}{T(t)}\right)+\frac{W_{\mathrm{A,s}}}{R_{\mathrm{m}}}\left(\frac{1}{T_{\mathrm{ref}}}-\frac{1}{T_{\mathrm{f}}(t)}\right)} = e^{\frac{W_{\mathrm{A}}}{R_{\mathrm{m}}}\left(\frac{1}{T_{\mathrm{ref}}}-\frac{x}{T(t)}-\frac{1-x}{T_{\mathrm{f}}(t)}\right)}. \tag{6.6}$$

It is also assumed that a part of the activation energy is related to stress relaxation $W_{\mathrm{A},\sigma}$ and another part to structural relaxation $W_{\mathrm{A,s}}$, where:

$$W_{\mathrm{A}} = W_{\mathrm{A},\sigma} + W_{\mathrm{A,s}}$$

and

R_{m}	Universal gas constant;
T_{ref}	Reference (or base) temperature [K];
T	Temperature [K];
T_{f}	Fictive temperature (structural relaxation) [K].

The stress relaxation rate depends on the time that has elapses after the change in strain has occurred. Because of the change in temperature during the process, the real time in the above equations must be converted into a so-called reduced time, ξ, which represents the time integral of a shift function:

$$\xi(t) = \int_0^t \phi(t')\mathrm{d}t'. \tag{6.7}$$

The constitutive relation from Eq. 6.3 can now be rewritten substituting the real time, t, with the reduced time, $\xi(t)$:

$$\begin{aligned}\sigma_{ij}(t) =& \delta_{ij} \int_0^t K\left(\xi(t)-\xi(t')\right) \frac{\partial\left(\varepsilon_{ii}-3\varepsilon^{\mathrm{th}}\right)}{\partial t'}\mathrm{d}t' \\ &+ 2\int_0^t G\left(\xi(t)-\xi(t')\right)\frac{\partial e_{ij}}{\partial t'}\mathrm{d}t'.\end{aligned} \tag{6.8}$$

The fictive temperature in Eq. 6.6 governs the structural relaxation which can be interpreted as a reorganisation of the microstructure in the material (long range order), which deals with the time changes. At temperatures below the transition temperature, the glass is in an unstable state where reorganisation of the microstructure has been prematurely halted by the transition to a solid. The response function, M_{p}, of the material properties taking into account the structural relaxation is expressed using the real and the fictive temperature:

$$M_p(t) = \frac{p(t) - p_2(\infty)}{p_2(0) - p_2(\infty)} = \frac{T_f(t) - T_2}{T_1 - T_2} \tag{6.9}$$

In the response function, p, is a property change; index 1 is the state before the temperature change and index 2 represents the state after the temperature change. With Eq. 6.9 the fictive temperature, T_f, can be calculated by integrating over time:

$$T_f(t) = T(t) - \int_0^t M_p(t - t') \frac{\partial T}{\partial t'} dt'. \tag{6.10}$$

The response function of the material properties can be set in analogy to the relaxation functions as a Prony series:

$$M_p(t) = \sum_{i=1}^{P} c_i \cdot e^{-\frac{t}{\lambda_i}} \tag{6.11}$$

where λ_i is a relaxation time and c_i is a weighting factor. If the temperature profile is known in each time step, the stress state can also be determined in each time step. Values for the material parameters and Prony series parameters are given in Tables 6.1 and 6.2.

Table 6.2 Experimental Prony parameters for shear, compression and structural relaxation at a reference temperature of $T_g = 869$ K [38]

	Shear relaxation		Bulk relaxation($K_\infty/K_0 = 0.18$)		Structural relaxation	
i	g_{1i}	τ_{1i}	k_{2i}	τ_{2i}	c_i	λ_i
1	0.05523	$6.658 \cdot 10^{-5}$	0.0222	$5.009 \cdot 10^{-5}$	0.05523	$5.965 \cdot 10^{-4}$
2	0.08205	$1.197 \cdot 10^{-3}$	0.0224	$9.945 \cdot 10^{-4}$	0.08205	$1.077 \cdot 10^{-2}$
3	0.1215	$1.514 \cdot 10^{-2}$	0.0286	$2.022 \cdot 10^{-3}$	0.1215	0.1362
4	0.2286	0.1672	0.2137	$1.925 \cdot 10^{-2}$	0.2286	1.505
5	0.2860	0.7497	0.3940	0.1199	0.2860	6.747
6	0.2265	3.292	0.3191	2.033	0.2265	29.63

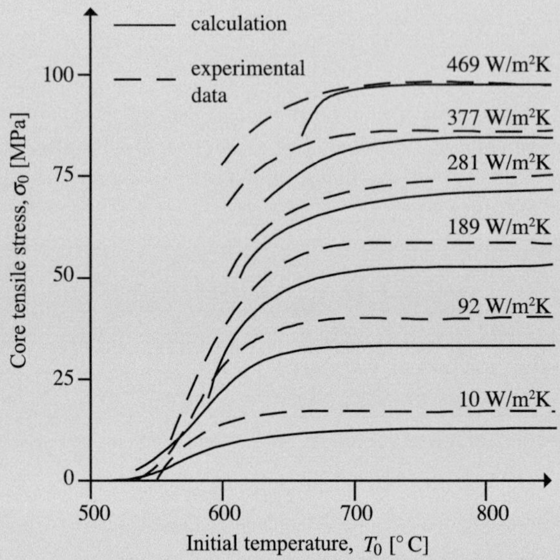

Fig. 6.12 Tensile stress in the core as a function of the pre-quench temperature for various heat convection constants and a glass thickness equal to $h = 6.1$ mm [5]

6.2.5.4 Examples

Taking into account the transient temperature distribution and the material model for the glass from the previous Sects. 6.2.5.2 and 6.2.5.3, calculations can be performed to quantify the sensitivity of the individual parameters. In a simple case, such as the cross-section of a glass panel far from edges and holes, one-dimensional calculations are sufficient. Figure 6.12 illustrates the residual tensile stress, σ_0, in the core of the panel, depending on the pre-quenching temperature, T_0, for various heat transmission coefficients, $\bar{h}$, with a glass thickness of $h = 6.1$ mm [5].

In Fig. 6.13, the resulting surface compression stress and maximum transient surface tension stress during the quenching for a $h = 3$ mm thick glass is shown as isolines in a heat convection constant, $\bar{h}$, versus pre-quench temperature, T_0, diagram.

Figure 6.14 plots measured residual stresses and calculated residual stresses for a plate with thickness $h = 4$ mm [42].

In more complex two-dimensional and three-dimensional geometries, the calculations can be carried out using the finite element method [9, 33, 45]. Such models can be used for predicting residual stresses and transient stresses near edges or holes in complex geometries, see Fig. 6.15. Such information is useful for both the designing engineer who needs to determine the local strength but also for the manufacturer

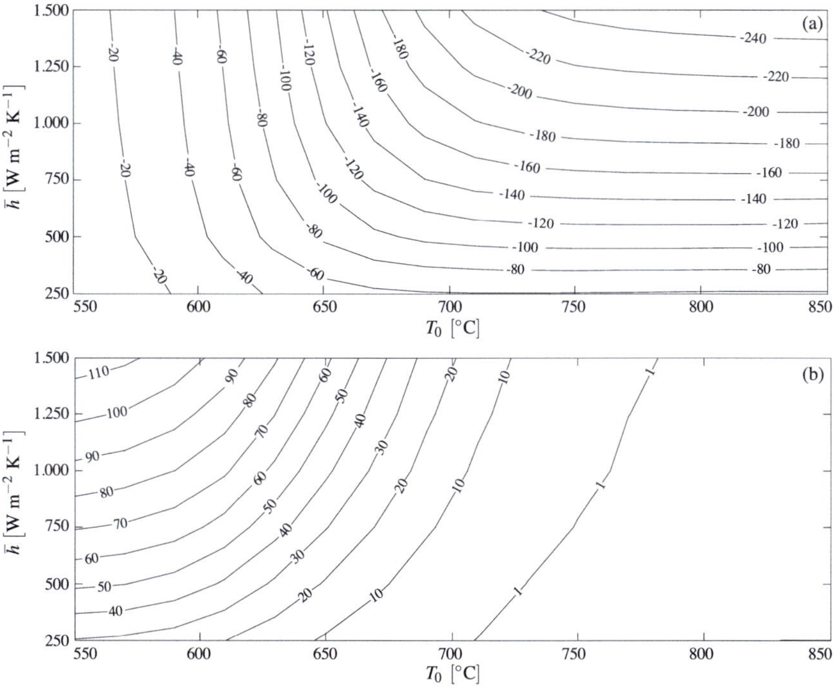

Fig. 6.13 **a** Resulting surface stress and **b** maximum transient surface tension during the quenching, depending on the pre-quench temperature, T_0, and the heat transfer coefficient, $\bar{h}$, for a glass thickness of $h = 3$ mm, from [39]

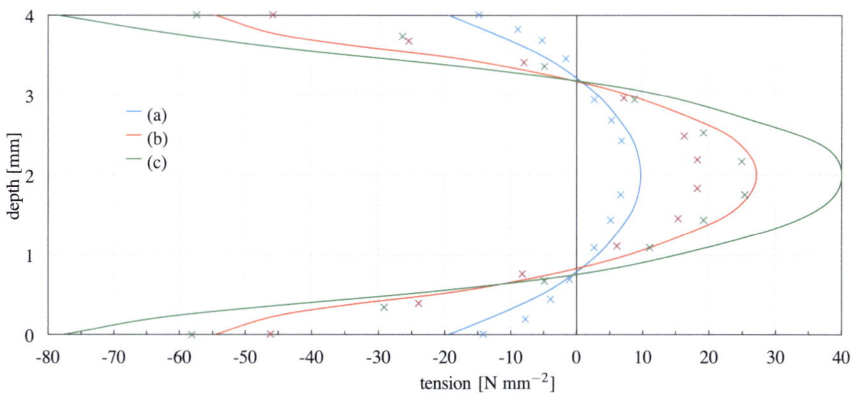

Fig. 6.14 Measured residual stresses (points) and calculated stress curves (solid lines) for heat transfer coefficients **a** $\bar{h} = 27\,\mathrm{Wm^{-2}K^{-1}}$, **b** $\bar{h} = 107\,\mathrm{Wm^{-2}K^{-1}}$ and **c** $\bar{h} = 190\,\mathrm{Wm^{-2}K^{-1}}$ [42]

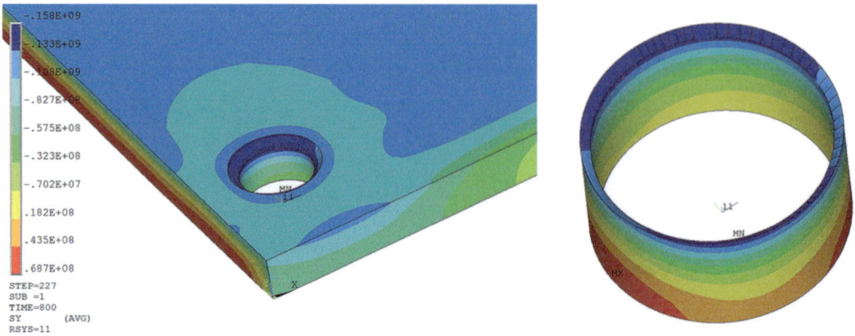

Fig. 6.15 Left: contour plot of the stress in the y-direction close to an edge and a hole. Right: contour plot of the stresses at the boundary of the hole

of thermally pre-stressed glass who needs to know if fracture occurs during the quenching of odd-shaped glass.

Another important factor to be taken into account in determining the strength of thermally pre-stressed glass is the influence of locally varying convection on the surface. This might lead to a significant reduction of the residual stress state, e.g. through shielding or insufficient movement during the thermal pre-stressing process.

As shown in [15], the residual stress state for complex geometries obtained from the above-mentioned material model can also be combined with the models of Crack Growth and Static Fatigue described in Sect. 5.3 to provide an accurate prediction of the fracture strength of the glass in these regions.

6.2.6 *Nickel Sulphide Inclusions and Heat-Soak Testing*

6.2.6.1 Phenomenon

It has been known since the 1960s that thermally toughened glass may fracture due to microscopic inclusions embedded in the glass during the production process. This is commonly referred to as *spontaneous fracture*. The fracture is caused by locally high tensile stresses due to a time/temperature dependent phase, and associated volume, change of small nickel-sulphide (NiS) inclusion in the glass. If these inclusions are sufficiently large and critically located, e.g. in a tension area such as the core of toughened glass, spontaneous fracture of the glass will occur. This induces fracture in the entire glass, in a seemingly spontaneous manner [46–52].

Fracture from nickel sulphide (NiS) inclusions often occurs in thermally toughened glass after installation [53]. Among other things, this is due to the fact that nickel sulphide inclusions are so small, with diameters of about 50–600 μm, that they are usually not found by standard detection systems. Fracture might still occur after years, or even decades, due to this delayed volume expansion. It has been shown

Fig. 6.16 Heat-soak oven for thermally pre-stressed glass

that both thermally toughened and heat-strengthened glass may be affected, although it occurs very infrequently in the latter [54]. Traces of NiS in the glass melt cannot be altogether avoided, therefore a proof-testing approach is adopted after manufacture in the form of a so-called heat-soak test, see Fig. 6.16. In this test, the glass is placed in an oven over a defined holding time e.g. from $t = 2\,\text{h}$ to $t = 4\,\text{h}$ [55] at high temperatures, usually at $T = 260(10)\,°\text{C}$, to bring about the conversion of the inclusions and trigger the fracture of the thermally toughened glass [56]. Glass that has survived the heat-soak test with a holding time of $t = 4\,\text{h}$ equates to a probability of failure $P_\text{f} < 10 \times 10^{-6}\,\text{a}^{-1}$, which is acceptable in the building industry [57, 58].

6.2.6.2 The Cause of Spontaneous Fracture by Phase Change of NiS

NiS is has a high-temperature phase (α-NiS) and a low-temperature phase (β-NiS). The low temperature phase increases by about 2.5 % in volume when it changes to its high temperature phase [46]. Up to 4 % increase in volume is reported in literature [47], but this value is not supported by experimental results and an associated fracture mechanical model [54]. The phase change temperature in the equilibrium state depends on the stoichiometric composition and is $T = 379\,°\text{C}$ for NiS, with $50\,\text{Wt} - \%$ nickel and sulphur, and may drop to $T = 282\,°\text{C}$ for NiS with about $48\,\text{Wt} - \%$ nickel and $52\,\text{Wt} - \%$ sulphur [59].

Around $T = 985\,°\text{C}$ (melting temperature), small NiS particles present in the glass melt, remain in suspension due to convection. In the glass melt, NiS is in its α-phase. The change of the NiS into the stable β-phase starts due to the slow cooling process in the annealing process (within the float process, see e.g. Sect. 2.2.1) at temperatures below $T = 379\,°\text{C}$. However, some of the NiS inclusions will remain in the α-phase.

In the thermal pre-stressing process, some of the NiS inclusions are thus initially in the α-Phase. These NiS inclusions are not fully converted into the β-phase due to the rapid cooling process (only by approximately 5 %). After the thermal pre-stressing process is completed, the glass contains some NiS inclusions in the unstable α-phase. These inclusions tend to change to the stable and larger β-phase over time. This change is time-dependent and might cause delayed fracture of the glass. The transformation time depends strongly on the temperature and the exact stoichiometric composition of the inclusion [59–61]. In case of normal operating temperatures in construction, this conversion can take up to decades to develop. In the heat-soak oven at $T = 290(10)$ °C the phase change is faster and usually takes place within a few hours. The heat-soak procedure does not guarantee that 100 % of the potentially critical NiS has fully developed into the β-phase.

6.2.6.3 Mechanical and Failure Hypothesis

The influence of inclusions on the fracture can be described using linear elastic fracture mechanics [47, 54]. Tangential stresses around a spherical inclusion in an infinite volume of glass can be found by a continuum mechanics approach [62] in which hydrostatic stresses due to changes in volume and differences in thermal expansion of the NiS and the glass, as well as residual stress are taken into account [54].

In addition to the material properties of the glass, the following information is required [47]:

- Young's modulus of NiS: $E_{\mathrm{NiS},\beta} = 70\,000\,\mathrm{MPa}$;
- Poisson ratio NiS (β-phase): $\nu_{\mathrm{NiS},\beta} = 0.20$;
- Thermal expansion coefficient NiS: $\alpha_{\mathrm{T,NiS},\beta} = 16.3 \times 10^{-6}\,\mathrm{K}^{-1}$;
- Heating temperature T_0;
- Ambient temperature before heating and after cooling down T;
- Maximum tensile stress in the centre of the glass σ_0;
- Diameter of the inclusion D_{NiS};
- Critical stress intensity factor K_{Ic};
- Stress crack corrosion limit for crack arrest K_{I0};
- Stress intensity factor at the crack tip for a circular crack around the inclusion K_{I}, see Fig. 6.17.

NiS inclusions may also result in high local stresses in float glass. However, unlike in thermally pre-stressed glass, the local stresses are insufficient to trigger large visible fractures, but will reduce the strength locally. The local stress occurs due to changes in volume of the spherical inclusion; however, the stresses diminish quickly with the distance from the particle and the crack stops propagating, see Fig. 6.17. Figure 6.18 plots the stress intensity factor for an inclusion with a diameter of $D = 200\,\mu\mathrm{m}$, which produces a hydrostatic stress state of about $\sigma_{\mathrm{h}} = 260\,\mathrm{MPa}$, resulting from a volume change of $\Delta V = 2.5\,\%$ and a cooling from $T_0 = 525\,°\mathrm{C}$ to $T = 25\,°\mathrm{C}$ and, for thermally toughened glass, superimposed by a tensile stress originating from an assumed residual stress state with $\sigma_0 = 40\,\mathrm{MPa}$.

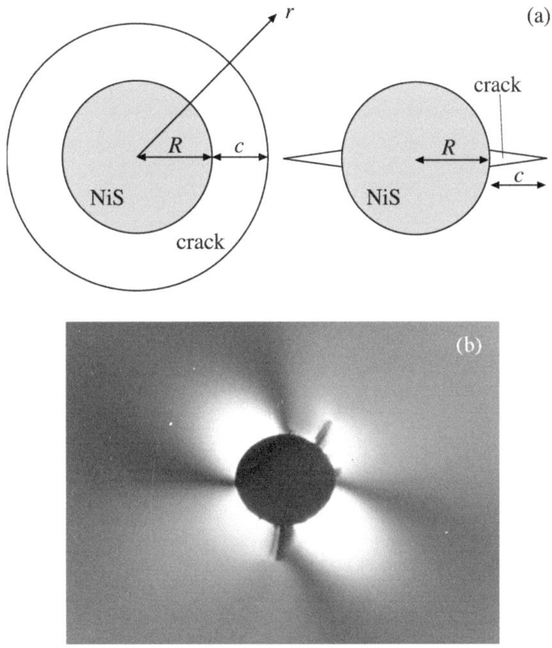

Fig. 6.17 **a** Model of a penny shaped (circular) crack around an inclusion [54] and **b** the recording of a fracture in thermally annealed glass by a NiS inclusion (image credit: Günther Mattes)

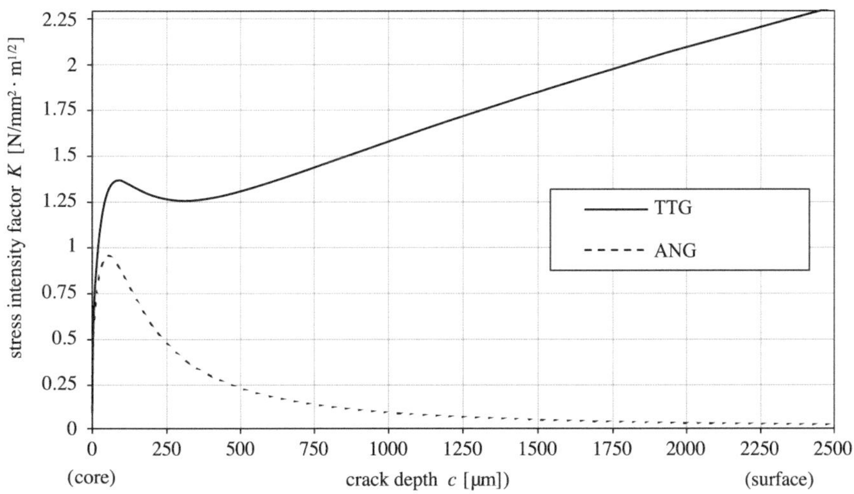

Fig. 6.18 Stress intensity K_I as a function of the crack depth c for a typical NiS inclusion in annealed glass (ANG) and thermally toughened glass (TTG) [54]

For annealed glass it is seen that the stress-intensity rises, but then rapidly decreases with the crack length (stable crack growth). However, if the residual stress state in thermally toughened glass is superposed an increase in the stress intensity is

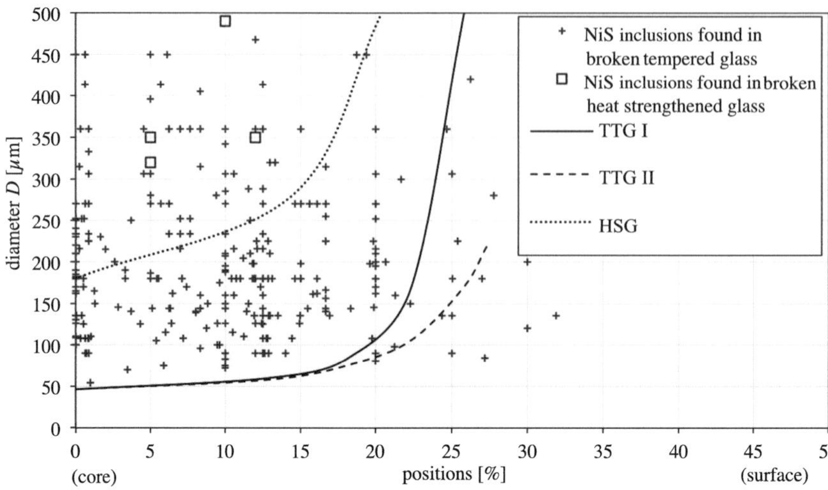

Fig. 6.19 Comparison between the diameter, D, and the relative position of empirically found inclusions relative to glass thickness, with limits for critical diameter and relative position for thermally toughened (TTG) and heat-strengthened glass (HSG) [54]

seen (unstable crack growth) leading to a complete fracture of the glass. The minimum diameter that leads to fracture can also be determined from the minimum of this function. This minimum is for a pre-stress of $\sigma_0 = 50\,\text{MPa}$ and a diameter of about $D = 48\,\mu\text{m}$ as seen from Fig. 6.19. This coincides with the empirical evaluation of 303 individual NiS fractures [63]. The fracture mechanics model can be used to determine the critical diameter and critical positions in thermally toughened glass, see Fig. 6.19. Figure 6.20 shows the statistical evaluation of the diameters of inclusions.

This data can also be used to estimate the frequency of NiS-induced fracture for pre-stressing levels typically found in heat-strengthened glass. In this case critical inclusions must be significantly larger than those found in thermally toughened glass: of all NiS fractures occurring in thermally toughened glass, up to 40 % can also be expected in heat-strengthened glass, depending on the pre-stressing level, see Fig. 6.21. The existence of NiS fracture in heat-strengthened glass has also been confirmed empirically by Stiell [95].

6.2.6.4 Fracture Pattern and Proof of Nickel Sulphide Inclusions

NiS-fracture is typically characterised by two larger fragments around the inclusion, forming the so-called "butterfly" pattern as illustrated in Fig. 6.22. However, such fracture pattern may also have other causes, such as local sharp indentations or other types of inclusions. The presence of NiS can be ascertained by means of optical microscopy and Scanning Electron Microscopy (SEM) accompanied by Energy-dispersive X-ray (EDX).

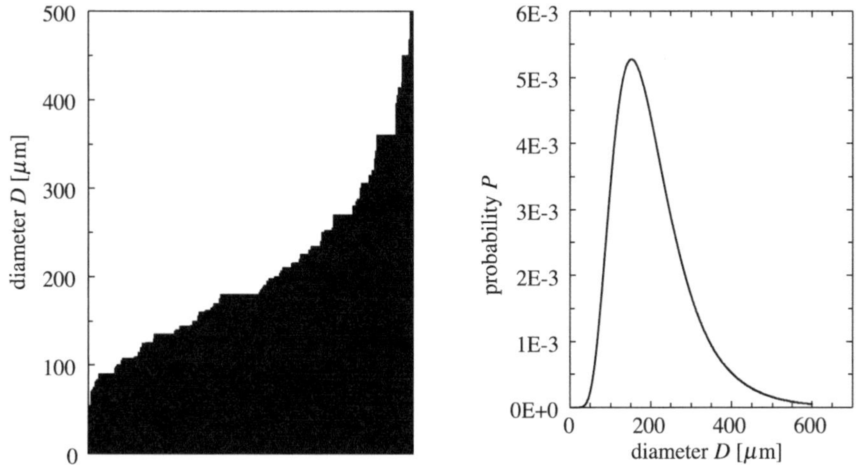

Fig. 6.20 Statistical evaluation of the diameter of 303 NiS inclusions [63]

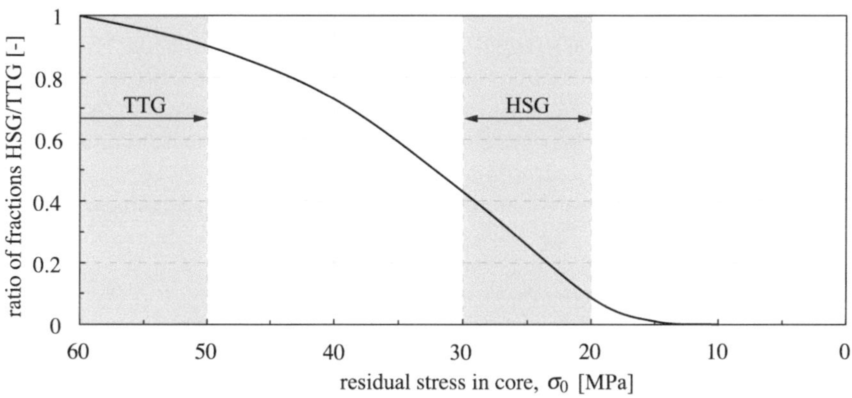

Fig. 6.21 Comparison between intrinsic stress and fracture frequency of heat-strengthened (HSG) and thermally toughened glass (TTG) [63]

6.2.7 Heat-Strengthened Glass

Heat-strengthened glass (HSG) is produced in the same process as thermally toughened glass, but cooled down at a lower rate. Consequently, the residual stress level and the bending strength is lower than for thermally toughened glass. The fracture pattern of HSG is somewhere in between that of annealed float glass and that of thermally toughened glass.

Due to difficulties in controlling the (lower) cooling rate, the production of heat-strengthened glass is currently only possible up to a nominal thickness of 12 mm. Specifications for minimum bending strength and fracture of heat-strengthened glass is outlined in the product standard EN 1863-1. According to this, the characteristic

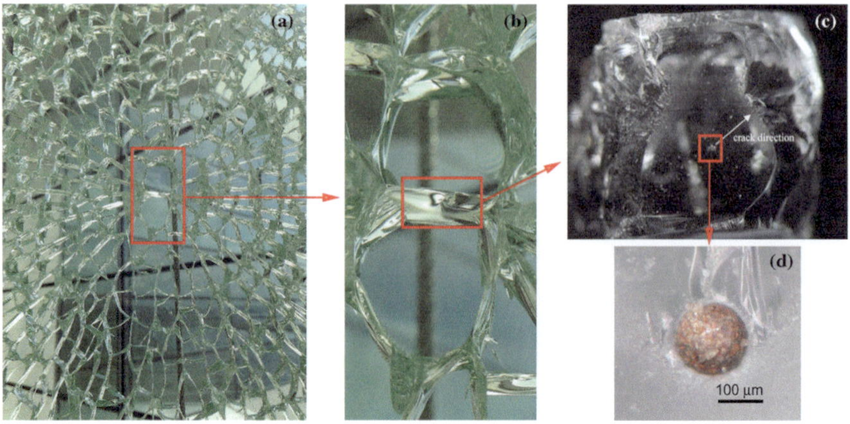

Fig. 6.22 **a**, **b** Typical fracture pattern caused by NiS inclusions; **c** NiS inclusion in the glass cross-section; **d** a NiS inclusion seen in an optical microscope (photo credits: Andreas Kasper)

bending strength, f_k, of heat strengthened glass is 70 MPa (5 %-fractile; 95 % probability). Technical approvals in Germany regulate this in dependence of the glass thickness to a bending strength, f_k, of up to 105 MPa [64]. Whereas in the US the surface prestress in HSG is between 3 and 52 MPa, whereas a surface prestress above 69 MPa is required for FTG according to [65].

The fracture pattern of heat-strengthened glass is determined in the same way as for thermally toughened glass. The allowed total area of the small fragments and *islands*, see Fig. 6.23, is limited. A distinction is made based on the area/mass equivalent, which is defined by the relationship area = mass / (thickness × density). An island is therefore a fragment of which the area/mass equivalent is $\geq 100mm^2$; the area/mass equivalent of a small fragment, on the other hand is $< 100mm^2$. At least one crack must reach the edge of the test specimen. To be classified as heat strengthened glass, four out of five specimens must meet the following requirements:

- The fracture pattern shall not have more than two islands;
- The islands must have an area/mass equivalent of less than 1 000 mm^2;
- The area/mass equivalent of all small fragments must not exceed 5 000 mm^2.

If one of the test specimens does not meet these requirements then it must meet the following requirements in order for the product to be classified as heat strengthened glass:

- The number of islands is less than three;
- The area/mass equivalent of all islands and small fragments is no larger than 50 000 mm^2.

Figure 6.23 shows an example of a representative fracture pattern of heat strengthened glass, which does not meet the requirements of EN 1863-1, since the island

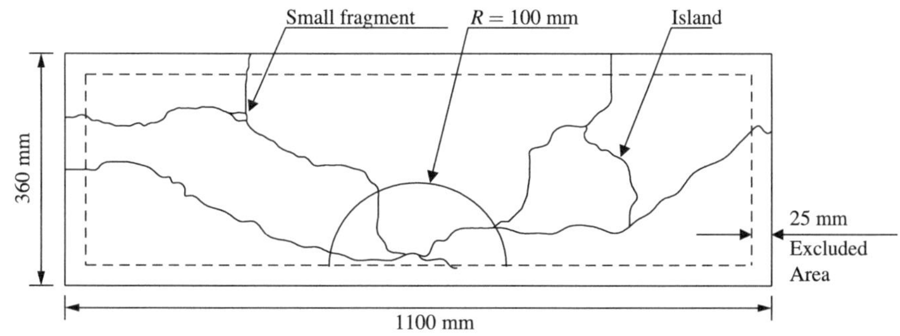

Fig. 6.23 Representative fracture pattern of heat-strengthened glass according to EN 1863-1

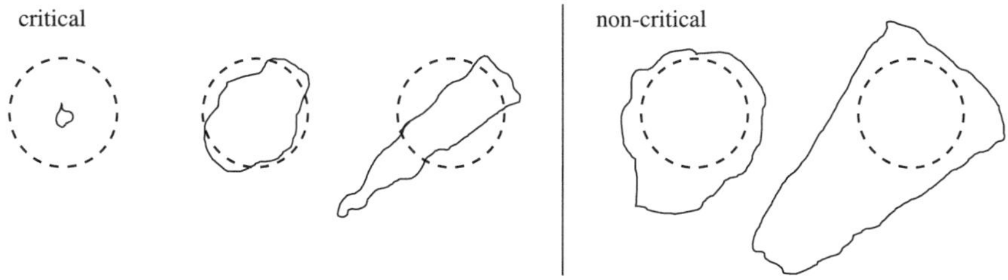

Fig. 6.24 Classification into *critical* and *non-critical* fragments for general building approvals of heat-strengthened glass in Germany. The diameter of the dotted circles $D = 120\,\text{mm}$

in this instance is larger than 1 000 mm^2. The fracture pattern of heat strengthened glass plays an important role in the post-fracture capacity of laminated safety glass using heat strengthened glass panes, see Fig. 6.25. In laminated glass panels the large fragments of the heat strengthened glass interlock through the interlayers thereby providing a degree of strength and stiffness to the fractured laminated glass. In obtaining technical approvals, additional requirements to the product standard may be requested regarding the type of fragments (*critical* or *uncritical*) and their area shares to ensure sufficient post-fracture load-bearing capacity. In Germany, for instance, the proportion of critical fragments must remain below 20 % of the total area, see Fig. 6.24.

6.2.8 *Fracture Patterns*

Elastic energy present in the glass originating from residual stresses and external loads have a significant influence on the fracture pattern. Thermally pre-stressed glass had a relatively high elastic strain energy which is stored in the glass. Consquently, thermally toughened glass generally fractures into small dice-shaped blunt fragments. External loading also influences the fracture pattern, which often reveals the origin

Fig. 6.25 Residual load-bearing capacity of a canopy glazing of laminated glass made from heat strengthened glass

of fracture. These blunt fragments can stick together in larger chunks. The elastic strain energy density, w, originating from the internal stress state, can be calculated by assuming linear elasticity and an equibiaxial stress state where the only non-zero stress is $\sigma = \sigma_\mathrm{x} = \sigma_\mathrm{y}$ as follows:

$$w = \frac{1}{2}\sigma_{ij}\epsilon_{ij} = \frac{\sigma^2\,(1-\nu)}{E}. \tag{6.12}$$

In a large plate it is reasonable to assume that the parabolic stress distribution shown in Eq. 6.1 is predominant. By further assuming that the stress state is homogeneous in plane, the stored elastic strain energy density can be calculated from a simple integration over the thickness of the glass:

$$W = \int_{-h/2}^{+h/2} \frac{\sigma^2(1-\nu)}{E} \mathrm{d}z\, \Delta x\, \Delta y = \frac{4}{5}\frac{\sigma_0^2(1-\nu)}{E} h\, \Delta x\, \Delta y. \tag{6.13}$$

The thickness is denoted h, σ_0 is the mid-plane tensile residual stress used to describe the parabola of the stress distribution. Δx and Δy are denoting the unit lengths in the x and y directions, respectively. Figure 6.26b is an evaluation of Eq. 6.13.

Relating Eq. 6.13 to the residual surface compression stress, the equation becomes:

$$W = \frac{1}{5}\frac{\sigma_\mathrm{c}^2(1-\nu)}{E} h\, \Delta x\, \Delta y. \tag{6.14}$$

Both the residual stress state, represented by e.g. σ_c and the thickness, h, influence the fracture pattern. Besides obtaining an increase in strength, the aim of thermally

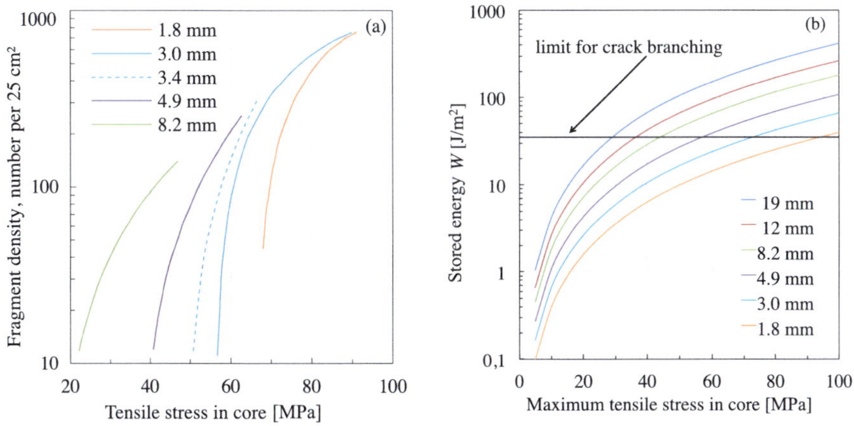

Fig. 6.26 **a** Correlation between the residual tensile stress in the mid-plane, σ_0, and the number of fragments per 25cm^2 according to [66]; **b** Correlation between σ_0 and the elastically stored energy, W, for different glass thicknesses; Crack branching threshold according to [67]

toughened glass is to provide a complete fragmentation upon fracture, which is considered safe in terms of risk of injury. In heat-strengthened glass, the aim is to have a higher fracture strength compared to annealed glass while maintaining large fragments upon fracture as illustrated in Fig. 6.28b thereby increasing the post-fracture strength of laminated glass.

There are only a few studies on the relationships between residual stress, crack branching and fracture pattern are available in literature. Figure 6.26a shows a correlation between the residual tensile stress, σ_0, and the fragment count (per 25cm^2) for varying thicknesses determined experimentally.

Fig. 6.27 Fracture image of laminated, chemically pre-stressed aluminosilicate glass with a thickness of $h = 1$ mm, fracture stress approximately $\sigma_\mathrm{f} = 700$ MPa (the origin of fracture is roughly in the middle of the bottom edge)

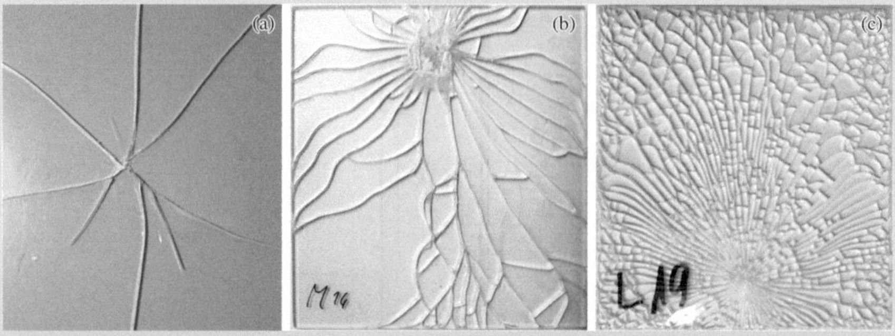

Fig. 6.28 Typical fracture image of **a** annealed float glass, **b** heat-strengthened glass and **c** thermally toughened glass; all punched locally with a sharp tool

Fig. 6.29 Typical fracture of: **a** annealed float glass, **b** heat-strengthened glass and **c** thermally toughened glass; all tested in a double ring bending test

The correlation between the residual tensile stress, σ_0, and the elastically stored strain energy for different glass thicknesses is evident. This also shows the threshold energy for the crack branching according to Fineberg [67]. Above an energy of $W = 35\,\mathrm{J/m^2}$ the cracks in soda-lime silica glass travel at 40–45% of Rayleigh's wave velocity ($v_R = 3.390\,\mathrm{m\ s^{-1}}$) and leads to crack branching [67–69]. High-speed recording of the fracture process for thermally toughened glass supports this [70].

Figure 6.26 also explains why chemically pre-stressed glass, which has a high residual stress level at the surface but a small amount of stored energy, does not exhibit a high fragmentation density, see also Sect. 6.3.

Due to high stresses induced by external loading, specimens can build up a large amount of strain energy, in addition to that generated by the pre-stress. This energy is released at the moment of fracture and may lead to a dense fracture pattern in annealed, heat strengthened or chemically pre-stressed glass as illustrated in Fig. 6.27 and in [71].

Thin glass must therefore be tempered to a higher degree than thick glass to ensure a dense fracture pattern in safety glass applications. The European product

Table 6.3 Minimum values for the number of fragments according to EN 12150-1

Nominal thickness h [mm]	Minimum number of fragments counted
2	15
3	15
4–12	40
15–25	30

standard for thermally toughened glass, EN 12150-1, prescribes a minimum number of fragments within an area of 50mm $\times$50mm within a test specimens having at total area of 360 $\times$ 1100 mm^2. However, the glass thickness, and the associated number of fragments, see Table 6.3, is not always consistent with theory, as one could expect larger thicknesses to correspond to a larger number of fragments.

During the test, specimens are impacted with a spring loaded centre punch at the mid-length of the long edge at a distance of 13 mm from the edge. The assessment of the fracture pattern is carried out by counting the number of fragments within a square (edge length: 50 mm) in the region of coarsest fracture, see Fig. 6.28c. The area around the impact origin and near the glass edges are neglected. Depending on the thickness of the glass, the minimum limit values for the number of pieces are defined in EN 12150-1. An additional and necessary condition for a glass pane to be classified as thermally toughened is that the maximum length of the longest fragment does not exceed 100 mm. The blunt-edged fragments often stick together in groups, especially for thick or enamelled glass.

Typical fracture patterns of annealed float glass, heat-strengthened glass and thermally toughened glass are shown in Figs. 6.28 and 6.29.

6.3 Chemical Pre-stressing

A chemical pre-stressing processes for glass was first successfully introduced by Kistler [72]. In this process compressive residual stresses of up to $\sigma_c = 1\,000$ MPa can be generated in the glass surface. While the high compressive residual stress provides clear advantages, a disadvantage is that the process takes considerably more time compared to thermal pre-stressing. Furthermore, the compressive zone is often only 40–50 μm deep, which is significantly less than in thermally pre-stressed glass, see Fig. 6.30. The thickness of the compressive surface pre-stress is also known as "depth of layer" (DoL) or case depth. Scratches deeper than the case depth can occur in the built environment due to transport, cleaning, etc., thereby leading to a chemically strengthened glass that is no stronger than annealed glass in that specific location.

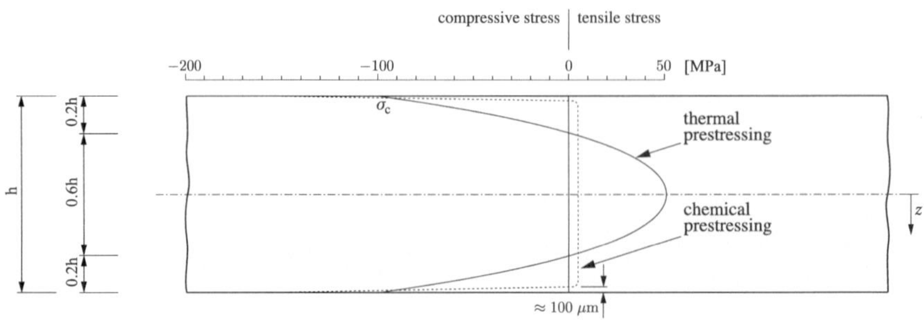

Fig. 6.30 Comparison of the stress distribution through the thickness of the glass for thermally pre-stressed and chemically pre-stressed glass

The visibility of scratches and the formation of lateral cracks around the scratches on the surface of pre-stressed glass depend heavily on the respective strengthening process. Here, chemically pre-stressed glass, with its high residual compressive stresses, usually yields a better scratch resistance than thermally pre-stressed glass. This is due to higher density of the surface layers caused by the ion exchange process which results in increased surface hardness [3, 73].

6.3.1 *Pre-stressing Processes*

The chemical pre-stressing process is based on the principle of ion exchange of the alkali ions in the area close to the surface [74, 75]. In the case of soda-lime silica glass, Na^+ ions (atomic radius: 153.7 pm) are replaced by K^+ ions (atomic radius: 227.2 pm) that are approximately 30 % larger, as shown schematically in Fig. 6.31. In commercial pre-stressing facilities, the ion exchange takes place in a bath of molten KNO_3 salt (potassium nitrate), see e.g. [73, 76]. The glass panels are stored in racks and completely immersed, usually for no longer than 24 h, see Fig. 6.32.

Initially, the surface compressive stress is proportional to the glass volume in which the ion exchange has taken place. This correlates roughly to the square root of the process time. The ion exchange is a diffusion process, and therefore time and temperature dependent. Ultimately, surface compression stresses decrease the longer the panels stay in the molten salt bath, while the depth of layer increases. Chemical pre-stressing of glass takes place at temperatures between 330 °C and 470 °C with a processing time of 1–12 h, due to the high temperature, significant stress relaxation will occurs [71, 77, 78].

The chemical composition of the glass also has significant influence on the chemical pre-stressing process. The ion exchange process is particularly effective in Aluminosilicate glass wherein compressive surface stresses increase proportionally to the aluminium content. In soda-lime silica glass, the ion exchange rate decreases with increasing calcium content [73]. However, it increases with an increasing number of Na^+ ions.

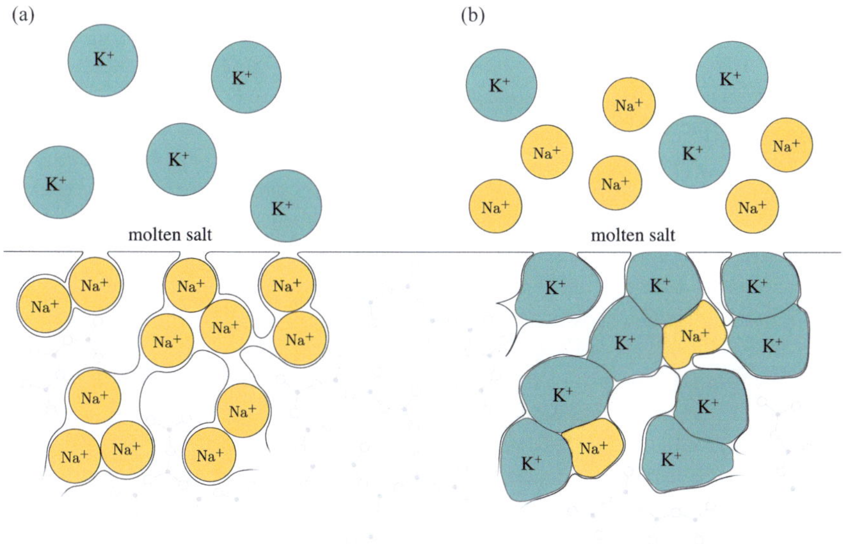

Fig. 6.31 Ion exchange process for chemically pre-stressing glass in the low temperature process: Na^+ ions stored in the glass are replaced by larger K^+- ions: **a** before ion exchange; **b** after ion exchange [3]

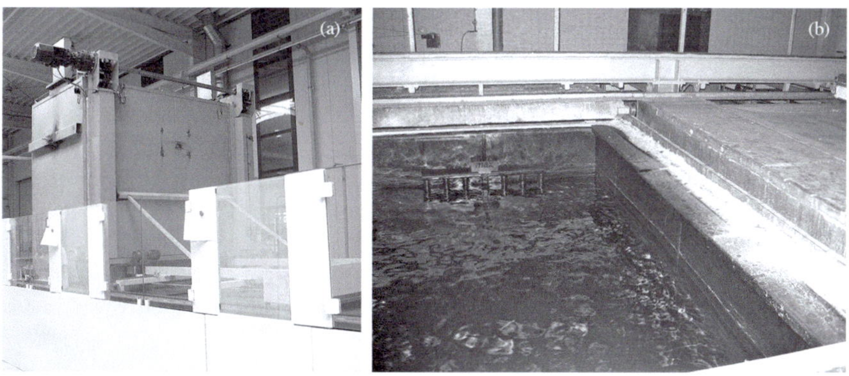

Fig. 6.32 Commercial plant for the chemical pre-stressing of glass: **a** Exterior view; and **b** view of the salt bath (picture credits: Yachtglass GmbH & Co. KG)

6.3.2 *Properties of Chemically Pre-stressed Glass*

The depth of the compression zone in chemically pre-stressed glass is 20–300 μm [79]; for commercial applications, it is generally $< 50\,\mu m$ [80]. In soda-lime silica glass, the practically achievable compressive surface stress is about 600 MPa. Commercial aluminosilicate glass used as chemically pre-stressed display glazing with a thickness of $h < 2$ mm is available with surface stresses $\sigma_V < -900$ MPa. [76] predict a theoretical bending tensile strength of $\sigma_f = 1450$ MPa for chemically pre-stressed soda-lime silica glass ($Na^+ \rightleftarrows K^+$) [76]. However, other authors, were only able to determine significantly lower bending strengths ranging from 255 MPa to 855 MPa [9, 72, 81–83]. EN 12337-1 specifies, for chemically pre-stressed soda-lime silica glass, a typical strength (5 %-fractile value) of $f_k = 150.0$ MPa.

Due to the small compressive zone, the practical strength of chemically pre-stressed glass significantly depends on the surface condition (size and distribution of flaws). Deep flaws may penetrate the compressive zone and thereby significantly lower the tensile strength of chemically pre-stressed glass to that of an annealed glass. Therefore, chemically pre-stressed glass is only occasionally used in structural applications, see e.g. Fig. 2.28a. Its use is more common in naval and aerospace applications, such as yacht glazing and aircraft cockpits, in which curved and/or thin[1] glass is used. Chemically pre-stressed glass may also be preferred over thermally pre-stressed glass in applications where optical anisotropy may be amplified by polarised light, e.g. in seaside locations. The limited market share for standard building components is also attributed to the fact that the thermally pre-stressing process is more cost-efficient due to a high level of automation and shorter processing time.

Even though high compressive residual stresses are found at the surface, the amount of strain energy stored in chemically pre-stressed glass is much lower than what is found in thermally pre-stressed glass and therefore the fracture pattern is similar to that of annealed glass, see Sect. 6.2.8.

Subsequent machining (e.g. cutting, drilling, grinding) of chemically pre-stressed glass is most often possible due to the usually relatively low strain energy stored in the glass.

6.4 Photoelastic Methods

Quality control of the residual stress state plays a decisive role in the manufacture of pre-stressed glass. Both the quantification of the stress and the homogeneity of the stress are of interest, especially for thermally toughened glass.

---

[1] In thin glass where the thickness is less than $h < 2$ mm thermal pre-stressing is difficult due to the need of very high cooling rates to establish a temperature gradient over the glass thickness resulting in reasonably high residual stresses.

Table 6.4 Photoelastic constants for different types of glass

Glass	Source	C [TPa^{-1}]
Soda-lime silica glass	[85, 85]	2.70
Soda-lime silica glass (float)	[86, 86]	2.75
Soda-lime silica glass TTG	[87, 87]	3.10
Borosilicate glass	[88, 88]	3.6–4.0
Aluminosilicate glass	[88, 88]	2.7–2.0
Acrylic glass PMMA	[89, 89]	4.0–5.0
Polycarbonate	[89, 89]	80

Glass is a photoelastic material and assessment of principal stresses can therefore be achieved by using polarized light. To do this, the specimen is positioned between two polarizing filters and illuminated through these. Due to the state of its own internal stress, a birefringence of the light beam occurs, which leads to a retardation. Different colours are generated, depending on the birefringence, according to Michel-Levy's colour chart [84]. In addition, a relationship between the retardation difference and the principal stress differences can be established:

$$\Delta s = C\,(\sigma_1 - \sigma_2) \cdot h. \tag{6.15}$$

where:

C is the photoelastic constant, see e.g. Table 6.4,
h is the glass thickness

The coloured lines (isochromats) are generated by polarization, and thus show the positions of the same principal stress differences, $\sigma_1 - \sigma_2$.

Furthermore, isoclines also become visible when using linearly polarized light. These black lines show the locations of the same principal stress directions. Since the isochromats can appear black at any difference in path retardation, even a slight one, it is essential to distinguish between isoclines and isochromats. This can be done by rotation of the sample, see Fig. 6.33, if the polarizer is linearly polarized or by direct use of circularly polarized filters.

A qualitative statement on the homogeneity of the residual stress state is thus possible using images taken through polarizing filters, see Fig. 6.10. A quantitative statement, for example, via the Michel-Lévy Table or variations thereof [90] is difficult, in that Eq. 6.15 is only valid if the principal stresses over the thickness remain constant. Otherwise, an integral is formed over the glass thickness with the (unknown) stress gradients in the integrand. In order to quantify anistropy due to inhomogeneous thermal pre-stressing, see Sect. 6.2.4, suitable quality management methods are still required [25]. However, important steps towards this has recently been reported in e.g. [23, 24, 91].

If, on the other hand, one of the two principal stresses is known, then a quantitative assertion can be made about the other principal stress by means of polarisation. The

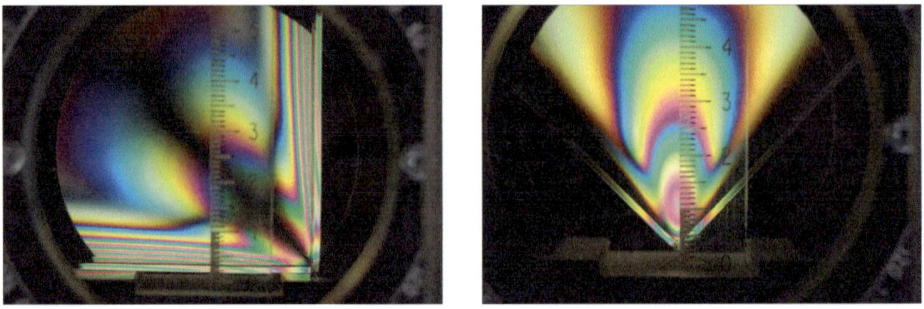

Fig. 6.33 Rotation of a sample in linear polarised light: isoclines change, isochromats remain the same

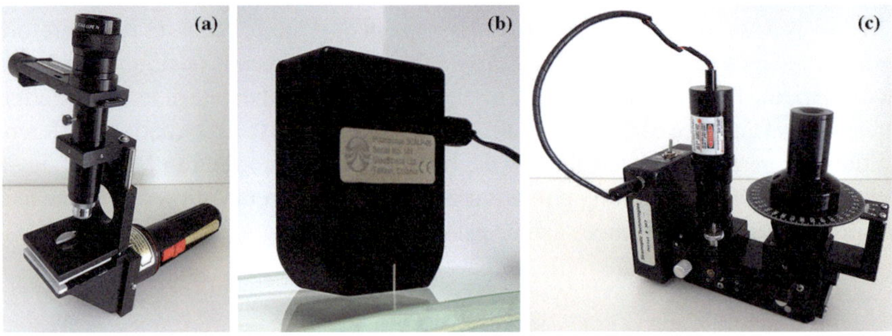

Fig. 6.34 Optical photoelastic instruments: **a** Edge stress meter, **b** SCALP and **c** Laser GASP

tension perpendicular to the free edge is at zero tension, to satisy static equilibrium, so that the counting of the isochromats found in the area of the edge due to thermal pre-stressing can be determined, see Fig. 6.34a.

A more complex photo-elastic measuring method is the scattered light method that uses a laser in order to quantify the internal stresses on a flat glass surface. In the scattered light method, a laser is directed diagonally into the glass via a prism. In order to avoid the refractive light from prism to glass, an immersion liquid with a refractive index similar to that of glass is used. The incident light is circularly polarized, wherein the polarization direction is varied during the measurement. The intensity of the recorded back scattered light by the camera and the difference in the change in path and stress distribution over the thickness of the glass would thereby be determined. A device that works with this method is SCALP, see Fig. 6.34b.

The Epibiascope (*GASP*, company Strainoptics, Fig. 6.34c) can be used for the Miragé effect. Since in the manufacturing process the glass floats on a liquid tin bath, see Sect. 2.2.1, tin ions diffuses about 10 μm deep into the glass structure and change the optical properties in the area close to the surface. The refractive angle changes from the surface to the centre and leads to a deflection of the light rays, so that they partially come out of the surface. The stress can be determined from the resulting interference effects. This method only shows the results of the surface pressure stress on the tin-bath side. An identification of the tin bath side is possible with short-wave UV light (UV-C): The tin bath side then appears milky, while the air side (fire side) remains clear.

ASTM C1279 [94] describes stress-optical methods commonly used in the US to determine the intrinsic stresses of flat glass. A more detailed description of the photoelastic measuring methods for glass can be found in, for example, [96].

References

1. H. Dannheim, H. Oel, W. Prechtl, Einfluß der Vorspannung auf die Oberflächendefektentstehung bei thermisch vorgespannten Flachgläsern. Glastechnische Berichte **54**(10), 312–318 (1981)
2. M. Zaccaria, M. Overend, Thermal healing of realistic flaws in glass. J. Mater. Civ. Eng. **28**(2) (2016)
3. S. Schula, Charakterisierung der Kratzanfälligkeit von Gläsern im Bauwesen. Dissertation. Technische Universität Darmstadt (2015)
4. J.H. Nielsen, K. Thiele, J. Schneider, M.J. Meyland, Compressive zone depth of thermally tempered glass. Constr. Build. Mater. **310**, 125238 (2021)
5. O.S. Narayanaswamy, Stress and structural relaxation in tempering glass. J. Am. Ceram. Soc. **61**(3–4), 146–152 (1978)
6. EN 1863-1, Glass in building—heat strengthened soda lime silicate glass—Part 1: definition and description (2012)
7. EN 12150-1, Glass in building—thermally toughened soda lime silicate safety glass—Part 1: definition and description (2015)
8. W. Laufs, Ein Bemessungskonzept zur Festigkeit thermisch vorgespannter Gläser. Dissertation. RWTH Aachen (2000)
9. J. Schneider, Festigkeit und Bemessung punktgelagerter Gläser und stoßbeanspruchter Gläser. Dissertation. Technische Universität Darmstadt (2001)
10. J.H. Nielsen, Tempered glass—bolted connections and related problems. Dissertation. Danmarks Tekniske Universitet (2009)
11. K. Lohr, B. Weller, Residual stress distribution in tempered glass with reground edges. Glas. Struct. Eng. **4**(1), 99–115 (2019)
12. J.H. Nielsen, M.J. Meyland, B.E. Thorup, A. Zugravu, J.F. Olesen, Investigating the strength effects of drilling in tempered glass: a first step towards a novel dowel joint. Glas. Struct. Eng. (2019)
13. J. Schneider, Glass strength in the borehole area of annealed float glass and tempered float glass. Int. J. Form. Process. **7**(4), 523–541 (2004)
14. J.H. Nielsen, J.F. Olesen, P.N. Poulsen, H. Stang, Simulation of residual stresses at holes in tempered glass: a parametric study. Mater. Struct. **43**(7), 947–961 (2010)
15. J. Watson, J. Nielsen, M. Overend, A critical flaw size approach for predicting the strength of bolted glass connections. Eng. Struct. **57**, 87–99 (2013)
16. M. Vandebroek, J. Belis, C. Louter, G. Van Tendeloo, Experimental validation of edge strength model for glass with polished and cut edge finishing. Eng. Fract. Mech. **96**, 480–489 (2012)
17. M. Vandebroek, J. Belis, C. Louter, R. Caspeele, Influence of the load history on the edge strength of glass with arrised and ground edge finishing. Eng. Fract. Mech. **104**, 29–40 (2013)
18. J. Kleuderlein, F. Ensslen, J. Schneider, Investigation of edge strength dependent on different types of edge processing, in *Engineered Transparency* (Düsseldorf, 2014), pp. 259–268
19. P.F. Nursey, Tempered glass, in *Popular Science Monthly*, vol. 7 (1875), pp. 554–563
20. F. Royer de la Bastie, *Patent des B.A. Royer de la Bastie in Richemont in Frankreich auf Vorrichtungen, Glas zu h ä rten* (1876)

21. R. Schultze, Literatur über Hartglas und unzerbrechliches Glas. Glastech. Ber. **9**, 267–268 (1931)
22. H. Frey, Vorspannen hat viele Namen, doch was steckt dahinter? Glaswelt **10**, 32–34 (2014)
23. S. Dix, P. Müller, C. Schuler, S. Kolling, J. Schneider, Digital image processing methods for the evaluation of optical anisotropy effects in tempered architectural glass using photoelastic measurements. Glas. Struct. Eng. **6**(1), 3–19 (2021)
24. Measurement and evaluation methods for optical anisotropic effects inthermally toughened glass. English translation. DIN (2022)
25. M. Illguth, C. Schuler, Ö. Bucak, The effect of optical anisotropies on building glass façades and its measurement methods. Front. Arch. Res. **4**(2), 119–126 (2015)
26. Y. Chen, D. Lochegnies, R. Defontaine, J. Anton, H. Aben, R. Langlais, Measuring the 2D residual surface stress mapping in tempered glass under the cooling jets: the influence of process parameters on the stress homogeneity and isotropy. Strain **49**(1), 60–67 (2013)
27. S. Pasetto, Anisotropy as a defect in U.K. architectural float heat-treated glass (University of Bath, Master-Thesis, 2014)
28. L. Adams, E. Williamson, The annealing of glass. J. Frankl. Inst. **190**(6), 835–870 (1920)
29. G.M. Bartenev, The phenomenon of the hardening of glass. Russian. J. Tech. Phys. **18**, 383–388 (1948)
30. E. Lee, T. Rogers, T. Woo, Residual stresses in a glass plate cooled symmetrically from both surfaces. J. Am. Ceram. Soc. **48**(9), 480–487 (1965)
31. O.S. Narayanaswamy, A model of structural relaxation in glass. J. Am. Ceram. Soc. **54**(10), 491–498 (1971)
32. G. Scherer, Relaxation in glass and composites (Wiley, 1986)
33. J.H. Nielsen, Tempered glass: -bolted connections and related problems. Ph.D. Technical University of Denmark (2009), p. 78
34. J.H. Nielsen, J.F. Olesen, P.N. Poulsen, H. Stang, Finite element implementation of a glass tempering model in three dimensions. Comput. Struct. **88**(17–18), 963–972 (2010)
35. C. Kurkjian, Relaxation of torsional stress in transformation range of soda-lime-silica glass. Phys. Chem. Glas. **4**(4), 128–136 (1963)
36. GANA Tempering Division, Engineering standards manual (Topeka, 1986)
37. H. Carré, L. Daudeville, Numerical simulation of soda-lime silicate glass tempering. J. Phys. IV **06**(C1), 175–185 (1996)
38. H. Carre, L. Daudeville, Load-bearing capacity of tempered structural glass. J. Eng. Mech. **125**(8), 914–921 (1999)
39. A. Aronen, R. Karvinen, Modeling of deformations and stresses during glass tempering, in *Proceedings of the ASME 2011 International Mechanical Engineering Congress & Exposition*. Fluids and Thermal Systems; Advances for Process Industries, Parts A and B, vol. 6 (Denver, USA, 2011), pp. 115–123
40. A. Aronen, Modelling of deformations and stresses in glass tempering. Dissertation. Tampere University of Technology (2012)
41. N. Pourmoghaddam, J. Schneider, Determination of the engine power for quenching of glass by forced convection: simplified model and experimental validation of residual stress levels. Glas. Struct. Eng. **4**(1), 117–125 (2019)
42. H. Oel, G. Helland, H. Dannheim, Thermisches Vorspannen von Glasscheiben. Glastechnische Berichte **57**(1), 1–6 (1984)
43. G. Hierl, E. Wartenberg, Principles of the polyliquid process for the thermal toughening of glass. Glastechnische Berichte **62**(5), 158–166 (1989)
44. J.H. Nielsen, Numerical investigation on optimizing the process for stronger tempered glass, in *Challenging Glass 4 & COST Action TU0905 Final Conference* (CRC Press, Nielsen, 2014), pp. 507–512
45. L. Daudeville, F. Bernard, R. Gy, Residual stresses near holes in tempered glass plates. Mater. Sci. Forum **404–407**, 43–48 (2002)
46. L. Merker, Zum Verhalten des Nickelsulfids im Glas. Glastechnische Berichte **47**(6), 116–121 (1974)

47. M. Swain, Nickel sulphide inclusions in glass: an example of microcracking induced by a volumetric expanding phase change. J. Mater. Sci. **16**, 151–158 (1981)
48. A. Kasper, *Nickelsulfid in Glas - Fortschritte der Glastechnik* (Verlag der Deutschen Glastechnischen Gesellschaft, Offenbach, 2003)
49. A. Kasper, Spontaneous cracking of thermally toughened safety glass. Part 1: properties of nickel sulphide inclusions. Glas. Struct. Eng. **4**(3), 279–313 (2019)
50. A. Kasper, Spontaneous cracking of thermally toughened safety glass. Part 3: statistic evaluation of field breakage records and consequences for residual breakage probability. Glas. Struct. Eng. **4**(3), 345–376 (2019)
51. A. Kasper, N. Pyeonglae, Z. Yuan, Spontaneous cracking of thermally toughened safety glass. Part 2: nickel sulphide inclusions identified in annealed glass. Glas. Struct. Eng. **4**(3), 315–343 (2019)
52. A. Kasper, F. Rubbert, Spontaneous cracking of thermally toughened safety glass. Part 4: a case study of isothermal breakages in a building, and conclusions thereof for the heat soak test. Glas. Struct. Eng. **5**(2), 211–229 (2020)
53. E. Ballantyne, Division of building research, in *CSIRO* (Melbourne, Australia, 1961), pp. 61–65
54. J. Schneider, J. Hilcken, Nickel sulphide (NiS-) induced failure of glass: fracture mechanics model and verification by fracture data, in *Engineered Transparency* (Düsseldorf, 2010), pp. 125–136
55. EN 14179-2, Glass in building—heat soaked thermally toughened soda lime silicate safety glass—Part 2: evaluation of conformity/Product standard (2005)
56. J. Schneider, J. Hilcken, A. Aronen, R. Karvinen, J.F. Olesen, J. Nielsen, Stress relaxation in tempered glass caused by heat soak testing. Eng. Struct. **122**, 42–49 (2016)
57. J. Hilcken, *Probabilistische und bruchmechanische Untersuchung Nickelsulfid-induzierter Spontanbrüche in Glas* (Diplomarbeit, Fachhochschule Frankfurt, 2010)
58. J. Schneider, J. Hilcken, A. Kasper, Ein Modell zur Bestimmung der Versagenswahrscheinlichkeit von heißgelagertem ESG. Glasbau **2012**, 171–184 (2012)
59. O. Yousfi, Y. Brechet, P. Donnadieu, A.M. Kasper, F. Serruys, Proposal to enhance the heat soak test following EN14179-1, in *Glass Performance Days* (2009), pp. 584–588
60. F. Bordeaux, A.M. Kasper, Reliable and shorter heat soak test to avoid spontaneous fracture of heat strengthened and tempered glass, in *Glass Processing Days* (1997), pp. 85–89
61. O. Yousfi, Transformations de phase des Sulfures de Nickel dans les verres trempés. Dissertation. Institut National Polytechnique de Grenoble (2009)
62. W. Becker, D. Gross, *Mechanik elastischer K ö rper und Strukturen* (Springer, Berlin, 2002)
63. J. Hilcken, J. Schneider, Nickel sulfide induced failure of thermally toughened glass. Glas. Struct. Eng. (2016)
64. Verglasungen aus teilvorgespanntem Glas "MONOPACT TVG plus". DIBt. (2016)
65. ASTM C1048, Standard specification for heat-treated flat glass—kind HS, kind FT coated and uncoated glass (2004)
66. K. Akeyoshi, E. Kanai, K. Yamamoto, S. Shima, Report of the research lab. Asahi Garasu Kenkyu Hokoku **17**(1), 23–26 (1967)
67. J. Fineberg, The dynamics of rapidly moving tensile cracks in brittle amorphous material, in *Dynamic Fracture Mechanics* (World Scientific, 2006), pp. 104–146
68. J. Fineberg, Universal features of the microbranching instability in dynamic fracture. Philos. Mag. B: Phys. Condens. Matter **78**(2), 243–251 (1998)
69. E. Sharon, G. Cohen, J. Fineberg, Propagating solitary waves along a rapidly moving crack front. Nature **410**(6824), 68–71 (2001)
70. J.H. Nielsen, J.F. Olesen, H. Stang, The fracture process of tempered soda-lime-silica glass. Exp. Mech. **49**(6), 855–870 (2009)
71. S. Nategh, M. Zaccaria, J. Missinne, J. Belis, Experimental strength characterisation of thin chemically pre-stressed glass based on laser-induced flaws. Glas. Struct. Eng. **7**(3), 471–486 (2022)

72. S.S. Kistler, Stresses in glass produced by nonuniform exchange of monovalent Ions. J. Am. Ceram. Soc. **45**(2), 59–68 (1962)
73. S. Karlsson, B. Jonson, C. Stålhandske, The technology of chemical glass strengthening a review. Glas. Technol.– Eur. J. Glas. Sci. Technol. Part A **51**(2), 41–54 (2010)
74. R. Gy, Ion exchange for glass strengthening. Mater. Sci. Eng.: B **149**(2), 159–165 (2008)
75. G. Marcelli, E. Poli, Chemically strengthened glass by ion exchange: residual stress profile and strength evaluation, in *Engineered Transparency* (Düsseldorf, 2014), pp. 231–240
76. I. Donald, Methods for improving the mechanical properties of oxide glasses. J. Mater. Sci. **24**(12), 4177–4208 (1989)
77. M. Zaccaria, Bi-tempered glass. Ph.D. thesis. University of Cambridge, Department of Engineering (2016)
78. T.M. Gross, Chemical strengthening of glass, in *Springer Handbook of Glass*, ed. by J.D. Musgraves, J. Hu, L. Calvez. Springer Handbooks (Springer International Publishing, Cham, 2019), p. 1841
79. D.R. Oakley, M.F. Green, Surface damage and the strength of chemically strengthened flat glass (1991)
80. A.Y. Sane, A.R. Cooper, Stress buildup and relaxation during ion exchange strengthening of glass. J. Am. Ceram. Soc. **70**(2), 86–89 (1987)
81. M.E. Nordberg, E.L. Mochel, H.M. Garfinkel, J.S. Olcott, Strengthening by Ion exchange. J. Am. Ceram. Soc. **47**(5), 215–219 (1964)
82. D. Connolly, A.C. Stockton, T.C. O'Sullivan, Large-deformation strength analysis of chemically strengthened glass disks. J. Am. Ceram. Soc. **72**(5), 859–863 (1989)
83. M.B. Abrams, D.J. Green, S. Jill Glass, Fracture behavior of engineered stress profile soda lime silicate glass. J. Non-Cryst. Solids **321**(1–2), 10–19 (2003)
84. A. Michel-Lévy, A. Lacroix, *Les min é raux des roches* (1888)
85. GlasStress, Scattered Light Polariscope SCALP-04 Instruction Manual Ver. 4.5.1 (2012)
86. U. Ankergren, A. Bak-Jensen, Structural glazing exposed to fire. MSc-Thesis, Technical University of Denmark (2008)
87. J.H. Nielsen, J.F. Olesen, H. Stang, Characterization of the residual stress state in commercially fully toughened glass. J. Mater. Civ. Eng. **22**(2), 179–185 (2010)
88. A.G. Schott, Schott technical glasses—physical and technical properties (Mainz, 2014)
89. H. Aben, C. Guillemet, *Photoelasticity of Glass* (Springer, Berlin, 1993)
90. B. Sørensen, A revised Michel-Lévy interference colour chart based on first-principles calculations. Eur. J. Mineral. **25**(1), 5–10 (2013)
91. S. Dix, *A Concept for Measuring and Evaluating Optical Anisotropy Effects in Tempered Architectural Glass* (Springer Vieweg Wiesbaden, 2024), p. 147
92. EN 12337-1, Glass in building—Chemically strengthened soda lime silicate glass—Part 1: Definition and description (2000)
93. EN 1288-3, Glass in building—Determination of the bending strength of glass—Part 3: Test with specimen supported at two points (four point bending) (2000)
94. ASTM C1279, Standard test method for non-destructive photoelastic measurement of edge and surface stresses in annealed, heat-strengthened, and fully tempered flat glass (2005)
95. W. Stiell, Aus Schäden lernen, in *Rosenheimer Fenstertage* (Institut für Fenstertechnik, Rosenheim, 2012), pp. 124–125
96. H. Aben, J. Anton, A. Errapart, Modern photoelasticity for residual stress measurement in glass. Strain **44**(1), 40–48 (2008)

Part IV
Glass Engineering Design

Chapter 7
Mechanical Connections

The brittle nature of glass calls for special attention to connections in glass structures. This chapter discusses mechanical connections for glass structures while the succeeding Chap. 8 discusses adhesive connections. Suitable contact materials for load transfer are discussed in Sect. 7.2. Subsequently, continuous and point support configurations are described in Sect. 7.3. Finally, mechanical connections for transfer of large forces between adjacent glass components and/or the underlying structure are described in Sect. 7.4.

7.1 General Considerations

The design of joints, supports and connections for glass structures, is governed by the brittle material nature of glass. In contrast to steel, glass does not have any plasticity at a macroscopic level, therefore stress concentrations, such as those caused by local contact with hard objects, cannot be absorbed by local yielding. Therefore, in principle, contact between glass and materials with a higher surface hardness or Young's modulus must be prevented in order to avoid glass surface damage or component failure. Consequently, aluminium or polymers are commonly used as intermediary contact materials.

However, care must be taken to avoid unwanted chemical or physical interactions between the materials used, such as contact corrosion or diffusion of plasticizers. Furthermore, the material behaviour (e.g. *Shore* hardness), durability (UV exposure, water, detergent, etc.) and behaviour under permanent load should be taken into account.

Several polymers have proven to be effective in glass connections including thermoplastics, elastomers and thermosetting polymers. The polymers are usually dyed black to increase their UV resistance.

J. Belis et al., *Engineering Glass Structures*,
https://doi.org/10.1007/978-3-032-03233-1_7

Finally, interfaces between and with glass should be designed and detailed carefully by accounting for (dimensional) tolerances, thermal expansions and movements thereby avoiding unnecessary stress concentrations.

7.2 Contact Materials for Load Transfer

7.2.1 General Information

Direct contact between glass and hard materials such as steel should be avoided, particularly when transferring large forces. To reduce stress concentrations, an intermediate contact material is placed between the glass and the hard material.

The contact materials must be as dimensionally stable as possible yet sufficiently flexible to minimise stress peaks in the glass. The typical materials used in this context are presented below. A distinction is made between thermoplastics, thermosets, elastomers and grouting mortar. In addition to the aforementioned polymers, aluminium, which has a Young's modulus comparable to glass is used as an intermediate contact material. An overview of typical stiffness ranges for suitable polymers is given in Table 7.1.

The characteristics of a good contact material are:

- large compressive strength;
- appropriate stiffness;
- relatively constant material properties in the operating temperature range from −20 °C up to 80 °C;
- hardness lower than that of glass;
- low creep under loading;
- resistance to ageing (e.g. stability against UV-light);
- resistance to cleaning agents;
- corrosion resistance;
- material compatibility with other materials in the local contact area.

In particular, a low creep under constant load is very important in the case of supports for structural glass components. In this respect, Polyamides (PA6 and

Table 7.1 Reference values of the calculated material stiffness of contact materials, taken from DIN 18008-3 [16]

	Elastomers	Thermoplastics	Grout	Aluminium
Young's mod. [MPa]	5–200	10–3 000	1 000–3 000	69 000
Poisson's ratio ν [-]	0.45	0.3–0.4	0.2–0.4	0.3

Table 7.2 Overview of different polymer contact materials, from [19]

Material[a]	Young's m. in tension [MPa]	Failure stress [MPa]	Elongation at break [%]	Service temperature [°C]	Suitability[b]
POM-C	3 000	70	30	−50 to +100	o
POM-C GF25	9 000	130	3	−20 to +100	+
PEEK	3 900	100	25	−60 to +250	++
PEEK GF30	9 700	156	2,7	−20 to +250	++
PEI	3 100	110	12	−50 to +170	+++
PA6	3 200	80	30	−40 to +85	–
PA6 GF30	9 500	180	3	−30 to +95	–

[a]The number behind GF indicates the glass fibre content in weight %
[b]Suitability for use as a load transfer material for large loads

PA66) outperform other polymers such as polyetheretherketone (PEEK), polyethylenterephthalat (PET) and Polyetherimide (PEI) [19]. While Polyoxymethylene copolymer (POM-C) provides an intermediate creep performance.

7.2.2 *Thermoplastics*

Various thermoplastics can be used as contact materials for large permanent loads. The most common are polypropylene (PP), (standard material for polymer blocks), Polyamide (PA6), polyether ether ketone (PEEK), polyetherimide (PEI), polysulfone (PSU) and polyoxymethylene (POM). The latter is also often used as a copolymer (POM-C) due to its higher stiffness. The polymers can be used in their unreinforced form, but many of the mentioned polymers are also used as matrix materials for short-fibre reinforced polymers to further increase their load capacity. The reinforcing fibres are typically made of glass. For example, unreinforced POM-C, has a maximum permanent compression stress capacity of about 10MPa, whereas a glass fibre reinforced POM-C GF can resist 30MPa without showing any relevant creep deformation [19]. The orientation of the fibres in the reinforced polymers is decisive for the achievable stiffness and compressive strength. It is advantageous if the glass fibres are oriented in the direction of the principal compressive stresses. An overview of different material properties with a rating of their suitability for large loads is shown in Table 7.2. For further reading see, for example, [12, 19].

7.2.3 Elastomers

Elastomers are the contact material of choice when it comes to permanent loads. They are normally made of silicone, Ethylene Propylene Diene Monomer rubber (EPDM) or polychloroprene (Neoprene). They are used, among other things, as gaskets, contact material for horizontal loads in point fixings and continuous edge supports for horizontal glazing. They have a relatively low stiffness and are therefore unsuitable for high loads. Silicone and EPDM used in close proximity can lead to material incompatibility issues and requires special attention.

7.2.4 Injection Mortars

Injection mortars are often used to fill tolerance gaps (e.g. assembly tolerances, misalignments, etc.) in bolted connections, other point-fixings in glass holes and continuous connections (e.g. for barrier glazing), see Fig. 7.1. A well-known product, which is primarily used as injection mortar for anchoring dowels in masonry, is a hybrid mortar based on methacrylate cement, such as fischer FIS V Plus, Hilti HIT®-HY 270 or its predecessor Hilti HIT®-HY 70.

After the two-component resin has been introduced, it hardens within a number of hours or even less than an hour, depending on the ambient temperature. For example, Hilti HIT®-HY 70 provides a compressive strength of $f_{c,k} = 59\,\text{MPa}$ with a Young's modulus of $E = 1760\,\text{MPa}$ after 24 h. These values do not change significantly even after long use [13].

Fig. 7.1 Compensation of tolerances inserting an injection mortar into the gap between a glass laminate and a steel bolt (image credits: Octatube)

7.2.5 Setting Blocks

Setting blocks are commonly used in vertical glazing to support the bottom edge of the panes from the window frame. These are inserted between the frame and the edge of the glass to directly support the self-weight and to stabilise the frame in its plane (Fig. 7.2). Essential functions of the setting blocks are:

- to provide an expansion gap for the glass in the frame;
- to distribute the glass weight in the frame;
- to protect the glass edge and edge seal;
- to provide ventilation and vapour pressure equalisation in rebate space;
- to stabilise the frame in operable windows.

Setting blocks, usually made of polymers, serve different functions. *Bearing blocks* are elements that transfer the dead load of the glazing onto the frame. The blocks are typically made of polypropylene (PP). *Bridge setting blocks* should be used in locations where bearing blocks coincide with drainage channels or electric cables. *Spacer blocks* maintain the gap between the glass edge and the frame. *Locking blocks* prevent excessive distortion of the glass casement frame during operation (e.g. due to opening of a door or window). *Glass rebate inserts* provide a smooth base for the setting blocks in various aluminium façade systems. For flat rebate areas without any profiling, blocks with a ventilation channel similar to bridge setting blocks must be used which allow for drainage and vapour pressure compensation and prevent the formation of trapped air in the rebate chamber.

Setting blocks used for insulating glass units are typically 80–100 mm long and should be approximately 2 mm wider than the glass unit. The blocks should usually not be placed directly at the corner of the pane to avoid overstressing the glass corner, but rather at a minimum distance of about 100 mm from the glass corners. For large glazing, this distance should even be increased to about 250 mm [18].

Glass panels must be free to expand in all directions to accommodate the inevitable thermal expansion caused by daily and seasonal temperature changes. This can be provided, for example, by lateral arrangements of the setting blocks as shown in Fig. 7.3. Inclined glazing must be mounted in the same way as vertical glazing using setting blocks along the lower edge and at the sides. Glazing with large inclinations should have a corresponding block placed in an inclined position (Fig. 7.4).

The transfer of large loads through setting blocks, e.g. supports for glass beams or glass fins, requires full surface contact between the block and the glass edge. Therefore, these areas must by detailed carefully to compensate for tolerances and movements of the glass under loading.

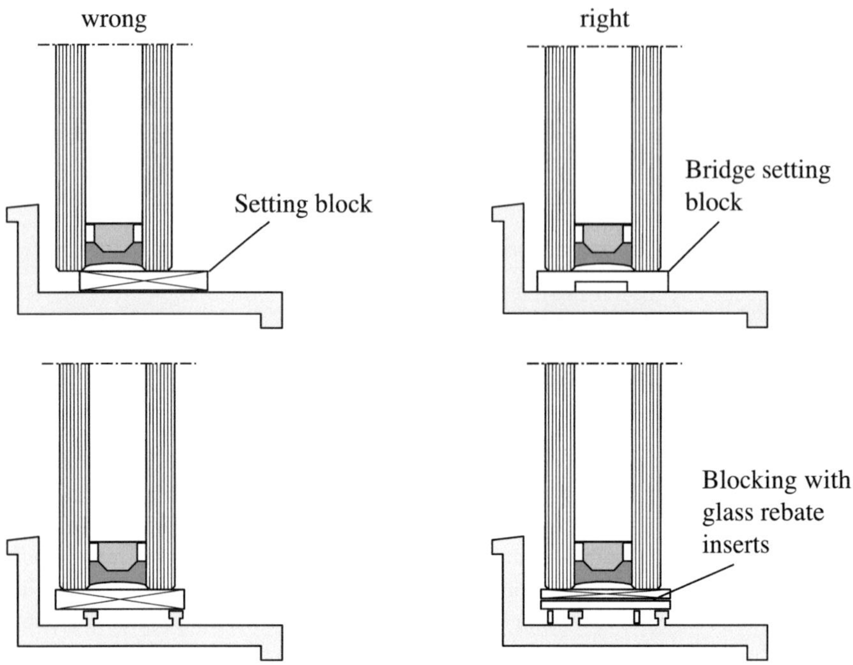

Fig. 7.2 Setting blocks for insulating glass units, according to [18]

7.3 Glass Support Configurations

This section describes support configurations that are predominantly used for façade and barrier glazing. A distinction is made between continuous supports and point supports. Combinations of continuous supports and point supports are also possible.

7.3.1 Continuous Support

The most common support used in glass construction is a *continuous support* perpendicular to the plate, see Figs. 7.5 and 7.6. For structural analysis, in most cases, a free rotation can be assumed for this type of support (simple support), which is achieved by the structural design. Continuously supported glazing is, for example, included in the regulations of DIN 18008-2 [10]. A minimum glass bite of 10 mm along the perimeter and a maximum allowable deflection of the substructure of $l/200$ is mentioned here to ensure a stable support. The continuous supports must be provided at both surfaces of the plate, thereby providing support against both wind pressure and suction, for example. It should be noted that in particular small, and therefore rela-

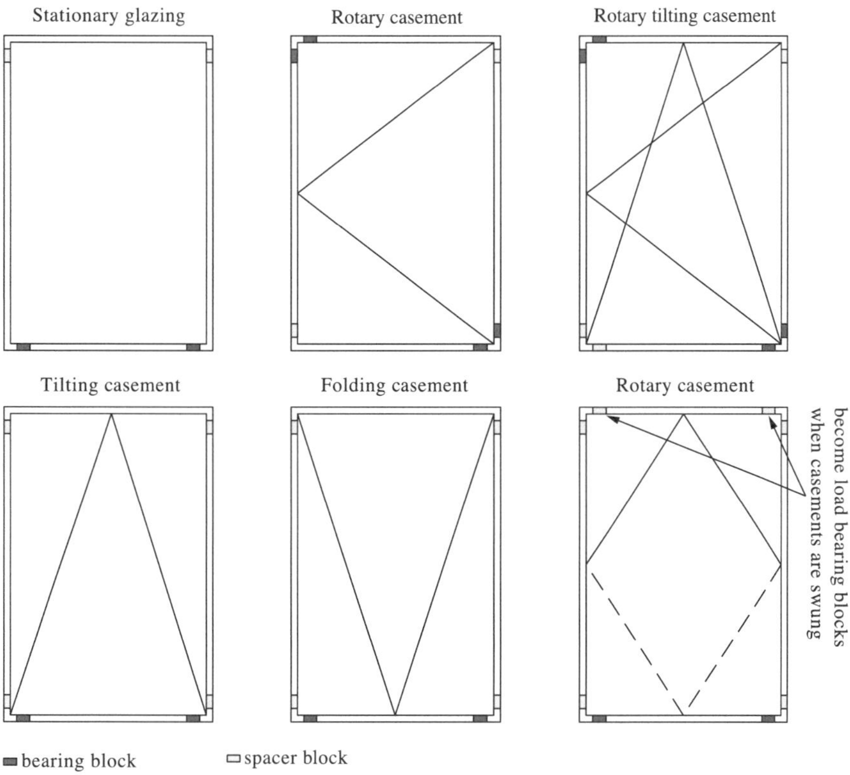

Fig. 7.3 Arrangement of setting blocks for different use cases, according to [18]

Fig. 7.4 Bearing block with strongly inclined glazing, according to [18]

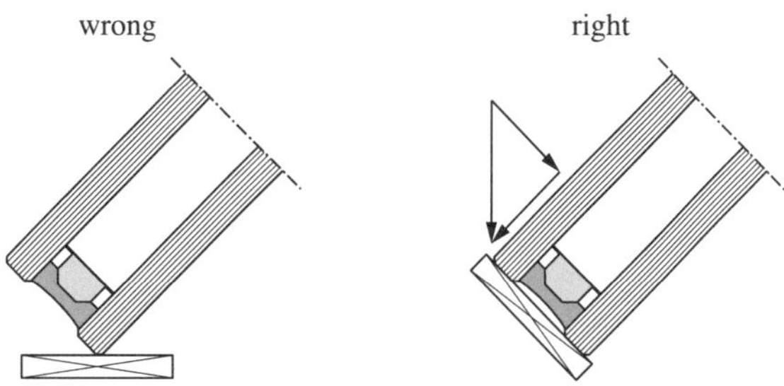

tively stiff plates are sensitive to deformations of the substructure, since the according deformations will result in considerable mechanical stress in the glazing.

A continuous support from a tape or elastomeric strip with a width of 10–15 mm can compensate for minor unevenness of the substructure and avoid undesirable stress concentrations in the glass. Elastomeric gaskets are commonly used for this purpose, but silicone gaskets or materials from seals used in high-pressure pipe connections are also possible. However, unintended mechanical stresses may arise in the assembly process or during the loading, which should be anticipated in the design and structural calculation of stress level in the glass. In the façade industry, various manufacturers offer a variety of extruded aluminium profiles (e.g. from EN AW 6060-T5), which meet the structural support conditions as indicated in the codes and also fulfil the relevant building physical requirements (Fig. 7.7a).

Clamped glazing is used where it is desirable or necessary to restrain the rotation of the glass edge. Such continuous, clamped supports are, for example, used for all-glass balustrades where the glass panels are supported at the bottom edge and are intended as barrier glazing, see Fig. 7.7b. The clamped height in a suitably formed substructure is typically $\geq$ 100 mm.

7.3.2 *Point Supports*

7.3.2.1 General

Mechanical point supports can be divided into clamped supports and bolted point fixings. In the case of bolted fixings, intermediate polymer layers, see Sect. 7.2, or, less frequently used, annealed aluminium alloys (e.g. AlMgPbSi) are preferably used in the area of holes to avoid direct contact between the steel bolt and the glass. Whereas the stress level in continuously supported plates can generally be calculated using simple structural mechanics, the verification of point-supported glass panes requires more elaborate techniques such as numerical modelling. The maximum principal tensile stress usually occurs close to the holes due to the stress concentration around them. Therefore edge treatment and edge surface characteristics in the area of the hole as well as the properties of the polymer layers can have a significant influence on the load capacity. Hence a detailed structural analysis of point supported glass is particularly important.

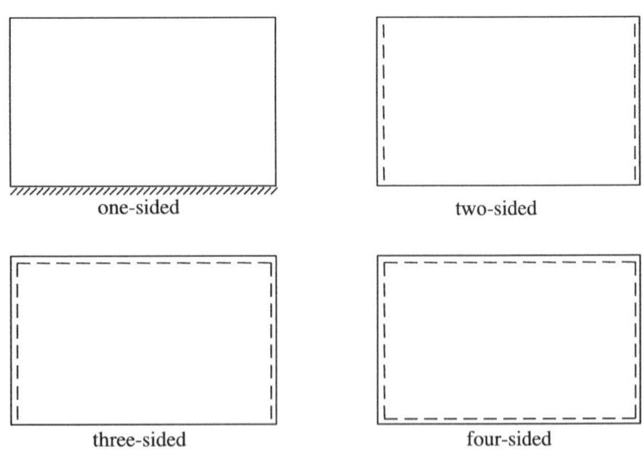

Fig. 7.5 Types of continuous support

Fig. 7.6 Continuously supported glazing at **a** the shopping center *MyZeil* in Frankfurt/ M. (photo credits: Schneider) and at **b** the *New Fair* in Milan (photo credits: Mero)

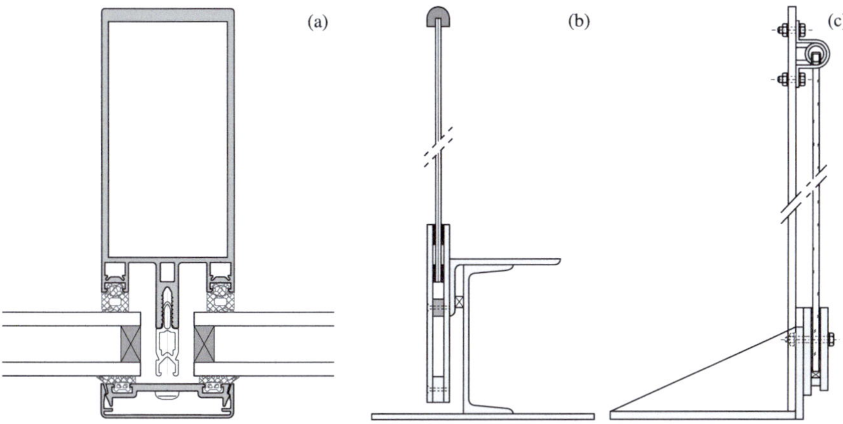

Fig. 7.7 **a** Typical support in a curtain wall system for insulating glass units (illustration after Schüco International), and **b**, **c** barrier glazing with locally clamped glass plates

The stress distribution at the edge of the hole is strongly influenced by the following parameters:

- type of support (centric articulated joint, eccentric articulated joint, clamped joint);
- hole characteristics (diameter, edge conditions, geometry, level of pre-stress from the thermal tempering process);
- geometry of the glass fitting;
- polymer material (thickness and stiffness);

In the case of thin polymer layers, the parameters of the materials used must be usually determined experimentally, since their mechanical behaviour can be significantly influenced by friction and confinement of the transverse expansion. Due

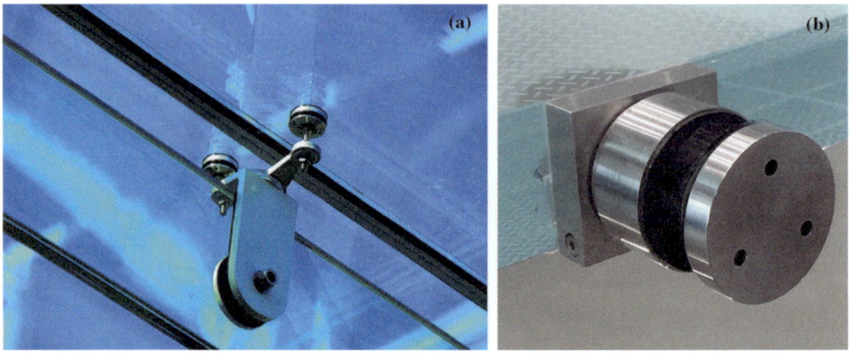

Fig. 7.8 Bolted point supports: **a** glass beams in *Sparkasse Senftenberg* (photo credits: Mero), **b** glass staircase at the *Apple Store* in Frankfurt am Main (photo credits: Schneider)

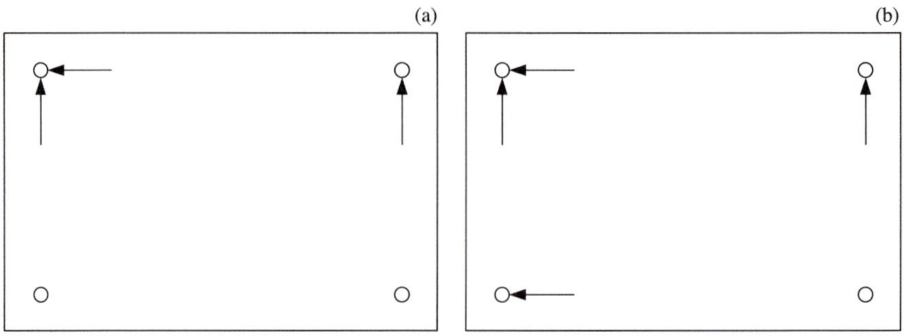

Fig. 7.9 Two alternative configurations for in-plane support of glass panels, using: **a** two out of four point-fittings (free thermal expansion possible) and **b** three out of four point-fittings (stress from thermal expansion of the glass can occur)

to the stress concentration around the holes, point supported systems typically use (thermally) pre-stressed glass (Fig. 7.8).

Undesirable stresses from thermal dilatation of the glass can be avoided in point-supported plates by providing a combination of fixed and sliding supports while providing the necessary support to the glass. This results in a statically determinate support condition where the lines of action do not intersect at one point: this type is shown in Fig. 7.9a or Fig. 7.9b, but the latter is more sensitive to geometric tolerances than the former. The in-plane sliding bearing required in these connections can be achieved by using slotted or oversized holes in the supporting substructure.

To achieve the configuration shown in Fig. 7.9a, a fitting hole, a slotted hole and two oversized holes can be provided in the steel parts (not in the glass), to which point connectors are attached. The in-plane sliding of the point connectors can be made possible by the use of intermediate materials of lower friction resistance (e.g. sliding films). Alternatively, the desired in-plane movement can also be achieved by the special structural design of the point connector. If statically determinate mounting

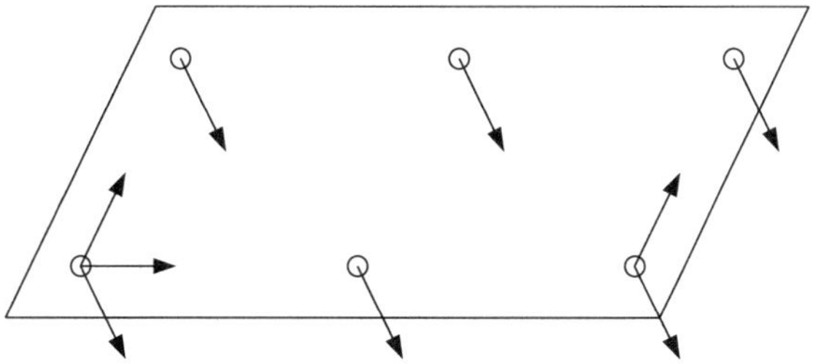

Fig. 7.10 Statically indeterminate out-of-plane support of a glass plate with six fittings

cannot be achieved, the stress resulting from the constraint thermal expansion and the deformation of the substructure must be taken into account in the structural analysis.

Out of plane, glass plates supported by local fittings are always statically indeterminate Fig. 7.10.

7.3.2.2 Clamp Brackets

Clamp brackets (Fig. 7.11) are mounted along the perimeter or at the corners of a glass plate and rely on frictional load transfer between the glass and the bracket rather than by bearing forces on the bolts. When clamping laminated safety glass with soft contact material such as EPDM, the contact material and the glass interlayer should not be squeezed out during mounting. Typically, the clamped surface should be at least 1 000 mm^2 and the glass rebate should be at least 25 mm. A lower rebate and a smaller clamped surface are permissible, if it is shown that, with a rebate of at least 8 mm, the load bearing capacity can still be ensured, taking into account tolerances and the movement of the glass plate in the support in its deformed state.

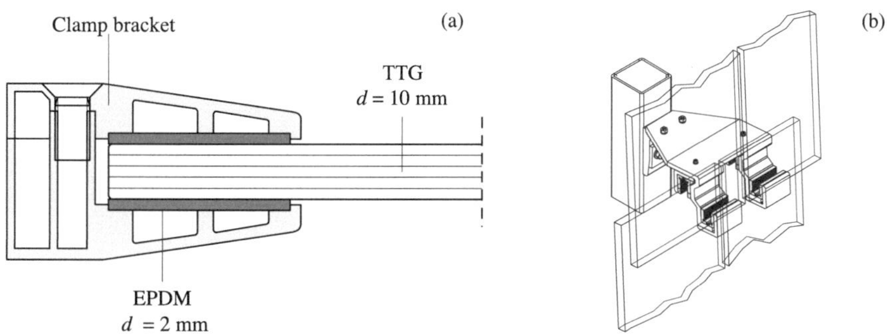

Fig. 7.11 **a** Clamp bracket for outward leaning vertical glazing and **b** support system for shingled glass façades (photo credits: Aweso Systemtechnik GmbH)

Fig. 7.12 Detailed view of the clamp brackets on the cable net façade of *Kempinski Hotel* at Munich Airport (photo credit: Schlaich Bergermann Partner)

7.3.2.3 Clamp Brackets for Cable Net Façades

Clamp brackets are also often used for cable net façades, see Fig. 7.12. For this purpose, clamps at the intersection of the vertical and horizontal cables are used to support the laminated safety glass. The prototype of such construction was built at the Kempinski Hotel at Munich Airport, see Fig. 7.13 [4]. The exceptional transparency of the façade construction system has since been adopted worldwide, see Fig. 7.14.

Out-of-plane façade deflections from wind loading are kept within acceptable limits by the level of pre-tensioning of the steel cables. Furthermore the twisting of the glass panels between adjacent brackets must be small enough to avoid excessive stress peaks in the glass close to the bracket. For the design of cables under uniform transverse load, see Appendix A.

7.3.2.4 Mechanical Point Fittings

Mechanical point fittings (or: point fixings) can be divided into fittings requiring cylindrical holes in the glass and fittings requiring countersunk holes, see Fig. 7.15. A point fitting must resist both positive and negative out-of-plane loads and support the dead load of the glass.

Fittings for cylindrical holes typically use metal support discs with a diameter of at least 50 mm, e.g. DIN 18008-3 [16]. Appropriate design shall ensure that the support discs provide a minimum support to the glass of at least 12 mm around the hole.

Countersunk fittings are more aesthetically pleasing as they provide a flush surface, but they are more sensitive to tolerances of the substructure and possible ply misalignment in laminated glass. However, a large number of manufacturers provide countersunk connection systems that overcome these challenges.

Fig. 7.13 Cable net façade on the *Kempinski Hotel* at Munich Airport (Architects: Murphy/Jahn, Engineers: Schlaich Bergermann Partner, photo credit: Schlaich Bergermann Partner)

The walls of the bores in the glass should always have a chamfer (at least a seamed edge), to make them less vulnerable to edge defects and to provide a continuous pre-stress level around the hole from the tempering process. In general the depth of the cylindrical part of the hole should be no less than 8 mm. The diameter of the hole should generally be at least equal to the thickness of the glass. The position of the holes in a glass pane and its clearance from edges and corners is specified in terms of the glass thickness, h, from the product standards for thermally toughened glass (EN 12150-1 [21]) and heat-strengthened glass (EN 1863-1 [14]) as follows:

- The distance between the hole perimeter and the edge of the pane should not be less than $2h$ (Fig. 7.16a);
- The distance between the perimeters of adjacent holes should not be less than $2h$ (Fig. 7.16b);
- The distance of a hole perimeter to the glass corner must not be less than $6h$ (Fig. 7.16c).

However, it should be noted that these standards are intended to regulate the glass quality from the production process of thermally pre-stressed glass. A design check of the load bearing capacity of the glass (based on the specific hole configuration) for a given connection is required even if the above-listed clearances are implemented. Published research shows that the residual stresses, governing the strength near the hole are dependent on these distances and the cooling process itself [11, 23].

In DIN 18008-3 [16] a clearance of at least 80 mm from edges and corners shall be maintained. Recent research shows that the regulations in DIN 18008-3 [16] are over-conservative [24] and smaller clearances are possible. Holes near corners should

Fig. 7.14 Cable net façade *Time Warner Center*, Columbus Circle, New York City (Architects: James Carpenter, Engineers: Schlaich Bergermann Partner, photo credit: Schlaich Bergermann Partner)

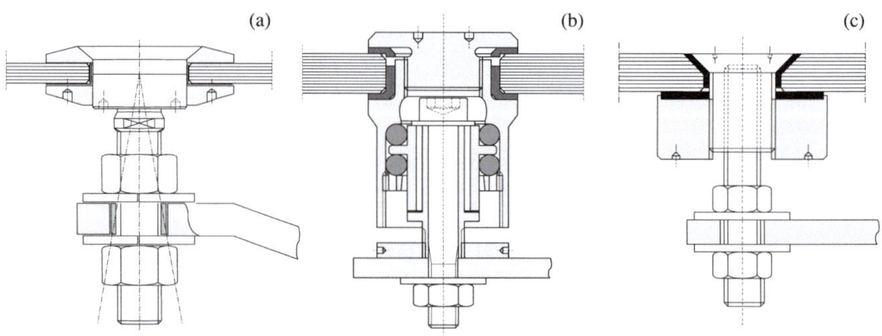

Fig. 7.15 Point fittings: **a** articulated (swivel) fitting (type: FXR 1003B, Sadev), **b** flexible fittings (type: Manet®EV 50/50, former Dorma), **c** countersunk fitting (type: Litewall, former Eckelt Glas)

be positioned with an offset of a least 15 mm to the diagonal of the glass pane to avoid stress concentrations originating from plate corner forces DIN 18516-4 [1].

The diameter of the holes in the glass, depends on the respective point fixing system and typically varies between 16 and 40 mm. The diameter of the support discs varies between 45 and 70 mm. Usually, all systems allow a large degree of rotation ±10° to ±20° (Fig. 7.15a–c) which is not really required for the final structural function but rather to avoid mechanical stresses from the assembly process. Exceptions to this are rigid (fixed) point fittings (Fig. 7.15b). Another important function is the out-of-plane adjustability, which accommodates unavoidable tolerances in the substructure during assembly. Finally, the sealant between adjacent glass panes must be flexible enough to avoid transmitting local forces from one pane to another and to ensure its barrier function. The structural detailing must prevent moisture penetration or the formation of corrosion from condensation. Especially for insulating glass units supported by point fittings, the air tightness of the seal around holes requires careful attention.

7.3.2.5 Undercut Anchors

Undercut anchors (also known as partial penetration anchors) are a type of point supports based on a mechanical expansion anchor where a blind hole is milled in the glass allowing to fasten the fitting (Fig. 7.17). The glass is therefore not fully penetrated, which makes the cleaning on the outside of the pane easier and the air/water tightness in the connection area is guaranteed (see Fig. 2.18). The holes with typical diameters of 16 mm are manufactured in a computer numerically controlled (CNC) milling machine. After milling, the glass is thermally pre-stressed. This is required due to the stress peaks around the holes from the mechanical loading. It should be noticed that due to the blind hole, the local pre-stressing stress level at the bottom of the hole is significantly lower than at the chamfer. The anchors, consisting of studbolts, spreading sleeve, polymer cap, compensation washer and retaining nut are inserted into the undercut anchor hole and tightened using a special torque wrench. Direct contact of steel to glass is avoided by the polymer cap. When using laminated

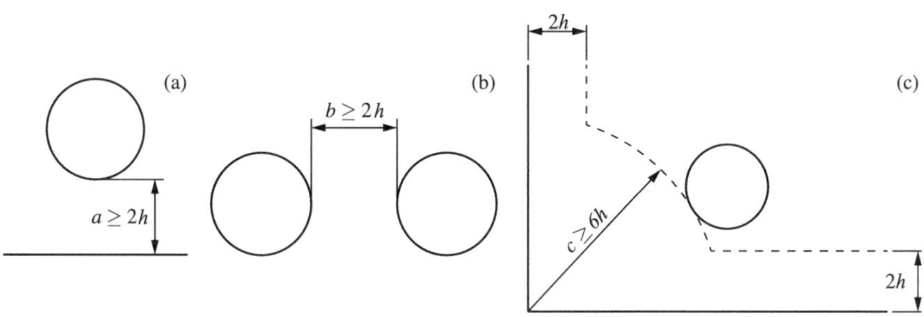

Fig. 7.16 Minimum hole clearances **a** to the glass edge, **b** between holes and **c** to the corner of the pane according to EN 12150-1 [21] and EN 1863-1 [14]

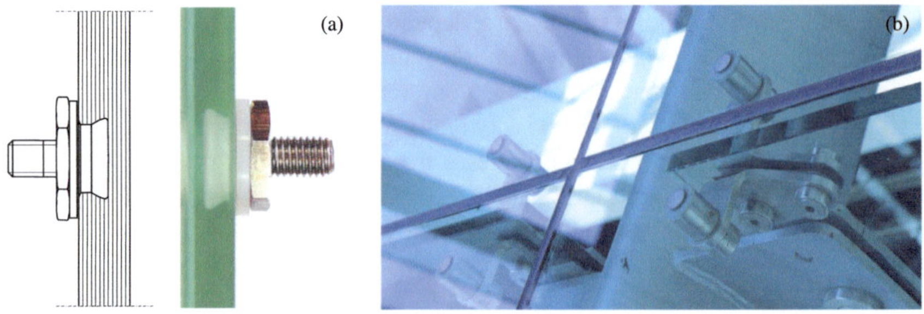

Fig. 7.17 Undercut anchor: **a** Schematic representation and detail of the anchor, **b** application at Century House in London (photo credit: fischerwerke GmbH & Co. KG)

safety glass, the glass pane situated towards the substructure is typically penetrated by a cylindrical hole and the undercut anchor applied on the outside glass.

The undercut hole geometries typically require a glass thickness of 10 mm or 12 mm. For laminated safety glass, the glass thickness combinations 10 mm/10 mm and 8 mm/10 mm can be applied, wherein the undercut is always positioned in the 10 mm thick glass pane. With a glass thickness of 10 mm, the anchoring depth of the fitting is 6 mm; and with a glass thickness of 12 mm, the depth is 7 mm. The edge distance should be at least 60 mm [3].

7.3.2.6 Toggle Fittings

Toggle fittings are special point fixings used to mount insulating glass units in a curtain wall façade system (Fig. 7.18). In this case, a local slot is integrated within the secondary edge seal of an insulating glass unit (Fig. 7.18). This slot provides space for placing a toggle fitting. This screw-in toggle fitting is first fixed in a parallel position to the substructure. After installation of the insulating glass unit, the toggle fittings are rotated by 90° and fit into the edge slots. While the toggle fittings are usually made of stainless steel, the slots can also be made of polymer or (injection-moulded) aluminium. The inner pane of the insulating glass unit is thus mechanically secured against suction loads, which act normal to the pane. The outer glass pane, on the other hand, is held exclusively by the bond strength of the secondary seal in the insulating glass unit. In some design codes, a mechanical retaining device shall be provided for structural glazed façades beginning at 8 m above ground level in accordance with building codes ([20] part II, appendix 2/1) for a scenario in which the polymer bonding fails. This mechanical retaining device could be realised, for example, by means of an external clamp. An alternative retaining device is depicted in Fig. 7.18b. In this case, a narrow slit is milled into the edge of the outer glass pane, which is thermally pre-stressed afterwards. However, it should be noted that the pre-stress at the slot cannot be precisely defined since, technically, adequate cooling during the thermal pre-stressing process cannot be accomplished in the narrow slot.

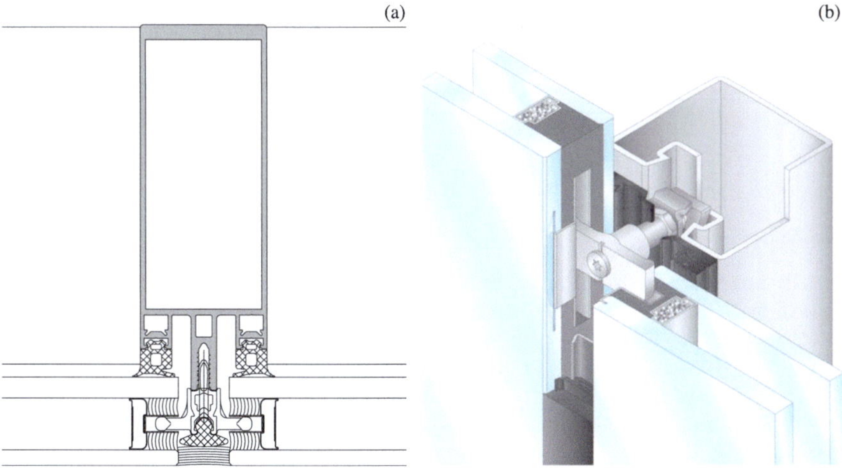

Fig. 7.18 **a** Schüco FW 50+SG (picture credits: Schüco International KG) and **b** Screw in mounting systems Vario S-FOR/DZ (picture credits: Eckelt Glas)

7.4 Structural Connections for Transfer of Large Forces

7.4.1 General

Mechanical connections can also be used in glass constructions to transfer large forces between adjacent glass components and/or the underlying structure. These are essentially different from the connections described in Sect. 7.3, which are predominantly suitable for cladding applications such those in façades, canopies, etc.

This section deals with high strength load bearing connections which are either in the form of bolted connections or friction-grip connections. For both connection types, it is crucial to avoid contact between glass and hard materials (e.g. steel, concrete, etc.). Direct glass-glass contact should also be avoided [2] unless a very high quality in glass milling and very small tolerances between the adjacent surfaces can be guaranteed.

7.4.2 Bolted Connections by Bearing Bolts

Bolted connections have been used successfully for many years in steel and timber construction. They provide a simple and adaptable connection system. In steel construction, bolts are commonly in a clearance hole and the stress peaks from the bolt bearing contact stresses are redistributed by local deformation and the elastic-plastic nature of the material. Since glass has no plasticity at a macro scale, an intermediate material, typically in the form of a sleeve and/or an injection mortar, see Sect. 7.2.4, is

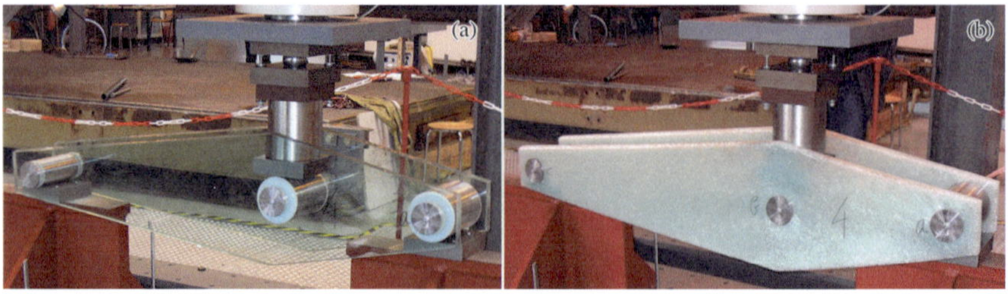

Fig. 7.19 Load test on a thermally pre-stressed glass beam with bolted connections, for the glass façade of the *Hauptbahnhof* in Berlin **a** before and **b** after glass fracture as a result of stress concentrations around the holes

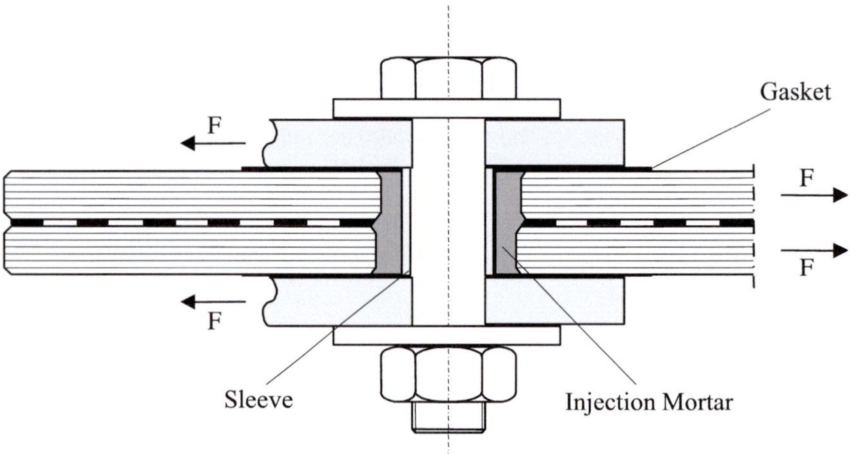

Fig. 7.20 Example of a bolted connection in glass construction

required to spread the contact force from the bolt onto the edge of the hole. The sleeve typically consist of aluminium or a polymer such as polyamide. An additional injection mortar, made of epoxy resin, polyester, polyurethane or urethane methacrylate cement, see Sect. 7.2, is particularly useful to accommodate possible hole misalignment in laminated glass by flowing into and filling the gaps, see Fig. 7.20. Direct contact between the other components of the bolted connection and the surface of the glass is prevented by means of a gasket, e.g. made of EPDM, nylon, silicone or PEEK. All materials used must be resistant to environmental factors (UV light, water), have strong fatigue resistant properties, e.g. to avoid creep effects and must be compatible to each other. In addition to investigation by means of component tests, a numerical evaluation of bolted connections is possible analytically or using FEM-analyses [5–9, 16]. An example of a bolted connection in a glass beam is given in Fig. 7.19.

7.4.3 Friction-Grip Connections

Friction-grip connections have a successful track-record as high-strength connections in steel construction. Steel plates are pressed against each other via preloaded bolts and friction is generated at the contact surfaces. However, the pre-treatment of the friction surfaces in steel construction is very complex. No such pre-treatments are used in glass construction. Instead, appropriate friction layers which are free of creep deformations are used (Fig. 7.21). For example high-pressure gaskets or sandpaper have been used successfully. The frictional resistance of the layer is to be determined in experiments.

The use of friction-grip connections in combination with laminated glass requires special attention, since the interlayer of the laminated glass can be squeezed out (creep deformation) due to high compressive forces from the preloaded bolt. The use of an aluminium insert at the compressed connection area provides a possible solution to avoid local creep deformation of the interlayer.

Similarly, an innovative yet early-stage concept is suggested in [15, 17, 22] where the possibility to establish a drilled connection in thermally pre-stressed glass is proposed. The method relies on (many) small blind holes drilled carefully into the compression zone of the pre-stressed glass used for transferring the load, see Fig. 7.22. While the presence of the holes cause a 3 times higher tensile stress when loaded, this is partly compensated by an increase in compressive residual stresses. The latter is due to a circumferential stress redistribution as a result of the post-tempering drilling process. It is shown theoretically and experimentally that for a small hole the residual compressive stress is doubled [22]. The hypothesis is that such a connection will not

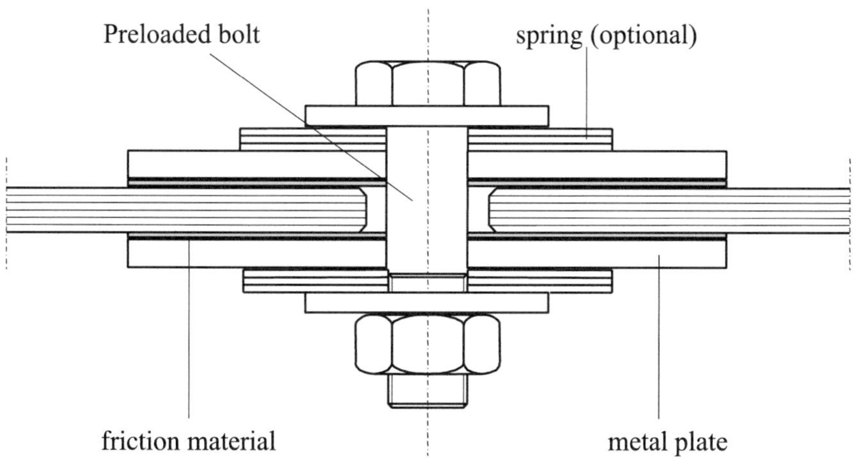

Fig. 7.21 Example of friction-grip connection in glass construction

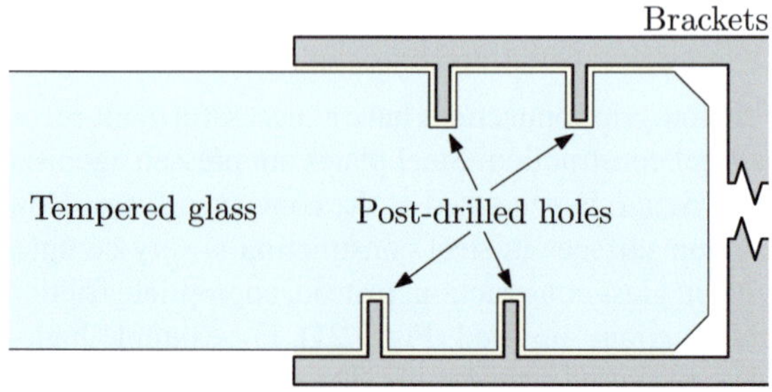

Fig. 7.22 Principal sketch of a possible way to attach a bracket to tempered glass with post-drilled holes [22]

only allow for post-drilling in thermally pre-stressed glass, but also provides some ductility in the connection due to the small metal pins.

References

1. DIN 18516-4. *Außenwandbekleidung, hinterlüftet – Einscheiben-Sicherheitsglas – Anforderungen, Bemessung, Prüfung* (1990)
2. T. Macfarlane, Construction details for structural glass assemblies, in *Glass Processing Days 1999* (Tampere, 1999), pp. 474–476
3. J. Schneider, P. Hof, C. Block, Fastening of glass panes with undercut anchors. Numerical tempering simulation, experiments, load-bearing capacity, in *Proceedings of the 2nd International Colloquium on Modelling of Glass Forming and Tempering (M.G.F.T.)* (2002), pp. 247–254
4. H. Schober, Gespannte Seilnetzfassaden. Stahlbau **73**(12), 973–981 (2004)
5. I. Maniatis, Numerical and Experimental Investigations on the Stress Distribution of Bolted Glass Connections under In-Plane Loads. Dissertation. Technische Universität München (2006)
6. M. Overend, I. Maniatis, The influence of finite element modelling parameters on the stress distribution in bolted glass connections, in *11th International Conference on Civil, Structural and Environmental Engineering Computing* (2007)
7. M. Feldmann, M. Pilsl, C.C. Segura, Tragfähigkeit von Lochleibungsverbindungen vorgespannter Glasscheiben zur Ausbildung von Anschlüssen im Konstruktiven Glasbau. Stahlbau **77**(1), 17–25 (2008)
8. J.H. Nielsen, Tempered Glass: -Bolted Connections and Related Problems. PhD. Technical University of Denmark (2009), p. 78
9. M. Baitinger, Zur Bemessung von SL-belasteten Anschlüssen im konstruktiven Glasbau. Dissertation. RWTH Aachen (2010)
10. DIN 18008-2. *Glas im Bauwesen – Bemessungs- und Konstruktionsregeln – Teil 2: Linienförmig gelagerte Verglasungen (mit Berichtigung 1 vom April 2011)* (2010)
11. J.H. Nielsen, J.F. Olesen, P.N. Poulsen, H. Stang, Simulation of residual stresses at holes in tempered glass: a parametric study. Mater. Struct. **43**(7), 947–961 (2010)
12. B. Weller, J. Schneider, S. Reich, J. Ebert, Kontaktmaterialien zur Einleitung von Druckkräften in Glas. Bauingenieur **85**(3), 118–127 (2010)
13. Ö. Bucak, A. Lorenz, *Gutachterliche Stellungnahme für die Verwendbarkeit von Hilti HIT-HY 70 im Glasbau* (2011)
14. EN 1863-1. *Glass in building - Heat strengthened soda lime silicate glass - Part 1: Definition and description* (2012)
15. J.H. Nielsen, P.N. Poulsen, Joints in tempered glass using glass dowel discs, in *Engineered Transparency* (2012)

16. DIN 18008-3. *Glas im Bauwesen – Bemessungs- und Konstruktionsregeln – Teil 3: Punktförmig gelagerte Verglasungen*. Beuth (2013)
17. J.H. Nielsen, Numerical investigation of a novel connection in tempered glass using holes drilled after tempering, in *COST Action TU0905 Mid-term Conference on Structural Glass* (CRC Press, 2013), pp. 499–505
18. Roto Frank AG, *Die Klotzfibel* (Leinfelden-Echterdingen, 2013)
19. J. Ebert, Einleitung hoher Lasten in Glaskanten – Ein Beitrag zum Einsatz von Kunststoffen als Klotzungsmaterial. Dissertation. Technische Universität Dresden (2014)
20. DIBt. *Muster-Liste der Technischen Baubestimmungen – Fassung Juni 2015* (2015)
21. EN 12150-1. *Glass in building - Thermally toughened soda lime silicate safety glass - Part 1: Definition and description* (2015)
22. J.H. Nielsen, M.J. Meyland, B.E. Thorup, A. Zugravu, J.F. Olesen, Investigating the strength effects of drilling in tempered glass: a first step towards a novel dowel joint. Glas. Struct. Eng. **4**, 243–256 (2019). https://doi.org/10.1007/s40940-019-00095-5
23. N. Pourmoghaddam, J. Schneider, Determination of the engine power for quenching of glass by forced convection: simplified model and experimental validation of residual stress levels. Glas. Struct. Eng. **4**(1), 117–125 (2019). https://doi.org/10.1007/s40940-018-0078-5
24. L. Efferz, S. Dix, C. Schuler, Punktgehaltene Gläser mit geringem Bohrungsrandabstand. ce/papers **5**(1), 268–279 (2022). eprint: https://onlinelibrary.wiley.com/doi/pdf/10.1002/cepa.1674

Chapter 8
Adhesive Connections

Chemical bonding by use of adhesives, offers an alternative means of connecting glass components to the mechanical connections discussed in Chap. 7. *The present Chapter describes the materials available for bonding and sealing in Sect.* 8.2 *and Sect.* 8.3, *followed by information on the necessary compatibility with other materials in Sect.* 8.4. *The use of adhesives for linearly bonded glazing and point supports is then described in Sect.* 8.5 *and Sect.* 8.6, *respectively.*

8.1 General

Adhesively bonded connections, such as those shown in Fig. 8.1, generally provide a relatively uniform stress distribution over a large area [6]. The surface pretreatment of glass contact surfaces is very simple in comparison to that of other materials, because the bonding surfaces typically only need to be dried and degreased. Polyurethanes, epoxy resins, silicones and acrylics have all been successfully used for bonding glass, albeit for different applications. Stiff interlayer materials such as ionomers may also be used for adhesive connections for glass, see Sect. 3.1.3.3. By changing the thickness of the adhesive layer and the bulk material stiffness of the adhesive it is possible to control the stress distribution and the deformation behaviour of the adhesive connection and to minimise stress peaks. The stress distribution in the adhesive and in the connected parts can be determined by FEM-analyses provided that the material properties are known and that suitable material models are used, see Sect. 4.1. An analytical calculation of overlapping bonded connections is also possible. Volkersen's approach [1], for example, also takes the deformation of the joining parts (adherends) into account (Fig. 8.2). For simplification, in practice the shear stresses, τ, are usually assumed to be uniformly distributed via the adhesive surface, A, and the shear force, F, and can be determined without solving the differential strain equation from [1] :

J. Belis et al., *Engineering Glass Structures*,
https://doi.org/10.1007/978-3-032-03233-1_8

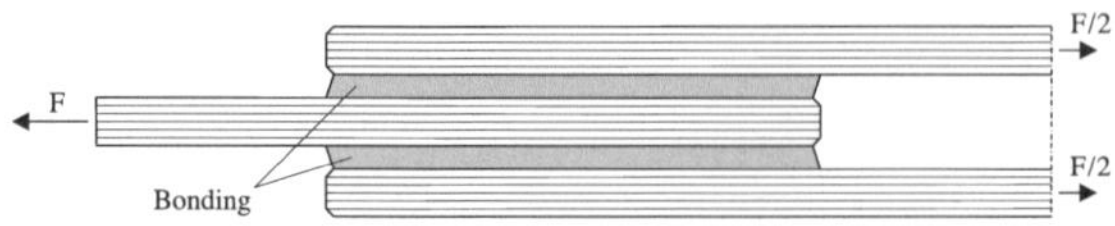

Fig. 8.1 Schematic representation of an adhesive connection in glass construction

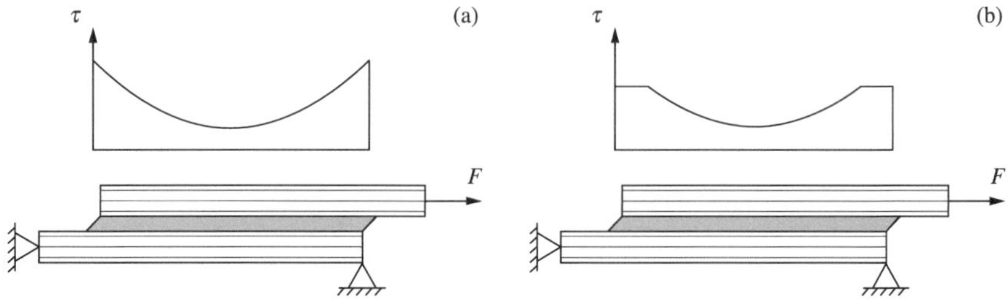

Fig. 8.2 Schematic representation of the shear stress gradient in adhesive joints for different material behaviour of the adhesive: **a** linear-elastic [1] and **b** linear-elastic-ideal-plastic [3]

$$\tau = \frac{F}{A} \tag{8.1}$$

This uniform shear stress is commonly referred to as the nominal shear stress.

It is a reasonable assumption for relatively flexible adhesive layers, but this simple formula is unable to capture the stress peaks near the ends of the adhesive layer that arise in relatively stiff and elastic adhesive layers, therefore this simple formula should be used with caution.

An overview of the analytical approaches is given by da Silva et al. [13].

Adhesives are often sensitive to moisture and temperature and therefore, the application of adhesives and forming of joints requires special attention and should only be carried out under controlled conditions, i.e. in climate-controlled premises. In addition, temperature fluctuations during service life can lead to considerable tension in the glass, since adhesive and glass usually have very different temperature expansion coefficients and relaxation is limited when using stiff adhesives. Figures 8.3 and 8.4, show the different types of stress and failure mechanisms for adhesive joints.

The long-term performance of adhesive connections is crucial, particularly for connections under sustained load, see Sect. 8.4. Creep and ageing effects should therefore be taken into account in the design of these connection, but data on these effects is not always available.

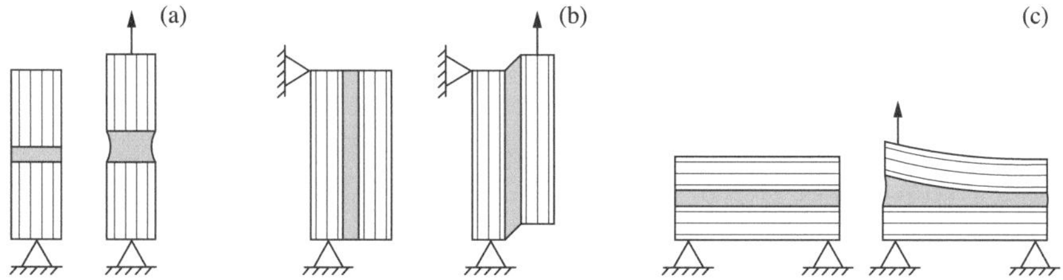

Fig. 8.3 Strain types of an adhesive joint according to [11]: **a** tensile stress, **b** shear stress and **c** peeling stress

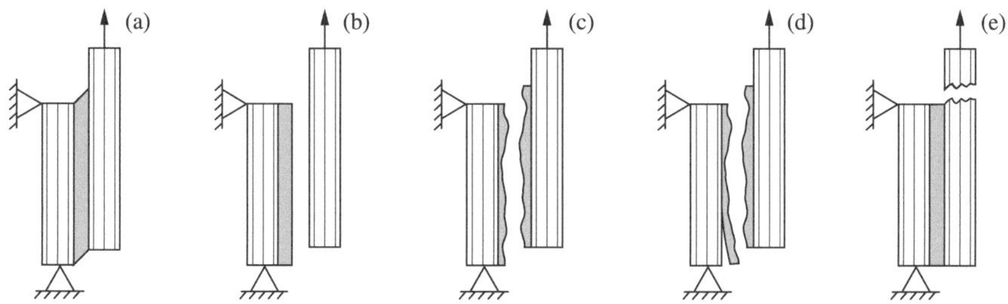

Fig. 8.4 Adhesion failure processes under shear stress according to [11]: **a** Intact bond, **b** adhesive failure, **c** cohesive failure, **d** mixed-mode adhesive and cohesive failure, and **e** adherend failure

8.2 Materials for Sealing and Bonding

8.2.1 General

Polymers are typically used for bonding and sealing in glass construction. The field of application ranges from adhesively bonded point supports and glass-glass structural bonding to weather seals for façades and hermetic edge seals for insulating glass units.

The following subsections list the salient adhesives and sealants in descending order according to their stiffness. The polymer structure, the properties of the hardened adhesive and the areas of application are described in turn. Silicones covering the largest range of applications are presented in more detail. Table 8.1 provides a limited overview of the adhesive and sealant products available on the market; many more suitable adhesives are available though. Figure 8.5 shows typical tensile shear strength and elongation at break of products used in glass construction.

Stiffness and cohesive strength are essential characteristics for selecting adhesives for a given application. Other important properties are adhesion and chemical compatibility with the parts to be joined, as well as the durability of the bond. In this respect, proper cleaning and degreasing of the bonding surfaces is essential. This requires the application of a cleaning agent (e.g. isopropylalcohol), but more advanced treatments are sometimes required such as ozone cleaning or plasma

Table 8.1 Mechanical properties of a limited selection of adhesives and sealants

Material	Tensile strength [MPa]	Elongation at break [%]	Shore Hardness [-]
Delo-Duopox®AD821[a2] [39]	10	60	50(D)
Delo-Photobond®4468[b1] [36]	14	200	83(A), 45(D)
SikaFast®-5211 NT[b2] [32]	10	200	90(A), 50(D)
Körapur 140[c1] [38]	4	400	55(A)
Sikaflex®-521 UV [c1] [33]	1,8	400	40(A)
SoudaSeal®2282[d1] [27]	2,8	>250	68(A)
Dow Corning®DC983[e2] [20]	2,0	400	35–45(A)
Dow Corning®TSSA[e2] [29]	8,5	250	75(A)
Ködiglaze S[e2] [30]	2,1	230	46(A)
Sikasil®SG-500[e2] [26]	2,2	300	45(A)
Thiokol®2282[f2] [24]	2,0	450–500	–
GD 116[f2] [23]	>0,8	–	38(A)

[a] epoxy resin, [b] acrylic, [c] polyurethane, [d] MS-Polymer, [e] silicone, [f] polysulphide
[1] one-component adhesive, [2] two-component adhesive

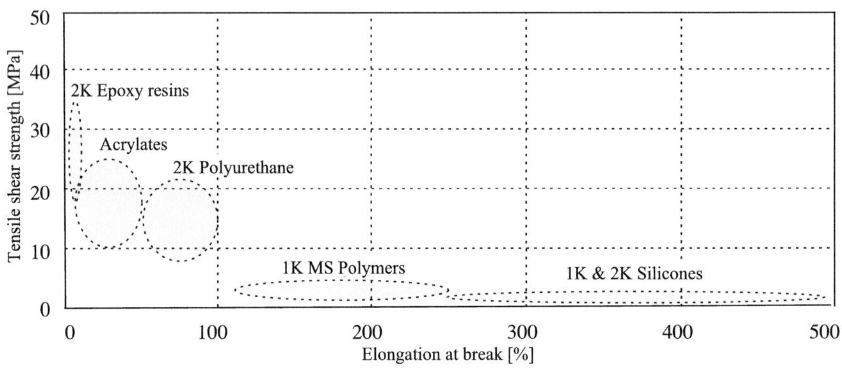

Fig. 8.5 Comparison of the mechanical properties of various adhesives used in glass construction [17]; 1K = one-component adhesive, 2K = two-component adhesive

cleaning. To further improve adhesion, the bonding surfaces may additionally be pre-treated with primers.

Adhesion science and technology constitutes a large body of scientific and practical knowledge, which cannot be entirely covered in this brief chapter. The works of [4, 22, 25] are particularly useful for a deeper insight into this topic.

8.2.2 Epoxy Resin

The monomers of epoxy resin adhesives are characterised by the epoxy group, which enables curing by means of polyaddition. The reaction requires a hardener as a second component. The final product can be classified as a thermoplastic characterised by a high stiffness, relatively high strength and a low elongation at break [11].

The epoxy resin adhesives typically used in glass construction harden at room temperature and are based on Bisphenol A [17]. The thickness of the adhesive layers is typically less than 0.5 mm.

8.2.3 Acrylate

Acrylates are adhesives which, in their uncured state, have monomers with a vinyl group (C = C double bond). One-component (1K or 1C) products are cured by chain polymerisation after activation by UV radiation, or in the case of two-component (2K or 2C) acrylates are cured by radical reactions [11].

Acrylates are classified as thermoplastics and exhibit high strength with low elongation at break. The thickness of the adhesive layers is usually 0.2 mm to 1 mm. If the bonding is visible, as with adhesively bonded point supports, transparent acrylates may be preferred.

8.2.4 Polyurethane

Cross-linked Polyurethane (PUR) is used as an adhesive and sealant in glass. It is characterised by the urethane group, and formation takes place by means of polyaddition. For one-component systems, the curing is driven by humidity and for two-component systems, by means of a hardener. Polyurethanes are also available as pre-formed end products such as elastomers, foam materials and thermosets [11].

Due to the great variety of products, PUR systems are available across a wide range of stiffnesses. The products used in glass construction are slightly more flexible than acrylates, but stiffer than silicones [17].

8.2.5 MS-Polymer

Polymers with modified silane (MS) groups, characterised by high elasticity down to low temperatures and high UV and weather resistance, have found growing application as adhesives and sealants [11]. Their strength and elongation at break places them between polyurethanes and silicones [17] and are applied with a minimum thickness of 2 mm.

8.2.6 *Silicone*

Almost all polymers have a structure based on carbon. Silicone is an exception. It consists of a helical silicon oxygen main chain with organic residues, see Fig. 8.6a. A strong adhesion is achieved between silicone and glass due to the chemical similarity between the two materials. In particular, covalent silicon-oxygen bonds are produced at the interface between glass and silicone [5]. Good adhesion is also achieved with other adherends such as aluminium or steel, so that a cohesive failure is always expected. This failure mode is used as a criterion for quality assurance tests.

Silicones can be classified as elastomers; the sparsely cross-linked network of one-component silicone is achieved by bonding with water from the air, whereby the duration of the curing process is no longer practical at larger joint depths. Consequently, two-component silicone is preferred for joints with deep aspect ratios. With these, cross-linking is instead generated by the reaction of two components, which are mixed shortly before the bonding process [11].

In addition to their excellent adhesion to glass and other joining parts, silicones are characterised by a high flexibility even at very low temperatures. The glass transition temperature is usually in the range of $T_g \approx -100\,°C$. Thus silicones are in

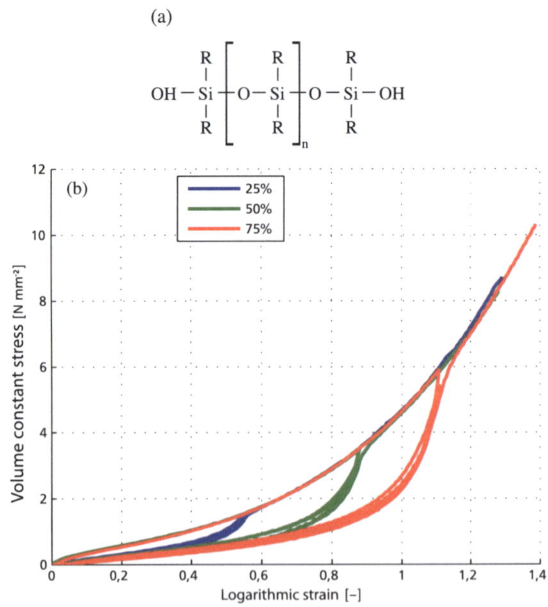

Fig. 8.6 **a** Chemical composition of silicone [11]; **b** Hyperelastic behaviour with Mullins effect for a silicone [34]

their rubbery state for most applications in the construction sector. Silicone adhesive compounds are also extremely durable against UV influences, high temperatures and humidity [11].

At room temperature, time dependence is generally negligible, and the material behaves almost as if it has a constant volume ($\nu \approx 0.5$). A good approximation can therefore be described with the aid of hyperelasticity. However, if after initial loading, the silicone is released, it will follow a different path from that during the loading. If it is loaded again, it follows the unloading path of the previous cycle up to the maximum initial load and then it continues along the initial load path. This (temporally reversible) stress softening is called the *Mullins effect* [2] and is illustrated in Fig. 8.6b.

Silicones in glass bonding applications are applied with a minimum thickness of 6 mm. Due to their long track record, they have become a standard structural adhesive in glass construction. The design of silicone-bonded glass façades is explained in Sect. 8.5. In addition, silicones are used as secondary sealing stages in insulating glass units and weather seals, especially for bonded glass façades.

8.2.7 Polysulphide

Polysulphides are used in addition to PUR and silicone as secondary stage sealants in insulating glass units. They are usually two-component adhesives and cure to an elastomer. The brand name Thiokol® is often used synonymously for polysulphide sealants.

8.2.8 Polyisobutylene

Polyisobutylene (PIB) is used as a primary stage sealant in insulating glass units. The common term for PIB is butyl. The term butyl rubber is commonly known, but it does correspond to the products used in insulating glass units. The primary sealing stages are amorphous thermoplastics with a plastic sticky consistency and are not elastomers.

PIB is characterised by a very low water vapour permeability and gas permeation, which are significantly superior to those of the secondary sealing stage.

8.3 Seals

Seals in a glass joint can be in the form of a dry seal (gasket) or wet seal (sealant) and have the task of permanently protecting the glass and frame against the ingress of air and moisture while simultaneously preventing sound transmission and heat

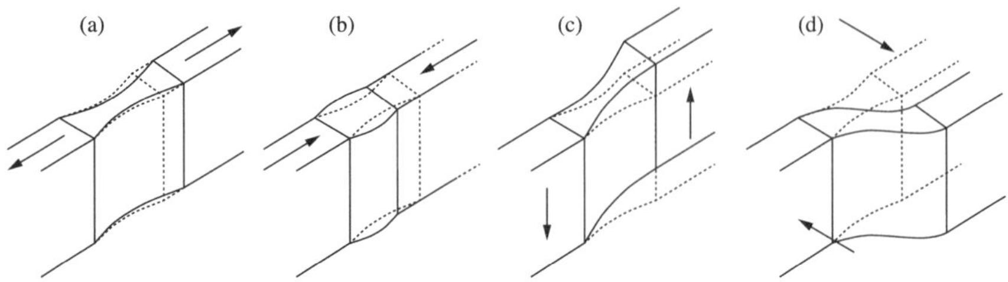

Fig. 8.7 Movements of joint seals (according to [9]): **a** elongation, **b** compression and **c**, **d** shear

transfer. The joint must also accommodate component and substructure movements, temperature expansions and moisture-related dimensional changes. Seals are generally exposed to the environment (e.g. gases, water vapour, UV radiation, heat and cold, etc.), chemical elements (e.g. acids, alkalines, solvents, etc.) and mechanical effects (e.g. change of shape and length, vibration, etc.). Therefore seals must, on the one hand, allow the movements of glass and the substructure that result from the effects of wind loads, movements, vibration and low-frequency oscillation, and on the other hand they must effectively prevent the flow of moisture and air. The sealing materials do not bear permanent loads e.g. support the self-weight of glass panes. The magnitude of differential movement between the glass and the substructure must be calculated explicitly to inform the design of the joint. For example, in practical applications, temperature fluctuations in a range from $-20\,°C$ until $60\,°C$ and thus temperature differences up to $\Delta T = 80\,K$ must be assumed. In case of highly heat-absorbing glasses, the body temperature of the glass can exceed $100\,°C$ in the presence of strong solar radiation. In this respect, a distinction is made between expansion/compression movements perpendicular to the longitudinal axis of the joint, shear movements perpendicular to the component surface and shear movements parallel to the longitudinal axis of the joint, see Fig. 8.7. These movements can also occur in combinations. Seals in glass construction are often considered as maintenance joints, whose performance must be checked at regular intervals and, if necessary, renewed to ensure functionality (e.g. prevent leakage due to seal embrittlement). Furthermore, care must be taken to ensure the chemical compatibility of all materials that come into contact (e.g. to polymer interlayers of laminated safety glass or edge sealing of insulating glass units) or to avoid contact between different incompatible materials, see Sect. 8.4.

Seals in butt joints and expansion joints should be at least 4 mm wide, but final dimensions of the joint is usually only possible after an exact estimation of all deformations of the supporting structure, see Fig. 8.8. For wet seals, reference values for the depth d_F can be determined as a function of the seal width a_F using the relationship:

$$d_F \approx \frac{a_F}{2} \quad \text{for } d_F \geq 6\,\text{mm} \tag{8.2}$$

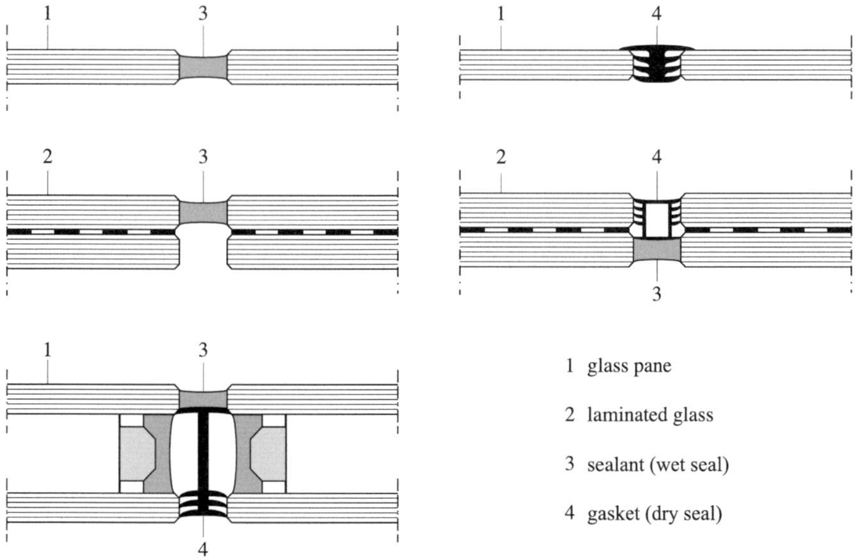

Fig. 8.8 Illustration of joint seals

In order not to constrain the necessary deformation behaviour of the sealants, they should only adhere to the two joint faces. Additional adhesion to the base of the joint (three-sided adhesion) or triangular joints must be avoided, to allow the deformability of the sealant which would otherwise be greatly reduced due to its high Poisson's ratio ($\nu \approx 0.5$) which leads to reduced durability. To avoid incorrect depths of the sealant and three-sided adhesion, a backing tape, backer rod or other backfill materials can be used, see Fig. 8.9. The backfill material should have an even, preferably convex boundary, be chemically compatible with the sealant, see Sect. 8.4, and waterproof. At the same time it should not interfere with desired deformability of the sealant and should not contain any substances which could impair adhesion. In practice, a closed-cell, rot-resistant, full round polyethylene foam rod is suitable for most sealant applications. The backfill material must not be damaged during installation and must be installed in a compressed state to ensure sufficient resistance when applying and smoothing the sealant. For this purpose, it is recommended to select the diameter of the backing rod material by a quarter to a third larger than the joint width. For joints with a shallow joint depth, polyethylene foils or a material equivalent in function and compatibility can be used to prevent three-face adhesion [37]. For all butt joints, the water drainage must be ensured by constructing the joint in such a way that water does not accumulate or become entrapped.

Sealants differ in their number of components.

One-component sealants are delivered ready for application. Multi-component sealants consist of at least two different components, which are mixed according to the manufacturer's instructions before application. Various materials such as silicone, polysulfides, polyurethanes or acrylic dispersions are suitable as sealants, see Sect. 8.2. In addition to wet sealing, where the sealants are inserted into the joints

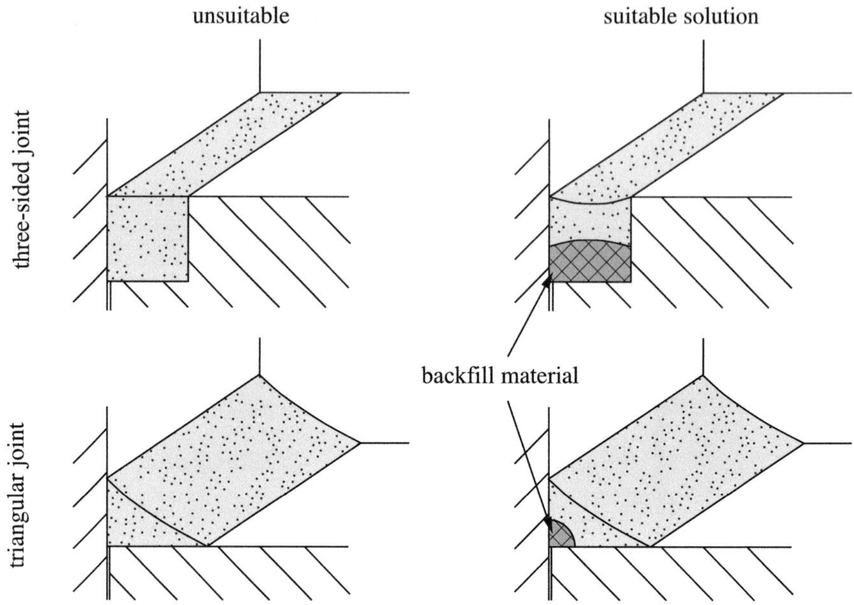

Fig. 8.9 Typology of joint seals when using backfill material to avoid three-sided adhesion, according to [9]

as an injectable or malleable mass, dry sealing with prefabricated silicone or rubber gaskets or combinations of the two sealing techniques are commonly used. However, joint sealing with materials whose UV resistance is not guaranteed, such as products based on polysulfides, may only be used in areas that are not exposed to direct sunlight.

Edge seals of insulating glass units, see Sect. 3.2, may be damaged if they come into contact with incompatible materials. Sealant materials usually contain plasticizers which can damage the edge compound of the insulating glass unit, if it comes into direct contact, thereby destroying the edge seal and causing fogging in the space between the panes which can damage the functional coating.

Similar compatibility issues arise between the wet or dry seals and the interlayers used in laminated glass. Compatibility must be checked with the manufacturers and is often proven by testing as described in 8.4.

Simple glazing, as shown in Fig. 8.10, is fixed within the frame structure preferably by means of glazing beads . Their support width, c, should be sufficiently large, e.g. 14 mm or more, when using a wooden frame. The glazing rebate is sealed with sealants to prevent water from penetrating into the glazing rebate.

The height of the glass rebate should typically be 20 mm or more. The glass bite is generally about 2/3 of the entire rebate height, although the glass inset

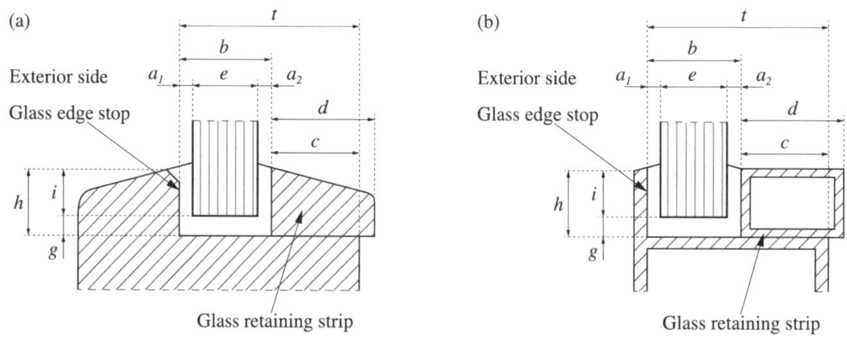

Fig. 8.10 Glazing with glazing beads according to DIN 18545 [41]; a_1 outer sealant thickness, a_2 inner sealant thickness, b glass rebate width, c support width of the glass bead, d width of the glass bead, e thickness of the glazing unit, g height of the glass rebate base, h glass rebate height, i glass bite, t total rebate width

should be less than 20 mm, to avoid thermally induced stresses at the edge of the pane. Depending on the length of the glass edge and the material of the frame, the sealant thickness should be between 3 and 6 mm; the height should be greater than 5 mm. In rooms where high humidity is to be expected, because of the risk of condensation collecting in the glass rebate, additional sealing films are bonded onto the glass and the base of the rebate. When using brackets for fixing the glass beads, the distance of the brackets should be less than 35 cm, the distance from the corners should be roughly 5 to 10 cm.

8.4 Compatibility and Durability

An engineered material must fulfil its assigned functions until the end of its service life. The following principal durability related questions arise when using polymers in construction, namely:

- Are the polymer materials used compatible with each other?
- How do the applied polymers age under given environmental influences and how does that affect their performance?

Figure 8.11a illustrates the diversity of polymers in a simple glass façade. It is evident that many different polymers are in contact with each other. Thus, a reaction can take place, which is usually of physical nature (material transport through diffusion processes) and less often of chemical nature (chemical substance conversion as a result of the diffusion processes). The consequences are usually a change in volume

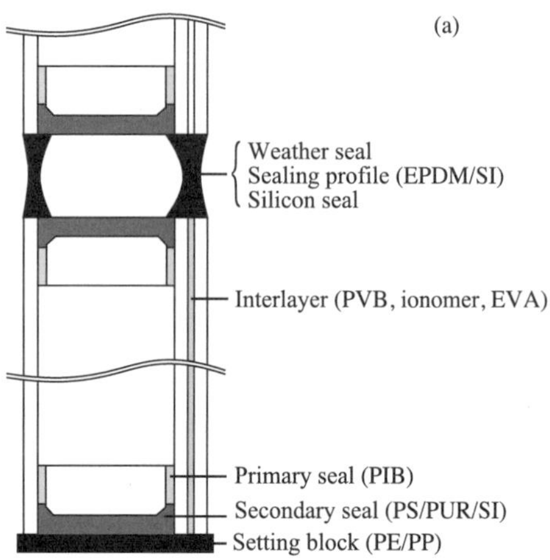

Fig. 8.11 **a** Polymers in façade glazing (according to [35]); **b** Butyl runners as a result of material incompatibilities (credit: Kömmerling)

and material properties such as stiffness, strength and viscosity [35]. In particular, the migration of plasticizers can cause damage, ranging from optical deficiencies to loss of load capacity. A typical damage pattern for insulating glass units is called butyl runners, as shown in Fig. 8.11b.

The compatibility of sealants in insulating glass units and laminated glass can be tested according to guidelines ift Richtlinie DI-01/1 [7] and ift Richtlinie DI-02/1 [12], or standard ASTM C1087-16 [48]. Tests on the compatibility of materials in contact with silicone bonding are proposed in ETAG 002 [21].

According to DIN 50035 [18], the term ageing can be defined as the totality of all chemical and physical processes irreversibly occurring in a material over time. These can be traced back to various causes, such as mechanical or thermal stress, radiation or oxidation. The consequences are, among others, discoloration, crack formation, volume change, leaching and changes in mechanical properties [19]. Therefore, an accelerated ageing test for polymers cannot be generally valid for all possible polymers, due to the different relevance of various influences. There are approaches for some applications, such as bonded glazing, see Sect. 8.5, or laminated glass EN ISO 12543-4 [16]. Overall, the determination of realistic ageing methods for the long design periods in the construction industry is still subject of ongoing research.

8.5 Bonded Glazing

Due to the architectural desire for unobstructed views through façades, there has been a worldwide trend since the 1980s towards bonded all-glass façades in which

Fig. 8.12 Bonded glazing at the *ETA-Fabrik* at the TU Darmstadt campus (Figure credits: HA Hessen Agentur GmbH, Jan Hosan)

the supporting structure is not visible from the outside. This is commonly known as bonded glazing or Structural-Sealant-Glazing-Systems (SSGS) according to ETAG 002 [21]. In bonded glazing, see Fig. 8.12, the bonding has a load-bearing function. In principle, all adhesives according to Sect. 8.2 are suitable for this, as far as sufficient durability and load-bearing capacity can be achieved. However, in practice, this so far only applies to UV-stable two-component silicones, see Sect. 8.2.6. The authoritative set of rules for SSGS façades is ETAG 002 [21], which, among other things, sets requirements for the construction, the mechanical fasteners and the adhesive with regard to load-bearing capacity, durability and compatibility. In addition, information is also provided on the dimensioning of the adhesive joint. Based on [21], an European Technical Approval (ETA) can be issued for adhesives or façade systems. The intended service life is 25 years.

Basically, glazed façade constructions are divided into the four types according to Fig. 8.13. However, not all four types are permitted in every country. For instance, in Germany only systems with mechanical self-weight support are permitted, see Type I and II in Fig. 8.13; a permanent stress on the adhesive joint is thus not permitted. From an installation height of 8 m above ground, mechanical retention devices must also be provided to resist wind suction in the event of bond failure, see Type I in Fig. 8.13 and 8.14, [40]. In addition to the vertical façades, a slightly inclined installation is permissible. In general, the glazing may be installed with an inclination to the vertical of up to 5° outwards and up to 10° inwards [40]. However, in individual cases an inward slope of the glazing up to 20° may be permitted.

The adherends regulated in ETAG 002 [21] are for bonding glass to anodised aluminium or stainless steel. The continuous silicone adhesive joint has a rectangular cross-section, widths between 6 mm $\leq a_F \leq$ 20 mm and a minimum height of $d_F \geq$ 4 mm. Cross-sectional dimensions of $a_F \times d_F$ = 12 mm × 6 mm are common. For approved silicone's, design stresses for dimensioning the adhesive joint are specified in their ETA [10, 28].

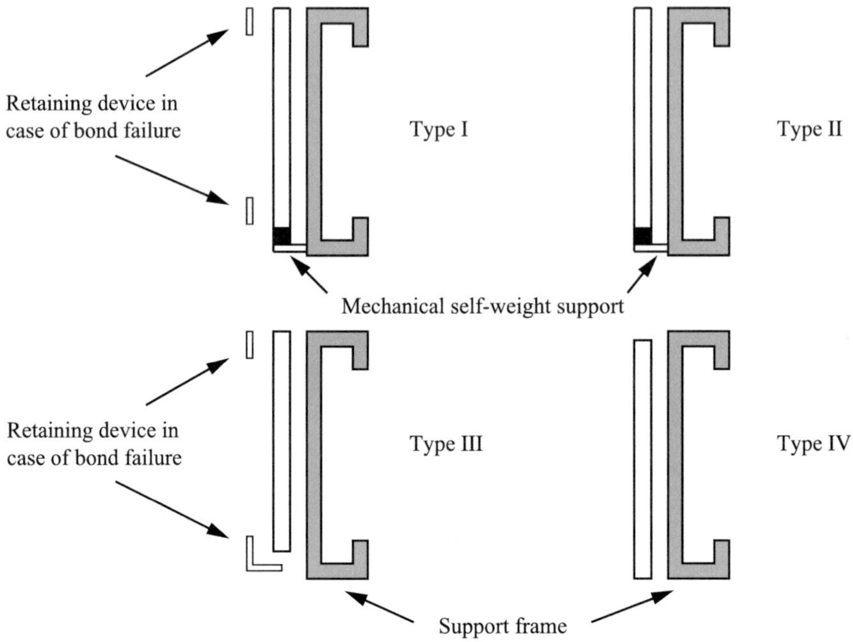

Fig. 8.13 Support configurations according to ETAG 002 [21]

Fig. 8.14 Structural-Sealant-Glazing-Façade with a mechanical retaining device at the outer pane

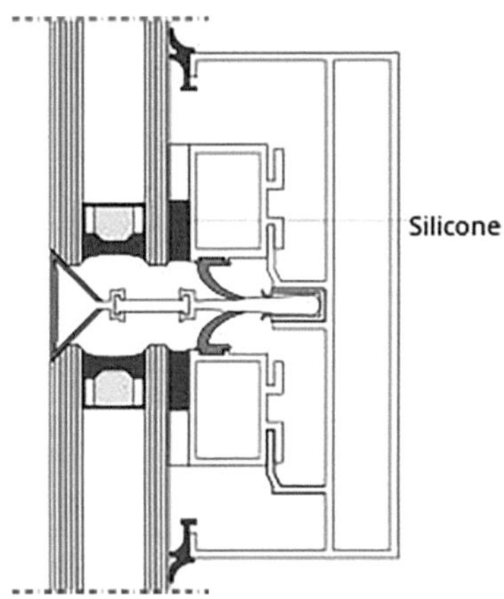

The dimensioning is carried out using a simplified approach in which the stress in the adhesive joint is determined by assuming a uniform stress distribution of the forces from the glass pane. Other essential aspects of the mechanical behaviour are neglected, for example the hyperelastic behaviour of the silicone adhesive. Among other things, this results in a high safety factor, which is $\gamma_{Q,glob} = 6$ for variable loads and $\gamma_{G,glob} = 60$ for permanent loads. The design approach is based on the now obsolete global safety concept (also known as the permissible strengh approach).

Accordingly, characteristic values of the actions without a partial safety factor are to be used in this calculation. This results in permissible stresses perpendicular and parallel to the adhesive bond of $\sigma_{adm} = 0.14$ MPa and $\tau_{adm} = 0.11$ MPa. These values are then used to determine the width a_F of the adhesive joint, as follows:

$$a_F \geq \left| \frac{a \cdot q_k}{2\sigma_{adm}} \right| \tag{8.3}$$

with

q_k is the characteristic load value,
a is the short edge of the glass panel.

The adhesive joint height d_F is determined from:

$$d_F \geq \left| \frac{G \cdot \Delta}{\tau_{adm}} \right| \tag{8.4}$$

with the shear modulus of the silicone G and the different thermal expansion of the parts to be joined Δ:
(a) The short edge a is supported (glazing is installed in portrait format):

$$\Delta = \left[(T_c - T_0)\alpha_{T,c} - (T_v - T_0)\alpha_{T,v}\right]\sqrt{(a/2)^2 + b^2} \tag{8.5}$$

(b) The long edge b is supported (glazing is installed in transverse format):

$$\Delta = \left[(T_c - T_0)\alpha_{T,c} - (T_v - T_0)\alpha_{T,v}\right]\sqrt{a^2 + (b/2)^2} \tag{8.6}$$

In each case this includes

T_c temperature of the metal frame when installed (typically 55 °C);
T_v temperature of the glass when installed (typically 80 °C);
T_0 temperature of all components during the bonding process (typically 20 °C);
$\alpha_{T,c}$ Coefficient of thermal expansion of the metal frame;
$\alpha_{T,v}$ Coefficient of thermal expansion of the glass.

The ratio between adhesive joint height and width (also known as the aspect ratio) should be in the range of:

$$d_F \leq a_F \leq 3d_F \tag{8.7}$$

The secondary sealing level in insulating glass units can be dimensioned analogously according to the load distribution of the glass panes. In addition, the deflection of the substructure is limited to l/300 and the deflection of the glass to l/100. The dimensioning of the adhesive bond according to ETAG 002 [21] provides conservative values due to the simplifications made. A more material-efficient verification requires a detailed consideration of the bonded joint. Simulation with the aid of FEM is a possible approach for this, whereby the selection of a suitable and validated material model is required in addition to a sufficiently refined mesh. Hyperelastic models are often well suited for silicones. However, the question of a suitable failure criterion for silicones has not yet been conclusively answered (See Sect. 4.1). In particular, local stress peaks (e.g. at corners) lead to difficulties in the evaluation of the results.

The visually obtrusive retention devices required for bonded façades in Germany limits the use of these façades. This is in contrast to most other European countries and the USA where bonded façades are commonly used. Various approaches are currently being pursued in Germany to dispense with retaining devices and still demonstrate sufficient safety of the construction during its service life. One important aspect is to carry out accelerated ageing (through UV radiation, water, surfactants, salts, etc.) under simultaneous mechanical load, which has not yet been taken into account in [21].

8.6 Other Bonded Connections

8.6.1 Bonded Point Supports

Adhesives or foils could also be used to produce point supports, as an alternative to mechanical point fittings. The benefit of doing so is that coring/drilling of the glass and the associated weakening are completely avoided. High demands are placed on the adhesives used in this manner. In particular, stiffness, strength and durability. For aesthetic reasons, transparent acrylics and epoxy resins are normally used for these applications. For vertical glazing, there may be additional stresses arising from the self-weight of the glazing that must also be resisted by the adhesive joint, which leads to a complex state of stress in the adhesive. Normally, the adhesive is not exposed to external environmental conditions due to concerns about the long term degradation of the adhesive.

Adhesive foils such as ionomer interlayer, or the transparent structural silicone adhesive (TSSA) foil (Dow®) (no longer commercially available) have been used in a limited number of case-studies for creating point supports, but their use has not been adopted more widely (Fig. 8.15).

When applying such connections on e.g. thermally pre-stressed glass, it is important to notice that the strength increasing residual stresses in the thermally pre-stressed glass are in-plane and that the same strength cannot be expected for local out-of-plane loading. Experiments have demonstrated that for a tensile load normal to the glass surface (part of) the compressive zone is simply peeled off (also known as “thin layer failure” or “plucking”) [8].

Fig. 8.15 **a** Adhesively bonded point fitting with Dow Corning®TSSA (credits: Sadev); **b** Embedded laminated point fitting with SentryGlas®

8.6.2 Embedded Connections

An alternative to bonding a connector onto the surface of a glass pane, as done for bonded point supports, the connectors can instead be integrated between the plies of a glass laminate, resulting in so-called embedded connections, see e.g. Fig. 8.15b. Typically, metal inserts (e.g. stainless steel or titanium) are used for creating embedded connections. The metal inserts can be bonded between the glass plies either by means of interlayer foils (e.g. ionomer or stiff PVB), or cast-in-place resins.

There are two types of embedded connections:

- *embedded connections with thin inserts*, where (metal) sheet connectors with a thickness in the range of 1–2 mm are embedded into a stack of foil interlayers or into a cast-in-place resin between the glass plies. In this type, the insert is accommodated within the thickness of the interlayer.
- *embedded connections with thick inserts*, where the inner glass ply (or plies) of a glass laminate is locally recessed to accommodate a (metal) connector of equal nominal thickness, e.g. in the range of 5 to 15 mm, which is bonded/laminated to the exterior plies. In this type, the insert is accommodated within the thickness of a glass ply.

Embedded connections have been used successfully in several real-world projects, e.g. for connections in glass staircases (Fig. 8.15b), facades [43] and other assemblies [15, 31, 46], and their structural performance and durability is studied in several research projects, e.g. [42, 44, 45, 47].

8.6.3 Spliced Connections

In spliced connections, multiple plies of glass are overlapped and bonded to create a shear-transferring connection. There are two basic types of spliced connections:

- *Spliced column-beam connections*, where a mortise and tenon joint is created between a laminated glass column and beam by overlapping the protruding outer plies of one laminated glass component with the protruding inner plies of the other laminated glass component. The cavity between the overlapping glass layers is typically filled with a transparent (UV-curing) adhesive or cast-in-place resin and subsequently cured to generate a bond between the glass components.
- *Splice-laminated glass components*, where multiple segments of glass are overlapped and laminated to create an oversized glass component that spans beyond the standard available length of glass panes. This technique is for instance used to create oversized glass beams, fins or façade panels. However, since the glass industry is gradually increasing the available lengths of glass panes, the splice-laminating technique is becoming less relevant.

8.6.4 All-Glass Corners

All-glass corners refer to specific corner configurations in glass façades without corner mullions, see Figs. 8.16 and 8.17. Aesthetically, this corner detail offers an emphasis on the lightness and transparency of the construction and a relatively unobstructed view. However, this type of corner detail is not free of static and structural challenges. The seemingly complete transparency of the building, which is suggested in architectural models, is only apparent when tall (e.g. floor-to-ceiling) glazing is used. Therefore, the all-glass corner should only be used where the advantages (transparency, usefulness of the all-glass corners as a design element) outweigh the disadvantages (structural issues, costs, execution, thermal performance).

The corners of a building are exposed to almost twice the wind load in comparison to the central area of a façade (EN 1991-1-4 [14]). These wind peak pressures generally govern the design of the glass panes. For protruding building elements it may be necessary to take even higher wind peak pressures into account. With suspended, ventilated façades, a reduction of the wind loads can be achieved by structural measures (e.g. by creating a compartment at the façade corners). Joints at the corner should be designed in such a way that the movements of the corner panes can be safely accommodated by the sealing material.

In contrast to bonded glazing where bonding is carried out on the glass surface near the edge, all-glass corners in insulating glass units are also bonded directly on the edge of the glass. The statically effective bonding of a glass edge with a glass surface is not yet regulated within the existing building standards. Special corner

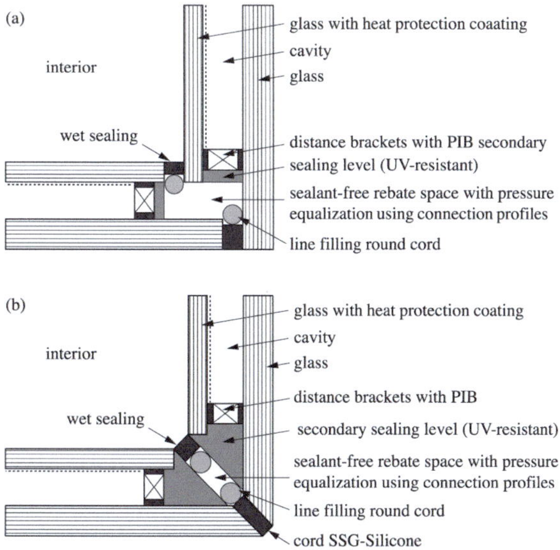

Fig. 8.16 Design of glass corners with **a** stepped insulating glass units and **b** mitre-cut insulating glass units

Fig. 8.17 All-glass corner on the façade *Three Pacific Place* in Hong Kong (architects: Wong and Ouyang; Engineers: Schlaich Bergermann and Partner)

solutions must therefore be approved on a project-by-project basis. Therefore a three sided bracket must be used to mount the glazing.

In addition, an all-glass corner creates a thermal bridge that can only be reduced with a sealant-free rebate space. It is important to compensate for the vapour pressure, for example, via the upper or lower profile to avoid condensation. A complete filling of the corner joint with silicone leads to an increased heat transfer. Figure 8.16 shows some design details for the execution of continuously mounted all-glass corners with insulating glass units. A version with stepped insulating glass units is generally more cost-effective and less sensitive to construction tolerances than mitre-cut glass edges.

Alternatively, all-glass corners can also be achieved by using point-fittings to hold the glass. Instead of corner profiles, perpendicular point fittings could be used to transfer horizontal and vertical loads to the substructure. By using point fittings, the design of all-glass corners is simplified, since the weight of the panes and the wind loads are transferred via the point fittings to the substructure. In this way, the corner of the façade design hardly differs from the main area and even complex geometries can be achieved by cutting the glass panes.

References

1. O. Volkersen, Die Schubkraftverteilung in Leim-, Niet und Bolzenverbindungen. Energie und Technik S. 68–71, 103–108 und 150–154 (1958)
2. J.A.C. Harwood, L. Mullins, A.R. Payne, Stress softening of natural rubber vulcanizates. Part II. Stress softening effects in pure gum and filler loaded rubbers. J. Appl. Polym. Sci. **9**, 3011–3021 (1965)
3. L.J. Hart-Smith, Adhesive-Bonded Single-Lap Joints. NASA CR 112236 (1973)
4. R. Adams, J. Comyn, W. Wake, *Structural Adhesive Joints in Engineering*, 2nd ed. eng (Springer, 1997), p. 372
5. D.E. Packham, *Handbook of Adhesion*, 2nd edn. (Wiley, New York, 2005)
6. T. Schadow, Beanspruchungsgerechtes Konstruieren von Klebverbindungen in Glastragwerken. Dissertation. Technische Universität Dresden (2006)
7. ift Richtlinie DI-01/1. *Verwendbarkeit von Dichtstoffen—Teil 1: Prüfung von Materialien in Kontakt mit dem Isolierglas-Randverbund* (2008)
8. A.B. Ølgaard, J.H. Nielsen, J.F. Olesen, H. Stang, Properties of an adhesive for structural glass applications, in *Challenging Glass—Conference on Architectural and Structural Applications of Glass*, pp. 263–272 (2008)
9. BIV TR 1. *Bundesinnungsverband des Glaserhandwerks: Technische Richtlinie des Glaserhandwerks—Nr. 1: Dichtstoffe für Verglasungen und Anschlussfugen, Bundesinnungsverband des Glaserhandwerks* (2009)
10. DIBt. *Europäische Technische Zulassung ETA-03/0038: Sikasil SG 500* (2009)
11. G. Habenicht, *Kleben: Grundlagen, Technologien, Anwendungen*. 6th ed. (Springer, 2009)
12. ift Richtlinie DI-02/1. *Verwendbarkeit von Dichtstoffen—Teil 2: Prüfung von Materialien in Kontakt mit der Kante von Verbund- und Verbundsicherheitsglas* (2009)
13. L.F.M. da Silva, P.J.C. das Neves, R.D. Adams, J.K. Spelt, Analytical models of adhesively bonded joints—Part I: Literature survey. Int. J. Adhesion Adhesives **29.3**, 319–330 (2009)
14. EN 1991-1-4, *National Annex—Nationally determined parameters—Eurocode 1: Actions on structures—Part 1-4: General actions—Wind actions* (2010)
15. J. O'Callaghan, Thinking big with structural glass, in *In The Future Envelope 3* (2010)
16. EN ISO 12543-4. *Glass in building—Laminated glass and laminated safety glass—Part 4: Test methods for durability* (2011)
17. B. Weller, M. Kothe, F. Nicklisch, T. Schadow, S. Tasche, I. Vogt, J. Wünsch, Kleben im konstruktiven Glasbau, in *Stahlbau Kalender 2011*. Ernst & Sohn, Chap. 10, pp. 585–646 (2011)
18. DIN 50035. *Begriffe auf dem Gebiet der Alterung von Materialien—Polymere Werkstoffe* (2012)
19. H. Domininghaus, *Kunststoffe—Eigenschaften und Anwendungen*, 7th edn. (Springer, Berlin, 2012)
20. Dow Corning. *Dow Corning 983 Structural Glazing Sealant–Product Information* (2012)
21. ETAG 002. *Guideline for European technical approval for structural sealant glazing kits (SSGK)—Part 1: Supported and unsupported systems, EOTA* (2012)
22. G. Habenicht. *Kleben—Erfolgreich und Fehlerfrei*, 6th ed. Vieweg+Teubner (2012)
23. Kömmerling. *Isolierglas-Dichtstoff GD 116, Product data sheet* (2012)

24. PolySpec. *Thiokol 2282, Product data sheet* (2012)
25. M. Rasche, *Handbuch Klebtechnik* (Carl Hanser Verlag, München, 2012)
26. Sika. *Sikasil SG-500, Product data sheet* (2012)
27. Soudal. *Soudaseal 270 HS, Product data sheet* (2012)
28. UBAtc. *European Technical Approval ETA-01/0005: DC 993 and DC 895* (2012)
29. Dow Corning. *Produktbroschüre: Dow Corning TSSA–Transparent Structural Silicone Adhesive* (2013)
30. Kömmerling. *Ködiglaze S, Product data sheet* (2013)
31. P. Lenk, F. Lancaster, Connections in structural glass, in *Glass Performance Days* (2013)
32. Sika. *SikaFast-5211 NT, Product data sheet* (2013)
33. Sika. *Sikaflex-521 UV, Product data sheet* (2013)
34. Y. Staudt, *Untersuchung des Materialverhaltens von Klebeverbindungen mit Silikon* (Technische Universität Darmstadt, MSc-Thesis, 2013)
35. W. Wittwer, Verträglichkeit von Kleb- und Dichtstoffen in der Fassade: Grundlagen, Prüfmethoden und Strategien, in *2. Darmstädter Ingenieurkongress*, Darmstadt, pp. 389–393 (2013)
36. Delo. *Delo-Photobond 4468, Product data sheet* (2014)
37. IVD. *Merkblatt Nr. 22—Anschlussfugen im Stahl- und Aluminium-Fassadenbau sowie konstruktiven Glasbau, Industrieverband Dichtstoffe e.V.* (2014)
38. Kömmerling. *Körapur 140, Product data sheet* (2014)
39. Delo. *Delo-Duopox AD821, Product data sheet* (2015)
40. DIBt. *Allgemeine bauaufsichtliche Zulassung Z-70.1-46: Geklebte Verglasungen Fa. Schüco* (2015)
41. DIN 18545. *Abdichten von Verglasungen mit Dichtstoffen—Anforderungen an Glasfalze und Verglasungssysteme* (2015)
42. A. Nussbaumer, C. Louter, M. Santarsiero, Laminated connections for structural glass components: a full-scale experimental study. eng., in *Glass Structures and Engineering*, vol. 2.1, pp. 79–101 (2017). https://doi.org/10.1007/s40940-016-0033-2
43. C. Teixidor, J. Torres, N. Guitart, Glass fins with embedded titanium inserts for the façades of the new Medical School of Montpellier. eng., in *Glass Structures and Engineering*, vol. 2.2, pp. 201–219 (2017). https://doi.org/10.1007/s40940-017-0049-2
44. C. Bedon, M. Santarsiero, Transparency in structural glass systems via mechanical, adhesive, and laminated connections—existing research and developments. eng., in *Advanced Engineering Materials*, vol. 20.5, p. 1700815 (2018)
45. E. Volakos, C. Davis, M. Teich, P. Lenk, M. Overend, Structural performance of a novel liquid-laminated embedded connection for glass, in *Glass Structures and Engineering*, vol. 6.4, pp. 487–510 (2021). https://doi.org/10.1007/s40940-021-00162-w
46. T. Fildhuth, P. Joos, T. Wüest, M. Haller, W. Stevels, Development and behavior of a thin fitting connection for lamination with structural PVB. eng., in *Challenging Glass Conference Proceedings* (2022)
47. J. Hänig, B. Weller, Integrated connections for glass-plastic-composite panels: an experimental study under tensile loading at +23, +40 and +60°C and different glass build-ups, in *Glass Structures and Engineering*, vol. 7.2, pp. 211–229 (2022). https://doi.org/10.1007/s40940-022-00174-0
48. ASTM C1087-16. *Standard Test Method for Determining Compatibility of Liquid-Applied Sealants with Accessories Used in Structural Glazing Systems* (2023)

Chapter 9
Glass Design Aspects

This Chapter covers various aspects that are important for designing glass components and structures. After an introduction in Sect. 9.1, general safety concepts and the salient features of current glass design standards used in different countries (including the European glass code CEN/TS 19100 "Design of glass structures") are described in Sect. 9.2. This is followed by key aspects, which have been addressed at a fundamental level in previous chapters, and are here implemented in practical engineering design contexts. In particular: designing with time dependent strength; designing with temperature-induced and deflection-induced loading; designing for blast and impact loading and structural analysis of plates, see Sects. 9.3, 9.4, 9.5 and 9.6, respectively. Also the design of specific glass components, including laminated glass and insulating glass units are discussed, including a section on edge protection, see Sects. 9.7, 9.8 and 9.9, respectively. Finally, Sect. 9.10 provides an insight and some design guidance on the environmental sustainability of glass.

9.1 General

Load bearing structures must meet several requirements simultaneously. Therefore, depending on the type of structure and its intended function, various aspects, such as:

- safety;
- serviceability;
- aesthetics;
- durability and
- efficiency

J. Belis et al., *Engineering Glass Structures*,
https://doi.org/10.1007/978-3-032-03233-1_9

need to be considered in the design process. The basic requirement is safety. This is particularly important due the inherent brittle nature of glass and the potentially high consequence of human injury in the event of fracture.

A good design should strive to meet all the performance requirements, but it may not be possible to satisfy all requirements simultaneously to their full extent. The design engineer is therefore responsible for setting priorities and designing accordingly. In all cases the safety aspect must not fall below an acceptable threshold.

Establishing this threshold is not trivial and requires careful consideration of all safety-relevant aspects at material, component and overall system levels. The risk, defined as the product of the failure probability and the associated consequences, is acceptable when it is sufficiently small for the individual and society.

It is difficult to define an acceptable risk level in an objective manner. For instance, an individual is quite willing to take relatively large risks in daily life, e.g. in playing sports, especially if the participant considers that the risks can be mitigated through their own actions. On the other hand, risks, even when demonstrably small, are not readily accepted if the individual feels a certain helplessness (e.g. fear of flying). One way of quantifying the permissible failure probability occurring and the permissible risk is provided by the (semi-)probabilistic approach developed in the past decades, which have since been enshrined in engineering design standards for the construction industry.

The implementation of the above-mentioned definition of risk as the product of the failure probability and the extent of failure leads to a number of rather complicated circumstances. The practising engineer can usually avoid complex risk analyses by referring to the design rules laid down in standards and regulations. However, the semi-probabilistic approach adopted by the Eurocodes still requires a significant effort by the engineer to identify the relevant factors for actions and resistances. The process has been simplified by the use of software, but it has also become more difficult for the engineer to validate the results, and results from computational methods often suggest a false level of accuracy which is not necessarily achieved in practise. For instance, the statistical data for the actions is generally rather poor. In particular, the determination of the distribution functions and the derivation of the associated fractiles are very questionable, especially in glass construction. The reason for this deficiency is that many influences, which can lead to glass fracture, are rarely recorded statistically. Examples of this are the effects of local impacts accumulated during the service life or micro-cracks arising during the production process which remain undetected. Therefore, a degree of post-fracture performance is essential for most applications.

This chapter will first introduce the fundamental design concepts considered in design and will explain how they are embedded in existing building regulations for the design of glass components. Further details are then provided on how salient factors, such as time and temperature and other actions (e.g. blast and impact loading) are considered in design of glass components. Finally, the chapter provides an overview of emerging methods for assessing the environmental sustainability of glass components, which ensures that glass is designed in a safe and sustainable manner.

Further information on the design of glass components can be found in literature [45, 47, 74, 83].

9.2 Design Concepts

9.2.1 General Safety Concepts in Construction

The topic of safety assessment in building designs has already been introduced in Section 9.1, including the definition of the anticipated risk as the product of the failure probability and the consequences of failure. However, selecting an acceptable level of risk is problematic, because it implicitly puts a value on human life and human injury. Therefore, the safety requirements in the Eurocode (EN 1990) are based on the various damage consequences of directly permissible failure probabilities. The failure consequence classes shown in Table 9.1 correspond to the reliability classes in Table 9.2. Unfortunately, the classes are not consistently implemented under building law. Furthermore, the low probabilities in Table 9.2 implicitly require a large quantity of statistically significant data (e.g. data on impact actions on glass). But there is a lack of such data for glass construction, which raises some questions on the validity and accuracy of the statistical methods used in contemporary standards.

Therefore, instead of reducing the probability of failure, it is advisable to also mitigate the risk by limiting the consequence of damage; i.e. to limit the consequence of failure with expert knowledge and human judgement, for example by means of fail-safe design features. The sensitivities of the various factors can be evaluated through risk assessment. The same procedure is used if post-fracture capacity or redundancy is required in glass construction, and glass types such as thermally prestressed or laminated safety glass are prescribed according to the type of design

Table 9.1 Damage consequences classes according to EN 1990

Damage consequence classes	Description	Examples
CC 1	Minor consequences for loss of human life and small or negligible economic, social or environmental consequences	Agricultural buildings without regular passenger traffic (e.g. barns, greenhouses)
CC 2	Medium-scale consequences for loss of human life, economic, social or environmental impact	Residential and office Building, public buildings with mediumscale impact (e.g. office buildings)
CC 3	Major consequences for loss of human life or serious economic, social or environmental consequences	Grandstands, public buildings with high failure consequences (e.g. concert halls)

Table 9.2 Reliability categories for ULS according to EN 1990

Reliability category	**Probability of failure P_f (Reliability index β)**	
	Reference period 1 year	Reference period 50 years
RC 1	10^{-5} (4.2)	$5 \cdot 10^{-4}$ (3.3)
RC 2	10^{-6} (4.7)[a]	$7 \cdot 10^{-5}$ (3.8)[b]
RC 3	10^{-7} (5.2)	$9 \cdot 10^{-6}$ (4.3)

[a]For the SLS applies: $2 \cdot 10^{-3}$ (2.9)
[b]Ror the SLS applies: $7 \cdot 10^{-2}$ (1.5)

and installation situation. Alternatively, more recent methods of uncertainty analysis provide a promising means of making the problem more accessible to engineers in future [70].

However, in spite of the various gaps and limitations of probabilistic approaches in glass design, the idea of a purely deterministic approach is even more questionable and a probabilistic approach that takes the consequences of failure into account is in general the more appropriate approach.

Further detail on the both approaches is provided below.

Deterministic Design Concept

If the action on a glass component and the corresponding resistance each have a fixed (deterministic) value, it would be sufficient to ensure that the magnitude of the action does note exceed the magnitude of the resistance in order to guarantee safety (Fig. 9.2).

However, this simple deterministic approach fails to consider the inherent statistical scatter of both the actions and the resistance. All actions, for example, self-weight, snow loads or wind loads, are subject to scatter (aka dispersion or spread) caused by inherent variabilities in manufacturing or in exposure conditions. Similarly, the scatter in the resistance values can be caused, for example, by inherent variations in the material. Additionally, further discrepancies arise from approximations in analytical or computational modelling in the design process. One approach to account for scatter and discrepancies is to adopt a global safety factor approach wherein the resistance value is reduced or the action is increased.

This classic and somewhat outdated safety concept has a long track record in glass construction (e.g. TRLV in Germany). Compliance is achieved by ensuring that the permissible stress did not exceed the stress caused by the action:

$$\sigma_{\text{action}} \leq \sigma_{\text{permissible}} \,. \tag{9.1}$$

The deterministic safety concept is no longer used in most countries because it is superseded by more recent design concepts.

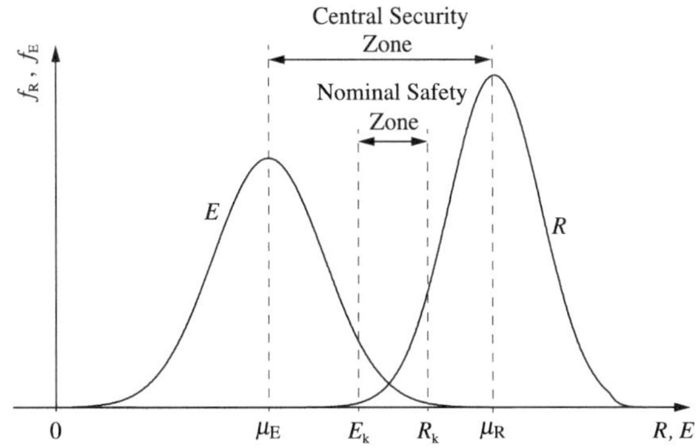

Fig. 9.1 Distribution function of effect of action E and resistance R

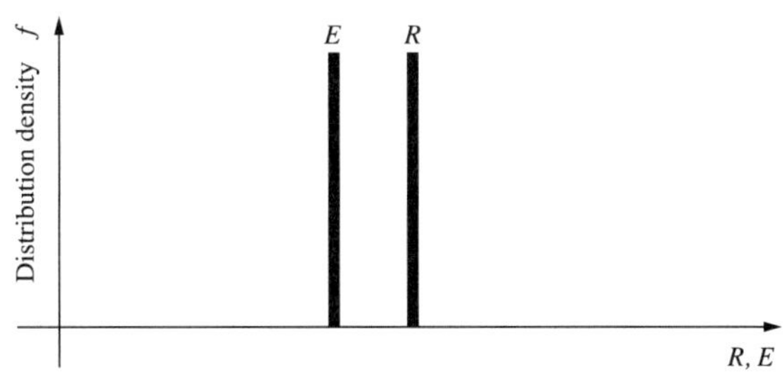

Fig. 9.2 Deterministic consideration $E < R$ (E: effect of the action, R: resistance)

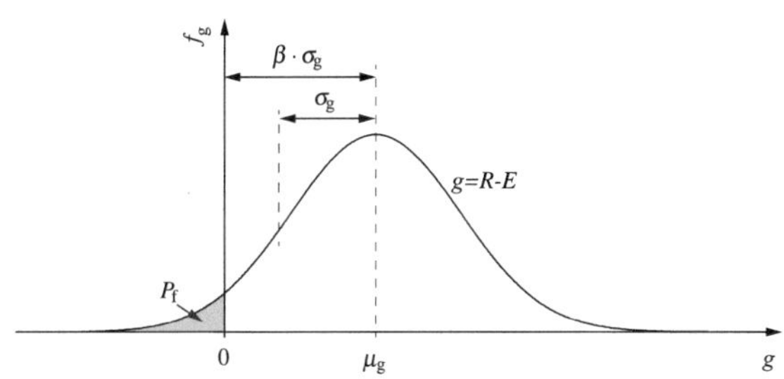

Fig. 9.3 Distribution Function $g = R - E$

Probabilistic Design Concept

Figure 9.1 shows a schematic plot of the scatters of the action and the resistance in the form of distribution densities. The two curves intersect, meaning that there are scenarios with a low probability wherein the resistance, R, is smaller than the action, E. Such scenarios must be limited to an acceptable failure probability P_f, according to Table 9.2, by providing a sufficient margin between the action and the resistance. This can be calculated with a convolution integral:

$$P_f = \int_{-\infty}^{+\infty} f_E(e) \left(\int_{-\infty}^{e} f_R(r)\, dr \right) de\,. \tag{9.2}$$

The resulting function $g = R - E$ is shown in Fig. 9.3.

Negative values of g indicate failure, so the probability P_{f} for this case must be kept to an acceptably low value. This is achieved by considering the the reliability index β of the distribution function g , such that, $\mu_{\mathrm{g}} = \beta \cdot \sigma_{\mathrm{g}}$, where σ_g is the standard deviation for g and σ_{g} is the mean. Values of β are given in Table 9.2. The required levels of β (or P_{f}), are achieved by applying safety coefficients (also known as partial factors) to the action and resistance. Thus specific characteristic values (also known as fractile or quantile values) for the action and resistances can be established.

As an example, at this point, action and resistance are assumed to be normally distributed random variables (see Sect. 5.4.3.2). This applies to the characteristic values of influence

$$E_{\mathrm{k}} = \mu_{\mathrm{E}}(1 + k_{\mathrm{E}} \cdot \sigma_{\mathrm{E}}) \tag{9.3}$$

and resistance

$$R_{\mathrm{k}} = \mu_{\mathrm{R}}(1 + k_{\mathrm{R}} \cdot \sigma_{\mathrm{R}})\,. \tag{9.4}$$

The factors k can be determined from the underlying fractile of the distribution function and are—when assuming the coverage of the population—specified for a normal distribution in Table 9.3. The limit state function g is then also distributed normally with the parameters

$$\mu_{\mathrm{g}} = \mu_{\mathrm{R}} - \mu_{\mathrm{E}} \tag{9.5}$$

and

$$\sigma_{\mathrm{g}} = \sqrt{\sigma_{\mathrm{R}}{}^2 + \sigma_{\mathrm{E}}{}^2}\,. \tag{9.6}$$

The reliability index, β, then becomes

$$\beta = \frac{\mu_{\mathrm{g}}}{\sigma_{\mathrm{g}}} \tag{9.7}$$

and the likelihood of failure to

$$P_{\mathrm{f}} = \Phi(-\beta) \tag{9.8}$$

with the cumulative distribution function of the standard normal distribution Φ.

In the partial factor approach, the action distribution is increased and the resistance distribution is reduced, thereby reducing the overlap between the two (Fig. 9.1) and achieving the desired reliability index β. This is done by introducing the sensitivity factors α_{E} and α_{R}, which account for the sensitivity of the distribution of the base

Table 9.3 Fractiles and related k values for the calculation of characteristic values, assuming normally distributed random variables

Fractile	99 %	98 %	95 %	50 %	5 %
k [−]	2.326	2.054	1.645	0.0	−1.645

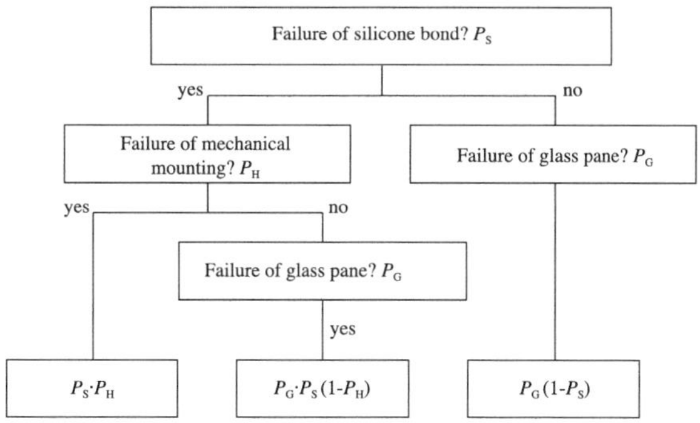

Fig. 9.4 Fault tree for bonded glass façades [18]

variables to the probability of failure. This leads directly to the partial safety factors for the effects of actions γ_E and the resistance γ_M. The general verification equation in the probabilistic design approach results in

$$E_d = \gamma_E \cdot E_k \leq R_d = \frac{R_k}{\gamma_M}. \tag{9.9}$$

In a real building, actions and resistances are not completely independent of each other and their inter-dependency cannot be described stochastically, therefore partial safety factors in standards and directives are determined through a practical, semiprobabilistic process. This is for example reflected in the definition of sensitivity factors as constant sizes. The partial safety coefficients only take into account the statistical uncertainties of the influences involved, such as a load or a material strength. Further information on the fundamentals of reliability theory is available in relevant specialised literature (e.g. [13, 15]).

For glass construction, the probabilistic design concept was first published by Shen [18]. Güsgen and Blank also proposed a probabilistic approach proof similar to the fatigue strength proof for steel structures [24].

However, for innovative designs or novel materials, it is still important to carefully consider the possible scenarios to ensure safety. Such a risk analysis (*fault tree*) is shown in Fig. 9.4 as an example for bonded glass façades. The overall probability of failure shall also meet the requirements of Table 9.2.

Challenges in the Implementation of Engineering Design Concepts in Practical Glass Construction

In real-world glass construction, several loads can be difficult to quantify. In practice, the most common causes of damage are thermal stresses, unforeseen constraints and impact loads. Examples of this are:

- Thermal stresses resulting from solar radiation on partially shaded glazing.
- Hard body impact caused by extreme hail storms, other flying debris in wind storms or falling objects.
- Soft body impact such as the humans stumbling into or onto the glass.
- Spontaneous fracture due to nickel sulphide inclusions in thermally toughened glass, see Sect. 6.2.6.

Therefore, a safe glass design is not guaranteed simply by applying the static dead and live loads, wind loads, and snow loads, which are typically sufficient for the structural design with other materials (e.g. steel, concrete, timber, etc.) .

In addition, the tensile strength of glass is governed by the presence of surface flaws that accumulate and grow over service life. The nucleation and morphology of the flaws is strongly dependent on the exposure to loading and environment of the glass component, but it is not yet possible to explicitly predict these flaws. Therefore the effect of the surface flaws can be mitigated by adequately low strength value for annealed glass and by thermal pre-stressing where needed, see Chap. 6. For high(er) consequences of failure, it is also important to consider the post-fracture performance, which is affected by e.g. the pre-stressing level, lamination and support conditions, see Chaps. 6 and 3.

In comparing the pros and cons of the deterministic and probabilistic design approach, one advantage of the deterministic method is its ease of use. On the other hand, this approach does not accurately represent the material properties of glass. The advantages of the probabilistic method are primarily the clearer separation of effect of actions and resistance, which makes it possible to derive the partial safety factors explicitly. In its original concept, the system is based on the uniformity of the safety requirements for the building industry - regardless of the respective building material or actions. Often substantial simplifications have to be made for the practical implementation and determination of the partial safety coefficients. Already the specification of constant sensitivity factors α_E and α_R, which should consider the influences of possible fluctuations of the scatter of effect of actions and resistance, have a considerable influence on the result - especially with a large scatter of resistance values as commonly found in glass [7, 20]. Strictly, each associated effect of action, dependent on the relevant duration of the load, must be compared to a set value of the resistance and for each pairing of effect of action and resistance the correct distribution function must be developed, in order to determine partial safety coefficients.

For the most relevant cases, the partial safety coefficients on the resistance side are derived from extensive series of tests. On the other hand, some of the partial safety coefficients, for example related to different types of surface treatments are approximations and require further research.

The disadvantage of probabilistic design for practical application is the complexity of the calculation, In particular, the large number of mutually influencing parameters which result in a large number of partial safety factors, combination coefficients and sensitivity factors. The consistent implementation of this procedure initially seems simple, but it does make a significant additional effort in the calculation. This approach often requires computer support. For example even the static verification of a simple insulating glass pane in the overhead area by manual calculation becomes a very laborious task, in which a large number of load combinations must be compared with different design values of the resistance as a function of the different durations of the loads involved.

It is therefore sensible to use simple deterministic approaches and other rules of thumb during preliminary/early design stages [120] and to revert to the more complex and accurate probabilistic approach in the latter stages of design (e.g. during detailed design).

9.2.2 Harmonised European Standard

Over the years there have been various draft guidelines such as the prEN 13474-3 [48] and prEN 16612 [79] which can be considered as forerunners of the European glass code currently in development. The prEN 16612 and prEN 16613 are now fully-fledged standards EN 16612 and EN 16613, respectively, but their scope is limited to infill panels in common applications, and other simple applications of glass where there is no requirement to assess the fracture and post-fracture performance explicitly. There has therefore been little for the engineer to turn to for effective guidance to a standardised approach to design of load bearing glass components and structures.

This lack of harmonised codified guidance for glass structures constitutes challenges for consulting engineers who must often resort to researching the fundamental characteristics of the materials, to testing and to verifying basic design principles or to carry out desktop research of published papers, etc. This is not a feasible approach for those who want to a safe, practical, and time-efficient approach for the design of glass structures. With this in mind, the European glass code CEN/TS 19100 "Design of glass structures" [109–111] was released for public comment in 2022. It currently consists of 3 parts as follows:

Part 1: Basis of design and materials, which sets out the basic design rules for load-bearing glass components. In particular the underlying design rules for load-bearing capacity, serviceability, fracture characteristics and failure consequences in relation to human safety, robustness, redundancy and durability of glass structures.

Table 9.4 Design checks required depending on limit state scenario of glass components. Note that LSS-0 is not in the scope of

	Limit state Scenario			
	LSS-0	LSS-1	LSS-2	LSS-3
Design for unfractured glass state	SLS	SLS	SLS	SLS
	ULS	ULS	ULS	ULS
Design for glass fracture state (safe glass fracture)	-	FLS	-	FLS
Design for the post-fractured state (residual load capacity)	-	-	PFLS	PFLS

Part 2: Design of out-of-plane loaded glass components, provides basic structural design rules for typically flat or curved (and typically laminated) glass components subjected to out-of-plane (i.e. lateral) loading. Here, out-of-plane loads refer to loads acting normal (e.g. wind) to or having a component (e.g. dead load, snow, ...) acting normal to the glass plane.

Part 3: Design of in-plane loaded glass components and their mechanical joints, provides guidelines on the design of glass components primarily subjected to in-plane loading. It also covers construction rules for mechanical joints for in-plane loaded glass components. The three parts adopt the customary format of other Eurocodes whilst addressing specific issues which the material characteristics of glass present to the designer.

The most significant new feature in this standard is the introduction of two additional limit states in the design of glass structures as summarised in Table 9.4. The familiar Ultimate Limit State (ULS) and Serviceability Limit State (SLS) which consist of combinations of actions: ULS- persistent or transient (fundamental), accidental, seismic design situation; SLS- characteristic, frequent, quasi-permanent combination) will feature in this new standard and will be similar to the design of other structural materials. However, these two limit states are insufficient to ensure a safe glass structure because some glass structures must also be designed to fail safely in the event of glass fracture. For example, a hammer falling onto a concrete slab has no significant structural performance consequences, but a hammer falling onto a glass floor can have serious consequences on human injury and costs of repair / replacement. For this reason, the new draft standard includes a Fracture Limit State (FLS) and a Post-Fracture Limit State (PFLS).

The Fracture Limit State (FLS) and Post-Fracture Limit State (PFLS) are intended to ensure that at the instant of fracture the glass fails in a manner that does not cause

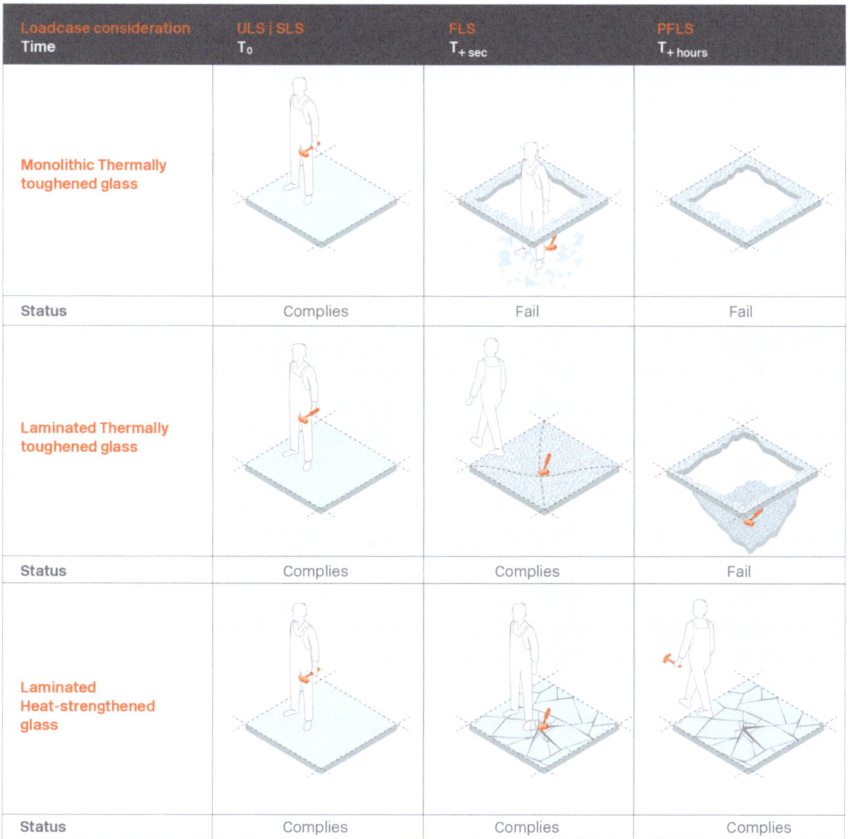

Fig. 9.5 Illustration of possible performance of different glass types at limit states [116]. Note that other cases and performances may apply in function of specific glass composition, size, etc

unacceptable risks of human injury or economic losses. In contrast, the Post-Fracture Limit State ensures that the fractured glass does not cause unacceptable risks of human injury or economic losses for a limited period of time (e.g., until the fractured glass can be replaced). Therefore if the consequence of failure of a glass component is low, it may be sufficient to design for SLS and ULS only. But for more critical load bearing glass elements e.g. glass floors, glass beams, etc. it will also be necessary to design for FLS and PFLS. To demonstrate this, it is useful to imagine a scenario where routine maintenance or cleaning is being carried out about a roof light, and the operative accidentally drops a tool on to the glass panel that has enough force to initiate a fracture, see Fig. 9.5.

Further information on the new harmonised European standard CEN/TS 19100 is available in [118].

9.2.3 Other Design Standards

9.2.3.1 ASTM E1300

The American standard Determining Load Resistance of Glass in Buildings [93] is based on the deterministic (global) safety concept. The standard describes procedures to determine the load resistance of glazing made of edge supported monolithic or laminated panels and exposed to a uniform lateral load of short or long duration (e.g. wind, snow, self-weight) for a specified probability of fracture. The standard does not apply to other applications such as balustrades, glass floor panels, aquariums, structural glass members and glass shelves. Design charts in the standard provide a means to determine the permissible non-factored loads and the maximum lateral deflections for a broad range of glass thickness and span. The permissible load charts are based on the probabilistic glass failure prediction model [8]. However, the underlying level of safety cannot be compared directly with the European standards. The probability of failure related to one year is $P_f = 8 \times 10^{-3}$ (eight lites per thousand) compared to $P_f = 1 \times 10^{-6}$ in EN 1990, but this value only refers to the resistance, while the scatter of the actions is not recorded in ASTM E1300 [93]. Further information on the US and Canadian design standards can be found in [45, 95]. The ASTM E1300 standard also provides an alternative analytical approach for determining the approximate maximum surface stress when performing independent stress analysis (e.g. when using computational finite element analysis instead of the design charts provided). Similarly to the design charts approach, this is also based on the probabilistic glass failure prediction model of Beason and Morgan, but with this approach it is possible for the design engineer to select a desired probability of fracture [8].

9.2.3.2 DIN 18008

DIN 18008 *Glass in Construction—Design and Construction rules* is divided into six parts:

- Part 1: Terms and general bases DIN 18008-1 [103],
- Part 2: Linearly supported glazing DIN 18008-2 [104],
- Part 3: Point fixed glazing DIN 18008-3 [75],
- Part 4: Additional requirements for barrier glazing DIN 18008-4 [76],
- Part 5: Additional requirements for walk-on glazing DIN 18008-5 [77],
- Part 6: Additional requirements for walk-on glazing in case of maintenance procedures and for fall-through glazing DIN 18008-6 [98].

This standard is based on the semi-probabilistic safety concept using partial safety factors. Since 2014, the Parts 1 to 5 were listed in the list of technical building codes [87] and by the end of 2015 they were introduced in all German federal states, thus completely superseding the older technical guidelines (TRLV, TRPV and TRAV). Part 6 was officially introduced in 2018 and the parts 1 and 2 were updated in 2020.

Table 9.5 Selected combination factors ψ according to EN 1990/NA and DIN 18008-1

Action	ψ_0	ψ_1	ψ_2
Imposed loads in buildings (types see EN 1991-1-1)			
– Category A: Areas for domestic and residential activities	0.7	0.5	0.3
– Category B: Office areas	0.7	0.5	0.3
– Category C: Areas where people may congregate	0.7	0.7	0.6
– Category H: Roofs	0.0	0.0	0.0
Snow and ice load (see EN 1991-1-3)			
– Locations up to sea level 1 000 m	0.5	0.2	0.0
– Locations above sea level 1 000 m	0.7	0.5	0.2
Wind loads (see EN 1991-1-4)	0.6	0.2	0.0
Climatic effects (change of temperature and change in	0.6	0.5	0.0
Meteorological air pressure) as well as			
Temperature induced loads			
Installation effects	1.0	1.0	1.0
Barrier loads	0.7	0.5	0.3

In *Ultimate Limit State* the glass is designed not to break under the expected actions or that any glass fracture does not have significant consequences. To this end, the actions according to *Eurocode 1* (EN 1991 series) are combined according to the specifications from *Eurocode 0* (EN 1990) with the corresponding combination factors and partial safety factors (Table 9.5). These effects of the actions are typically used to calculate stresses (e.g. the principal tensile stresses on the glass surface) and compare them to the glass strength (resistance) with a defined level of safety $E_\mathrm{d} \leq R_\mathrm{d}$.

The actions (loads) for the design with DIN 18008-1 are to be taken from the *Eurocode 1* (EN 1991 series series). In addition, actions on glass structures should be considered. For exmaple, *Barrier loads* are horizontal loads often to be considered for large glazing and are part of EN 1991-1-1 and the corresponding national annexes. Continuous line loads (of $0.5\,\mathrm{kN\,m^{-1}} \leq q_\mathrm{k} \leq 2.0\,\mathrm{kN\,m^{-1}}$, depending on the respective application and category), must be used as a barrier load for the railing or to design the glass itself and are typically applied up to a height of 1.2 m measured from the finished floor level.

In the case of IGU, the pressure differences between the ambient air and the hermetically sealed cavity must also be used for the design and can be considerable. These *climatic loads* are described in Section 9.8.1.

The glass resistance is determined from:

$$R_{\mathrm{d}} = k_{\mathrm{edge}} \cdot k_{\mathrm{LSG}} \cdot \frac{k_{\mathrm{mod}} \cdot k_{\mathrm{c}} \cdot f_{\mathrm{k}}}{\gamma_{\mathrm{M}}} . \tag{9.10}$$

where

k_{edge} factor to account for a reduced edge strength;
k_{LSG} factor to account for a reduced likelihood of failure for laminated safety glass;
k_{mod} factor to account for load duration;
k_{c} factor take account for the type of construction to account for a fail-safe fracture of the glass;
f_{k} characteristic strength of the glass quality used;
γ_{M} partial safety factor of the glass.

In the case of annealed glass with glass edges under tension the edge strength is reduced to 80 % ($k_{\mathrm{edge}} = 0.8$) irrespective of the edge processing, see Sects. 5.2 and 2.3.2). However, for thermally pre-stressed glass, $k_{\mathrm{edge}} = 1.0$.

If laminated safety glass (LSG) is used, the resistance may be increased by 10 % ($k_{\mathrm{LSG}} = 1.1$). The reason for this is not the shear coupling through the polymer interlayer, but a lower probability that both plates encounter the same (low) glass strength (statistical interference). Moreover, the glass surfaces in contact with the polymer interlayer are protected from external surface damage after lamination.

Due to the subcritical crack growth (see Sect. 5.3), the strength of annealed glass depends on the load duration. This is taken into account using the factor k_{mod}, which is described in detail in the Section 9.3. When combining different influences, for simplification, the action with the shortest duration is used to determine the k_{mod}. But all variations of combinations of actions have to be taken into account to ensure a safe design. For pre-stressed glass, generally $k_{\mathrm{mod}} = 1.0$ applies, as the level of pre-stress for most cases can be assumed to be large enough to avoid subcritical crack growth.

A fail-safe glass structure depends on its post-fracture performance and thus on the type of structure, its boundary conditions and the glass type (e.g. use of laminated safety glass and the fracture pattern of the glass). In DIN 18008-1, the influence of the type of construction can be considered using a construction factor k_{c}, which may be chosen to $k_{\mathrm{c}} = 1.8$ for the most typical glazing type, continuously supported annealed glass as part of an IGU. Otherwise $k_{\mathrm{c}} = 1.0$.

The characteristic strength of the selected glass types can be found in the respective product standards (see Table 5.2).

The partial safety factor in DIN 18008-1 depends on the glass quality. It is $\gamma_{\mathrm{M}} = 1.8$ for annealed glass and, due to statistical superposition of the glass strength and the pre-stress and thus a lower variation of the strength values of

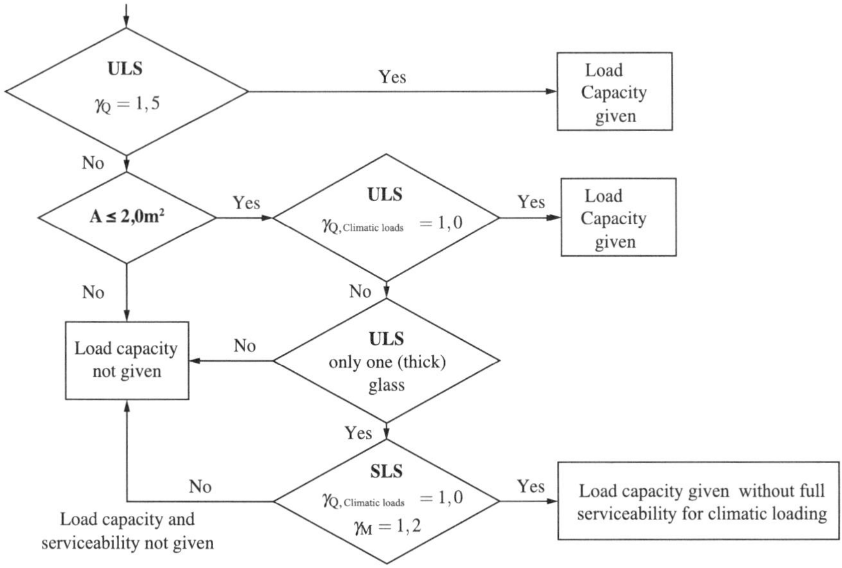

Fig. 9.6 Design concept for small format insulating glazing

thermally pre-stressed glass compared to annealed glass, it can be reduced to $\gamma_M = 1.5$.

Due to the low failure consequences of the fracture of a glass in an IGU, DIN 18008-1 also defines a specific design format for small sized IGUs ($A \leq 2.0\,\text{m}^2$) that ensures the full load capacity for wind and snow loads but cannot fully ensure the serviceability for climatic loading and the resulting pressure differences between the cavity and the ambience, see Fig. 9.6. The design concept here uses characteristic values for the climatic loads ($\gamma_{Q,\text{climatic}} = 1.0$).

For the *serviceability limit state*, typically only a check of the maximum deflection is required:

$$E_d \leq C_d. \tag{9.11}$$

The design value of the effect of actions E_d in this case is the deflection that results from the governing load case. Usually the *characteristic influence combinations* according to EN 1990 are applied, which do not stipulate an increase of the actions through partial safety factors i.e. $\gamma_F = 1.0$.

The target values for the SLS depend on the type of construction: for continuously or point-supported glazing, the maximum deflection is to be limited to $C_\mathrm{d} = l/100$, where l is the governing (smaller) span. In the case of continuously supported glazing, this verification can be dispensed if instead it is demonstrated that the glass does not slip from its supports under maximum load and a minimum edge support of 5 mm can be achieved for the glass. But stricter deflection limits for insulating glass that also depend on the respective manufacturing process and IGU components (typically $C_\mathrm{d} = l/200$), must be taken into consideration. For glass floors, the maximum deflection should be limited to $C_\mathrm{d} = l/200$.

Limitations regarding the deflection also have to be considered for the substructure that supports glass elements. The deflection of the substructure should be limited to $l/200$ for glazing which is continuously supported along the edge.

9.2.4 Other National and International Design Guidelines

Several countries have standards and guidelines for the design of load bearing glass in construction. The level of detail and information given in each of these standards varies and a comprehensive review of all of these is beyond the scope of this book. A useful overview is provided in Feldmann and Kasper, who describe how aspects of safety and serviceability are implemented in glass design requirements in different countries [82]. Guidelines on design methodology, specification, materials and techniques in the design and construction of glass structures are also provided in O'Regan [83] covering: design of glass elements such as floor plates, beams and columns; connection design; computational analysis of glass structures; design for extreme loading conditions and fire protection. This guidance does not replace national or international design standards, but is rather a useful source of information for initial design stages.

9.3 Designing with Time Dependent Strength

Due to crack growth effects, the tensile strength of glass is a function of the stress history (see Sect. 5.3.1). Consequently, load duration is often taken into account in glass design guidelines. In the European standard a step-wise modification factor, k_mod, is used to account for time dependent effects instead of a continuous function to simplify the design, see Table 9.6. The time dependent effects can be neglected

Table 9.6 Modification factors k_{mod} to take into account the load duration for constant loads

Type	Load duration	Action	k_{mod}	
			DIN 18008	CEN/TS
Permanent	Permanent	Self-weight, difference in altitude, permanent cold bending	0.25	0.29
Variable	Intermediate	Snow (3 to 4 weeks)	0.4[e]	0.43[c]
		Imposed vertical action (1 week)	0.4	0.45
		Temperature change and change in the meteorological barometric pressure (8 h)	0.4	0.58[d]
	Short	Maintenance load (30 min)	0.7	0.69
		Wind (10 min)	0.7[e]	0.74[a]
		Barrier personnel loads–crowds (5 min)	0.7	0.77
		Barrier personnel loads–normal duty (30 s)	0.7	0.89[b]
	Very short	Wind (3 s)	–	1.00
	Dynamic	Impact (100 ms)	1.8	1.20

[a]The value of $k_{mod} = 0.74$ is based on a cumulative equivalent duration of 10 min, considered representative of the effect of a storm which may last several hours. Higher values of k_{mod} can be considered for wind

[b]The value of $k_{mod} = 0.89$ is based on a personnel load of 30 s duration. Other values can be considered depending on the type of personnel load being evaluated and also the building use

[c]$k_{mod} = 0.43$ can be considered representative for snow loads lasting between 5 d ($k_{mod} = 0.49$) and 3 month ($k_{mod} = 0.41$). Other values of k_{mod} can be appropriate depending on local climate

[d]$k_{mod} = 0.58$ can be considered representative for climatic loads lasting between 6 h ($k_{mod} = 0.59$) and 12 h ($k_{mod} = 0.57$). Other values of k_{mod} can be appropriate depending on local climate

[e]From original table

in pre-stressed glass when the stresses due to bending do not exceed the residual compressive stress.

The modification factor is the equivalent constant stress load $\sigma_{f,s}$ due to the applied load relative to the characteristic static strength of glass:

$$k_{mod} = \frac{\sigma_{f,s}}{f_{g,k}}. \tag{9.12}$$

Figure 9.7 shows the modification factors k_{mod} for constant loads specified in and DIN 18008-1 as a function of the load duration.

Based on the crack growth constants specified in Table 5.1, it is possible to determine the load duration modification factors, k_{mod}, which take into account the load duration for soda-lime-silica glass, assuming a characteristic bending strength of $f_k = 45$ MPa for thermally pre-stressed glass, see Table 5.2, and the recalculation of

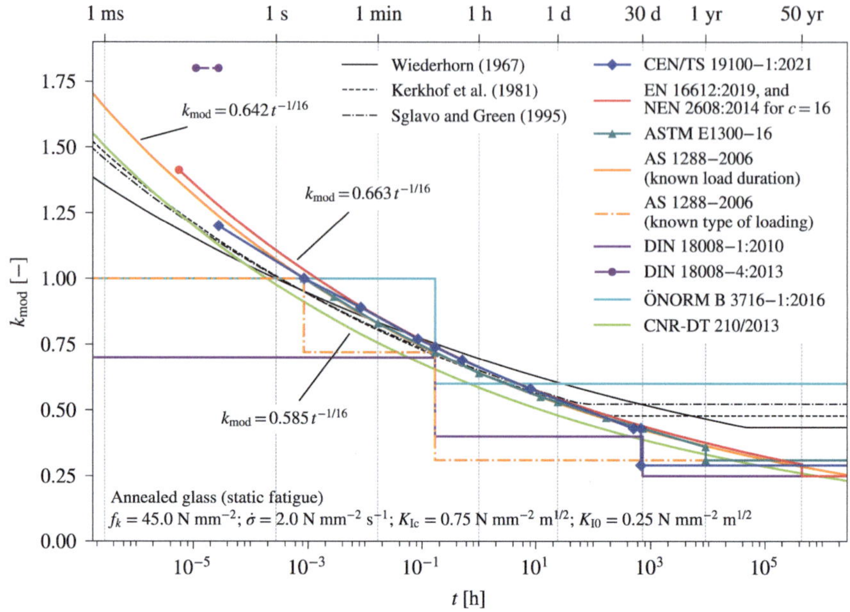

Fig. 9.7 Modification factors k_{mod} to take into account the load duration: contrasting the normative rules of DIN 18008-1 or in the European standard design modification factors for constant stresses are contrasted by material specific modification factors calculated on the basis of the crack growth constants referred to in Table 5.1. The data points shown on the curve correspond to the k_{mod} values shown in Table 9.6

the respective initial crack depth according to Eq. 5.14.[1] as well as the static creep strength for a constant load corresponding to Eq. 5.7.

It is striking that the diverse load duration relationships represented in Fig. 9.7 are very similar in the range of $30\,\mathrm{s} < t \leq 2\,\mathrm{h}$. From an effective duration in the range of $2\,\mathrm{h} < t \leq 1$ months or 3 years for crack growth constants according to Wiederhorn [3], the gradients show a levelling off in the constant modification factors k_{mod} due to the crack growth threshold σ_{I0}. The modification factors listed in and DIN 18008-1 are mostly below the material specific calculated values; therefore they provide a safe approximation of the actual crack growth behaviour.

In addition to constant static stresses, modification factors, k_{mod}, in accordance with are also useful for dynamic loads under soft body impact, see Table 9.7. The load duration during a soft impact is markedly shorter than the load

[1] The calculations are based on a stress rate of $\dot{\sigma} = 2\,\mathrm{MPa\,s^{-1}}$.

Table 9.7 Modifcation factors k_{mod} in construction for soft body impact loads according to DIN 18008-4

Glass type	Standard	Modification factor k_{mod}
[−]	[−]	[−]
Annealed glass	EN 572-1	1.8
Heat strengthened glass	EN 1863-1	1.7
Thermally toughened glass	EN 12150-1	1.4

duration in the bending test for the determination of the strength. In addition, relatively large tensile stresses from bending occur, which may exceed the surface compressive stresses in thermally pre-stressed glass. For these reasons, the strength of thermally pre-stressed glasses for the short-term dynamic load case during an impact is also higher than under quasi-static load and it may be subject to a modification factor $k_{\mathrm{mod}} > 1$. For DIN 18008-4, the modification factors were determined according to the following relationship:

$$k_{\mathrm{mod}} = \frac{\sigma_{\mathrm{f,impact}}}{f_{\mathrm{k}}} \tag{9.13}$$

The stress during a pendulum impact is close to inert strength. Using the mean bending tensile strength, Schula demonstrated that the short-term strengths resulting from a material specific consideration taking into account subcritical crack growth effects are higher than those in DIN 18008-4 specified values [92].

9.4 Designing with Temperature-Induced and/or Displacement-Induced Loading

In addition to the predictable, calculable stress effects, such as climate induced stress for insulating glass units, further unknown boundary constraints in glass components can lead to fracture, see Fig. 9.8. Glazing should therefore be mounted in a manner that minimises stresses caused by constraints. This applies in particular to differential thermal expansions and the movement of substructures.

While the absorption of solar radiation from uncoated façade glazing is 8 % to 20 %, it increases to values above 70 % for dark-coated, enamelled and coloured glass as well as for photovoltaic modules and often must be addressed explicitly during design and construction. Temperature variations through the thickness may lead to bending and stresses even for in-plane statically determinate glass panes. Finite

Fig. 9.8 Fracture of glass blocks as a result of external forces through the shrinkage of the surrounding concrete. The latter did not crack due to its high grade

element software is useful for assessing temperature-induced stresses by performing a multi-physics thermo-mechanical analysis. In the worst case scenario, a summer rain storm leads to a stress due to the shock-like cooling (ΔT) of the glass surface with respect to the temperature of the core, as follows:

$$\sigma_{1.2} = \alpha_{\mathrm{T}} \frac{E}{1-\nu} \Delta T \,. \tag{9.14}$$

Partial shading or partial printing can also result in temperature variations in the plane of the glass. If the perimeter of the pane is colder than the centre area (e.g. due to the shading from the frame), tensile stresses arise along the edge of the glass. The glass edge has a relatively low strength, see Sect. 5.2; therefore, such a load can lead to fracture with a characteristic pattern, referred to as *thermal fracture*, see Figs. 10.6 and 6.28.

The is a lack of specific guidance on the normative temperatures to be used in designs> For example, in Germany the temperatures have not been set, but EN 1991-1-5 gives recommended values for stationary temperature profiles. Hourly values for outside temperatures and radiation can be taken from test reference years of national meteorological institutes. A methodology to do so has been developed by Vandebroek [86]. The French glass design standard NF DTU 39 [21] contains some rules for thermal fracture calculations. For special constructions such as bonded glass façades, the testing temperatures can be obtained from the corresponding regulations (ETAG 002 [69]). Load combinations, which define the statistical occurrence of temperature and other actions over life cycles of 25 or 50 years, do not yet exist.

If temperatures are not explicitly taken into account in the design, it is essential to allow for the necessary expansion or contraction in the design details. In the design stage, it is therefore important to accommodate several tolerances e.g.

the misalignment of holes in thermally pre-stressed laminated glass, the long-term expansion/contraction and settlement of supports, etc.

9.5 Design for Blast and Impact Loading

9.5.1 General

When a structure of a given mass deforms under load its acceleration (and inertia) may either be negligible or significant. The former is referred to as a static or quasi-static response and a static analysis will provide sufficiently accurate results. The latter is known as a dynamic response, which often requires a dynamic analysis, e.g. in the event of impact or blast.

9.5.2 Soft Body Impact

In structural glass design, soft body impact testing is used to mimic impact from a human being. The principal test consists of a 50 kg double tyre pendulum and a tyre inflation pressure of 0.35 MPa as described in, see Fig. 9.9a, b.

The EN 12600 [30] test involves the swinging of the pendulum body from three drop heights into the test pane of glass held vertically in a standardised steel testing frame. The drop height of the pendulum and manner in which the glass fractures informs the classification as follows:

Class 3 (a 190 mm drop height) corresponds to an adult pushing hard against the glass or a child running into it;
Class 2 (a 450 mm drop height) corresponds to a significantly higher impact than an adult walking into glass, but less than an adult trying to force through by running into it;
Class 1 (a 1200 mm drop height) is the most demanding classification, applied in testing glass for critical applications.

Type A fracture, characterised by large sharp edged fragments;
Type B fracture, characterised by the fragments being held together; and
Type C fracture, characterised by the disintegration of the glass into small, less dangerous pieces.

Type A is typical of annealed glass; Type B is typical of laminated glass and mirrors with a safety film backing; Type C is typical of thermally toughened glass.

EN 12600 [30] uses a three-part classification system forming a three digit code representing (in this order):

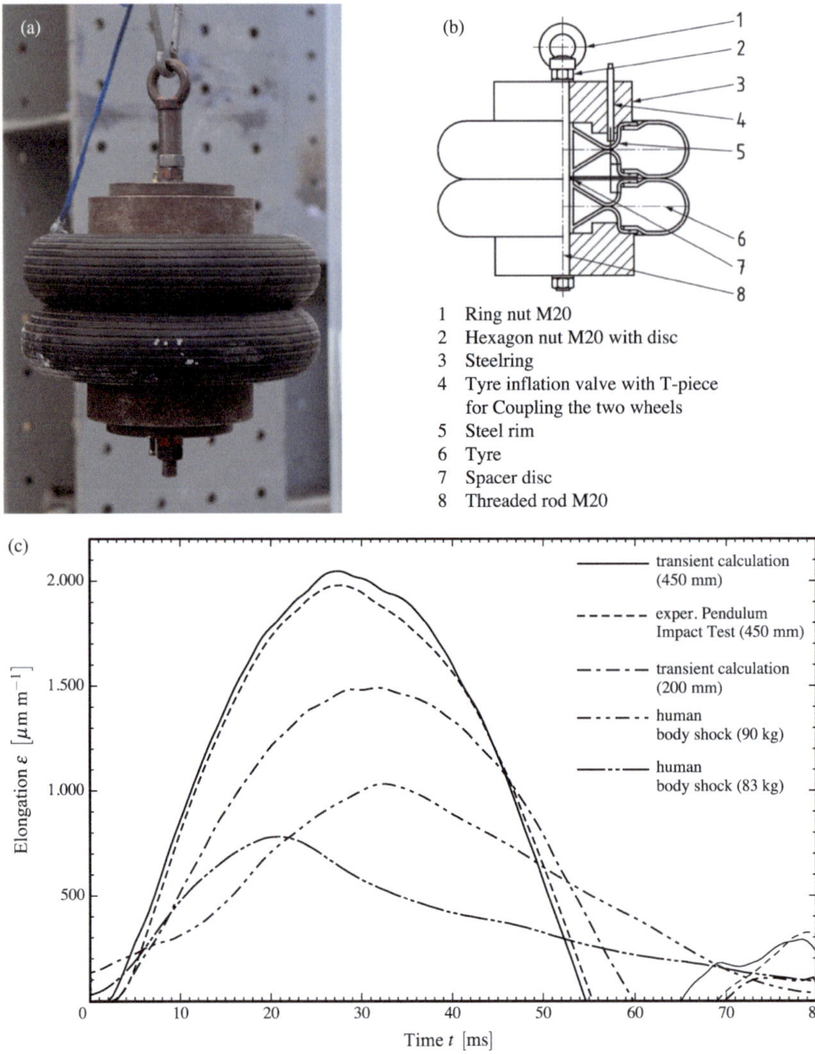

Fig. 9.9 **a**, **b** Double tyre pendulum test according to EN 12600 or DIN 18008-4 and **c** comparison of max. measured strains due to intentional human body impact with a run-up distance of 2.5 m in comparison with a pendulum impact test (drop height 450 mm, [26]) and transient calculations

The strength as measured by the aforementioned drop height Class 1, 2 or 3; the breakage type A, B or C; the highest drop height class at which the glass did not break or did not allow penetration (again classified as 1, 2 or 3).

Other soft body impact tests are available for overhead glazing. For example, the Centre for Window and Cladding Technology [50, 51] specifies tests to classify the fragility of overhead glazing. The classification is intended to assess safety for anyone walking on the glass (e.g. pedestrians or maintenance staff) as well as for anyone below the glass. The following 4 classifications are identfed by the CWCT standards:

Class 0 —for glass with unrestricted (walk-on) access by building occupants;
Class 1 —for glass that might be walked on for occasional cleaning/maintenance activities;
Class 2 —for glass that is not walked on, but where maintenance personnel could accidentally thread onto the glass;
Class 3 —for glass that is deemed fragile.

9.5.3 Hard Body Impact

A hard impact is defined as the impact of a hard (relative to glass) solid body at high speed (e.g. hail, falling tools, etc.). The immediate influence of a hard impact is usually limited to the local environment of the impact location, see Fig. 9.10a. While reliable computational methods exist for simulating the effects of soft impacts for different materials [11], the effects of hard impacts on brittle materials can only be ascertained experimentally. Thermally pre-stressed glass has a much higher surface resistance to hard impacts than annealed glass. For laminated glass, the polymer interlayers have an additional positive influence as they absorbs part of the impact energy. In addition, the very short load duration means that full shear transfer through the interlayer can be assumed during the impact event, see Sect. 9.7.1.

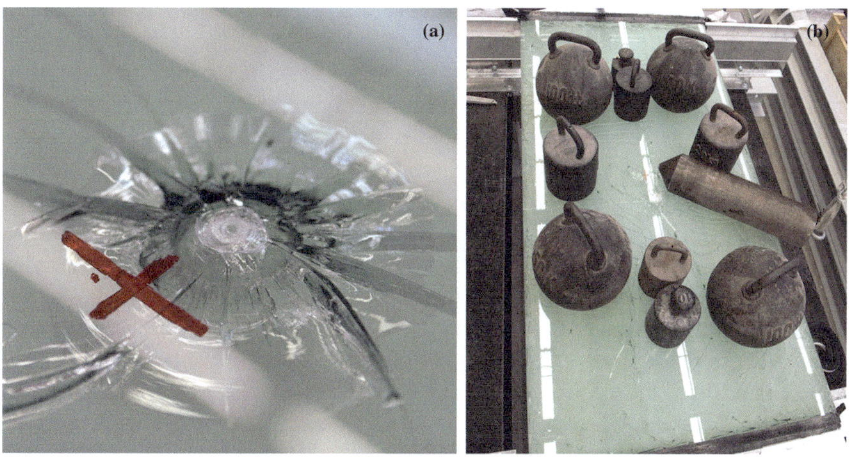

Fig. 9.10 Hard impact on glass: **a** impact location after a ball case test according to DIN 52338 and **b** after a collision of the impact body according to DIN 18008-5 for the detection of the shock-carrying capacity of walk-on glazing

The drop-ball test is used to check the behaviour of laminated glass in the event of an impact caused by a small hard object. In EN356:2002 steel balls with a mass of about 4110 g are used as the impact body. The drop height is between 1.5 m and 9.0 m depending on the intended category. This test is used for assessing the degree of resistance to burglary, see Sect. 3.5.2. In contrast, EN 14449 Annex C Section C.2.2 prescribes a ball drop test intended for the producer to check the quality of the laminate, see Sect. 3.1.4.

In the German standard [77], a so-called "torpedo" is used as the impacting body for the testing of walkable glazing, see Fig. 9.10b. With a weight of 40 kg, it simulates a half body mass, which impacts the glass surface with a hard small object: a steel M8 hexagonal bolt mounted at the contact tip. Before dropping the object on the glazing, an equivalent load of 1 kN per person is positioned on an area of 200 mm × 200 mm in an unfavourable load position, where the equivalent load of no less than one person is assumed.

9.5.4 *Blast Load*

An explosion represents a sudden physical or chemical change in state, in which a high amount of energy is released in a very short time. This creates a blast wave that spreads spatially and is transient. In structural design it is advisable to distinguish between low and high propagation velocities of the pressure waves [1].

Deflagration refers to the blast wave spreads at less than the speed of sound. The burned particles move opposite to the propagation of the blast wave; examples are dust or gas explosions.

In detonations, the propagation velocity of the pressure wave is higher than the speed of sound in the burning medium and the burnt particles move in the direction of the pressure wave. Examples are solid explosives or nuclear explosions. Since the pressure wave propagates faster at high pressure amplitudes, a sudden increase in air pressure occurs at a distance from the source of the detonation, independent of the original shape of the pressure wave, see Fig. 9.11. This process, which is due to non-linear relationships in wave propagation, is called steepening [1]. Blast waves, which spread at supersonic speed and have such a shape, are called shock waves.

In building design it is customary to consider shock waves resulting from detonations. The characteristic shape of the incident pressure-time curve at a certain distance from the source of detonation is shown in Fig. 9.11. The air pressure rises instantly from the ambient pressure, p_0, to the maximum pressure, $p_{\max}$. The difference is referred to as the maximum peak overpressure p_E . This is followed by an exponential decay, where the pressure drops from the peak to below the ambient pressure, known as the negative phase. The area below the pressure-time curve in the over-pressured area is called a positive impulse i_+; the duration of the positive pressure phase is referred to as t_+, see Fig. 9.11.

If the blast wave meets obstacles, it is reflected. Because of reflection and the accumulation of the air particles, the obstacle experiences at least twice the load

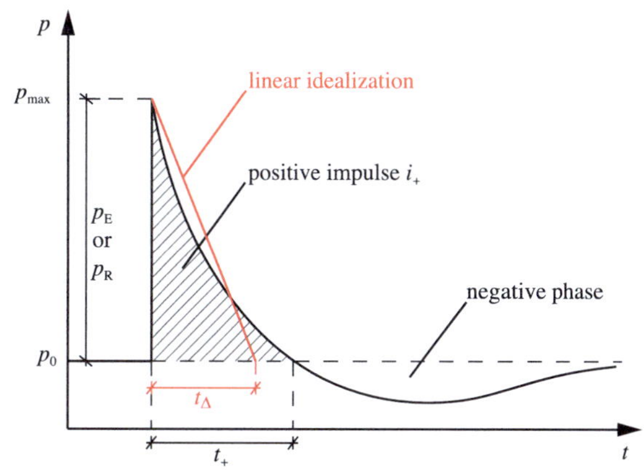

Fig. 9.11 Friedlander Curve: the pressure of a shockwave as function of time

compared to the incident pressure. This is referred to as reflected positive pressure, p_R. Building components must therefore be designed to resist this reflected positive pressure. Still, the characteristic shape of the pressure-time curve according to Fig. 9.11 is preserved. It can be described by the Friedlander equation:

$$p(t) = p_0 + p_R \cdot \left(1 - \frac{t}{t_+}\right) \cdot e^{-A\frac{t}{t_+}}. \tag{9.15}$$

For simple cases (spherical or semi-hemispherical expansion of the blast wave) the load can be estimated using analytical or empirical formulas. For complex situations with confinements and reflections in e.g. urban environments, a blast wave propagation can be simulated using explicit finite element software (Hydrocodes), see e.g. Fig. 9.12.

Once the the blast wave progression is known, the glass can be designed accordingly, for example by using a dynamic (transient) calculation with a Finite Element software. The mechanical response (i.e. the principal tensile stresses and deformations in the glass arsing from the blast load) can be computed in a relatively accurate manner up to glass fracture [78]. Beyond this, the simulation of post-fracture behaviour is relatively complex and reliable predictions are difficult to

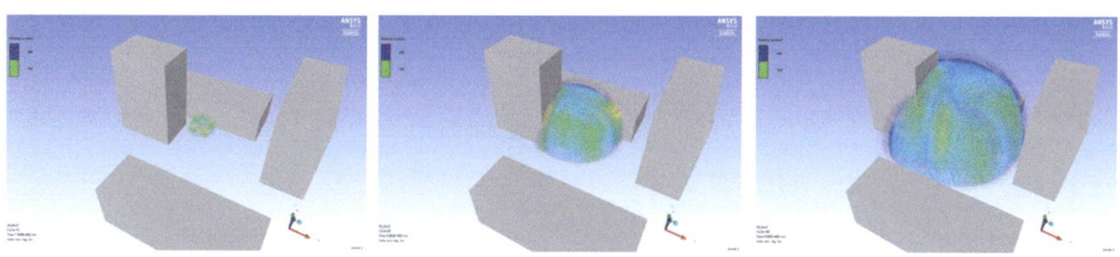

Fig. 9.12 Simulation of an explosion event in Ansys Autodyn

obtain [73, 91]. In fact, the post-fracture performance of laminated glass under high strain rates is the subject of ongoing research [108, 115].

A common simplification in calculations is the linearisation of the Friedlander curve while maintaining the positive impulse, which is equivalent to setting $A = 0$ in Eq. 9.15. Due to the short duration of the load, full composite action can be assumed in laminated glass during a blast event.

9.6 Linear and Non-linear Plate Analysis

A two-way spanning rectangular plate, simply supported along its perimeter (four edges supported) and subjected to a uniformly distributed load, is one of the most common configurations or static systems encountered in glass engineering, see Fig. 9.13. The out-of-plane deflections of glass plates are typically greater than the thickness of the plate. In such cases, a simple bending (Kirchhoff) plate theory is no longer valid and leads to an overestimation of the deflections and stresses. Depending on the aspect ratio of the plate, membrane stresses develop in addition to the bending stresses and in the case of non-developable surfaces (e.g. a 2-way spanning rectangular plate) radial tensile membrane stresses are also generated towards the

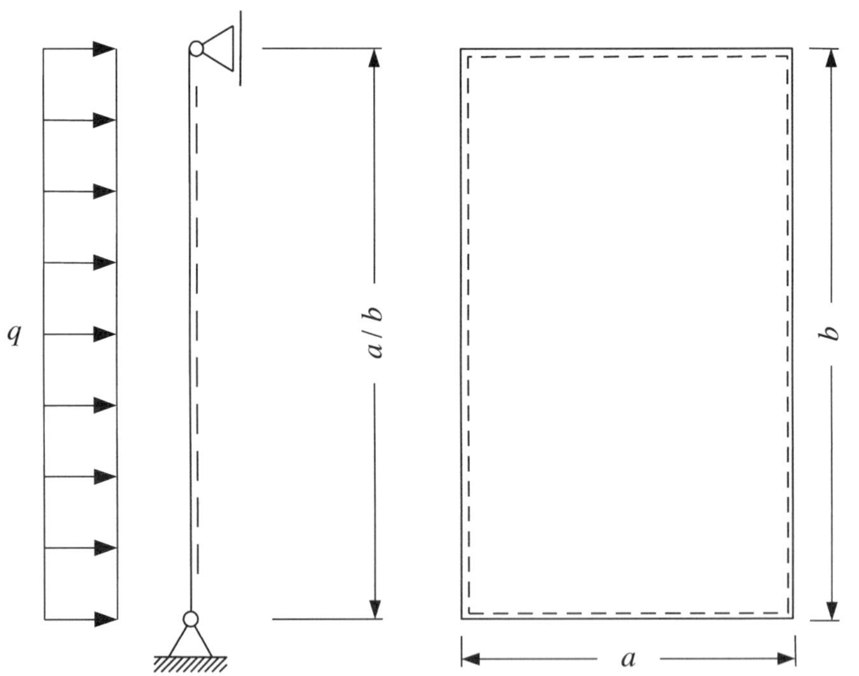

Fig. 9.13 Static system of a rectangular plate ($b \geq a$) with constant thickness d, simply (Navier-) supported along its perimeter under a uniformly distributed load q

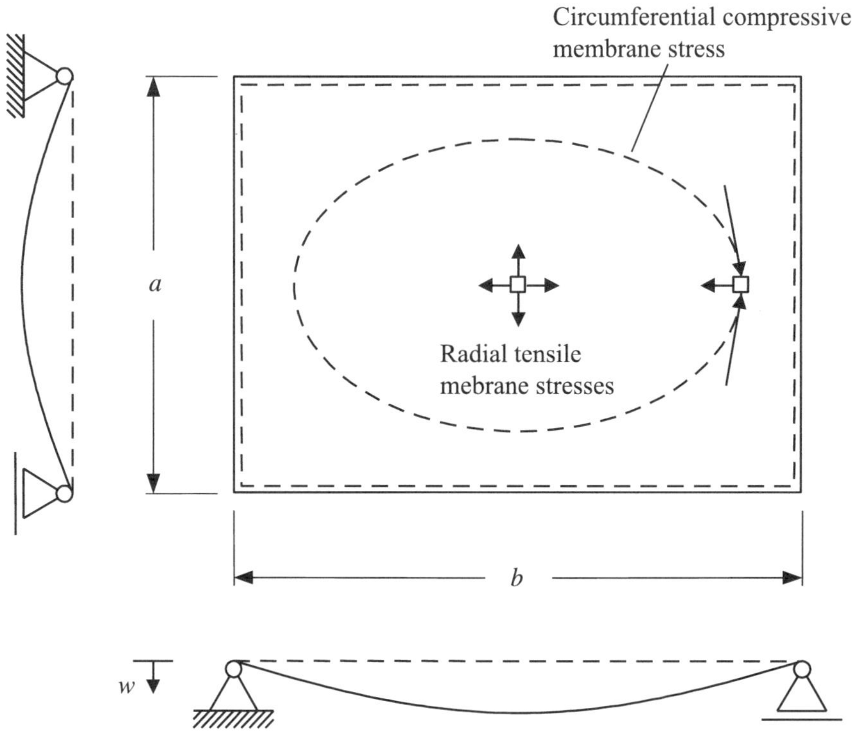

Fig. 9.14 Activating the membrane effect at large deformations in rectangular glass plates that are simply supported along their perimeter

centre of the plate, which are in equilibrium with circumferential (hoop) compressive membrane stresses, see Fig. 9.14.

For example, using linear Kirchhoff plate theory for a four side supported, square glass plate with a thickness of $d = 4\,\text{mm}$, edge length $a = b = 1\,000\,\text{mm}$ under a uniformly distributed pressure of $q = 2\,\text{kN/m}^2$ leads to an overestimation of about 100 % and 25 % in the deflection and principal surface tensile stress at the centre of the plate, respectively. In general Kirchhoff plate theory provides conservative estimates of deflections and stresses, however the location of the maximum stresses change when large deflections are taken into account. As a rule of thumb, non-linear effects become significant when the deflection is larger than half the thickness of the plate. It is therefore commonplace to undertake geometrically non-linear analysis even for relatively small plates.

The stresses and deflections in rectangular plates subjected to large deformations can be assessed by means of tables [6]. The maximum stress in the plate can be calculated from:

$$\sigma_{\max} = \frac{a^2}{t^2} \cdot B_\sigma \cdot q\,. \tag{9.16}$$

Table 9.8 Coefficients B_σ for the calculation of the maximum main tensile stress in a rectangular glass plate depending on the aspect ratio a/b supported along its perimeter [6]

a/b	q^*									
	0	5	10	20	30	50	100	200	300	500
1.0	0.272	0.271	0.268	0.258	0.245	0.227	0.207	0.188	0.178	0.165
0.9	0.323	0.320	0.314	0.293	0.269	0.243	0.222	0.203	0.193	0.180
0.8	0.383	0.378	0.365	0.329	0.294	0.262	0.240	0.221	0.210	0.198
0.7	0.451	0.442	0.421	0.368	0.322	0.282	0.261	0.241	0.230	0.217
0.6	0.526	0.514	0.485	0.417	0.362	0.305	0.284	0.263	0.252	0.239
0.5	0:603	0.590	0.560	0.485	0.424	0.342	0.309	0.289	0.277	0.264
0.4	0.673	0.665	0.643	0.580	0.519	0.429	0.377	0.317	0.306	0.292
0.3	0.725	0.722	0.714	0.687	0.650	0.575	0.444	0.349	0.337	0.323
0.2	0.748	0.747	0.746	0.744	0.739	0.724	0.671	0.561	0.481	0.348
0.1	0.749	0.749	0.749	0.749	0.749	0.749	0.749	0.748	0.746	0.739
0.0	0.750	0.750	0.750	0.750	0.750	0.750	0.750	0.750	0.750	0.750

The maximum deflection that occurs in the middle of the plate can be determined by the following equation:

$$w_{\max} = \frac{a^4}{E \cdot t^3} \cdot B_{\mathrm{w}} \cdot q\,. \tag{9.17}$$

For the design of insulating glass units, the volume, V, of the deflected shape of the plate is of interest:

$$V = \frac{a^4}{E \cdot t^3} \cdot B_{\mathrm{v}} \cdot A \cdot q \tag{9.18}$$

where:

A	Area of the rectangular plate	[m^2],
a	Short edge length of the plate	[m];
b	Long edge length of the plate	[m];
E	Modulus of elasticity of the glass	[MPa];
t	Thickness of the glass plate	[m];
q	Uniformly distributed load	[MPa].

The factors B_{w}, B_{v} and B_σ are dimensionless coefficients obtained from Table 9.8 to Table 9.10. In case of large plates, the deflections depend on the normalised load:

$$q^* = \frac{a^4}{t^4} \cdot \frac{q}{E}. \tag{9.19}$$

For a normalised load, $q^* < 10$, the principal tensile stresses in the centre of the plate are underestimated by less than 10% compared to a linear calculation. For small deflections, q^*, approaches zero.

Table 9.9 Coefficients B_w for the calculation of the maximum deflection of a rectangular glass plate supported along its perimeter dependent on the aspect ratio a/b [6]

a/b	q^*									
	0	5	10	20	30	50	100	200	300	500
1.0	0.046	0.046	0.045	0.041	0.034	0.032	0.024	0.017	0.014	0.011
0.9	0.056	0.056	0.054	0.049	0.044	0.036	0.027	0.019	0.015	0.012
0.8	0.068	0.067	0.065	0.057	0.051	0.041	0.030	0.021	0.018	0.014
0.7	0.083	0.081	0.077	0.068	0.059	0.048	0.035	0.025	0.020	0.016
0.6	0.099	0.097	0.092	0.081	0.071	0.057	0.041	0.029	0.024	0.019
0.5	0.115	0.113	0.109	0.097	0.086	0.070	0.051	0.036	0.030	0.023
0.4	0.131	0.129	0.126	0.116	0.105	0.088	0.065	0.046	0.038	0.030
0.3	0.142	0.142	0.140	0.135	0.128	0.114	0.088	0.064	0.053	0.041
0.2	0.147	0.147	0.147	0.146	0.145	0.140	0.126	0.101	0.085	0.067
0.1	0.148	0.148	0.148	0.148	0.148	0.148	0.147	0.146	0.143	0.136
0.0	0.148	0.148	0.148	0.148	0.148	0.148	0.148	0.148	0.148	0.148

Table 9.10 Coefficients B_v for the calculation of the volume resulting from the bent shape of a rectangular glass plate supported along its perimeter depending on the aspect ratio a/b [6]

a/b	q^*									
	0	5	10	20	30	50	100	200	300	500
1.0	0.019	0.019	0.019	0.018	0.017	0.015	0.011	0.008	0.007	0.005
0.9	0.024	0.024	0.023	0.022	0.020	0.017	0.013	0.009	0.007	0.006
0.8	0.029	0.029	0.028	0.026	0.023	0.020	0.015	0.011	0.008	0.007
0.7	0.035	0.035	0.034	0.031	0.028	0.023	0.017	0.012	0.010	0.008
0.6	0.042	0.042	0.040	0.037	0.033	0.027	0.020	0.014	0.012	0.009
0.5	0.050	0.050	0.048	0.044	0.040	0.033	0.025	0.018	0.014	0.011
0.4	0.059	0.058	0.057	0.053	0.049	0.042	0.031	0.022	0.018	0.014
0.3	0.068	0.067	0.066	0.064	0.061	0.054	0.042	0.031	0.025	0.020
0.2	0.077	0.077	0.076	0.076	0.074	0.071	0.062	0.048	0.040	0.031
0.1	0.086	0.086	0.086	0.086	0.086	0.085	0.084	0.081	0.077	0.068
0.0	0.095	0.095	0.095	0.095	0.095	0.095	0.095	0.095	0.095	0.095

A biaxial load transfer to the supports is necessary for a membrane stress state to develop in a simply-supported plate. The extent of the biaxial distribution is expressed by the coefficient, B_σ, which is a function of the edge length ratio, a/b, and the normalised load, q^*, as shown in Fig. 9.15 for three different cases: a square plate ($a/b = 1$), an infinite strip ($a/b = 0$) and a rectangular plate with $a/b = 0.5$.

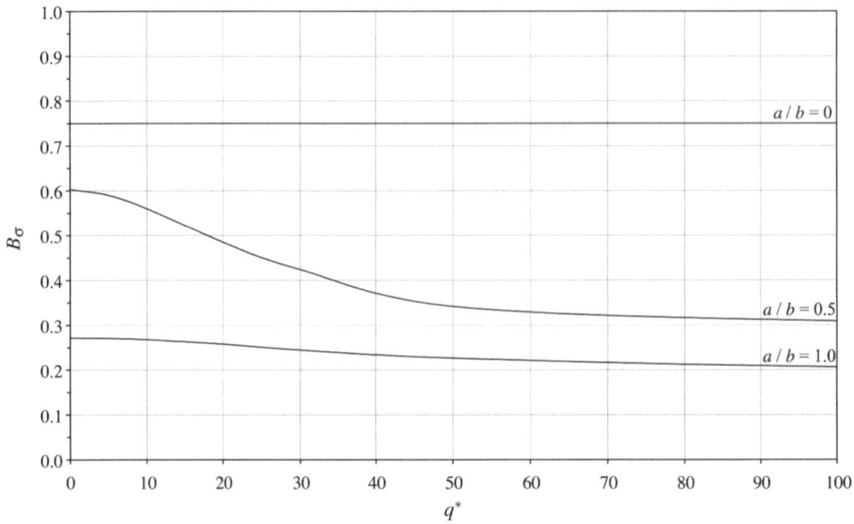

Fig. 9.15 Change of the maximum principal surface stress at the centre of a plate with different side length ratios

An alternative approach for calculating deflections and stresses with non-linear plate theory can be found in Kutterer. Based on an empirically determined scaling factor, the deflection in the centre of the plate can be defined with the linear solution. The maximum stress is then determined either by an approximation formula or by means of load-stress diagrams. Based on an empirically determined scaling factor, the deflection in the centre of the plate can be defined with the linear solution. The maximum stress is then determined either by an approximation formula or by means of load-stress diagrams.

It is also possible to perform geometrically non-linear analysis by means of finite element software, but this requires the use of the appropriate finite elements (e.g. plate or shells) and the correct solver functions to capture the membrane effects.

9.7 Design of Laminated Glass

As explained in Sect. 3.1, laminated glass consists of two or more glass plies bonded with polymer interlayers. The basic function of laminated glass is to provide safety in the event of glass fracture.

Whereas Sect. 3.1 mostly explains the production of laminated glass, the present section provides the principles of mechanical design of laminated glass.

9.7.1 Shear Transfer in Laminated Glass

In its unfractured state, the interlayer in laminated glass provides shear transfer between the glass plies. The ability to transfer shear forces through the interlayer is governed by the shear stiffness of the interlayer and therefore by temperature and load duration. Therefore, the actual composite behaviour of laminated glass lies between the two extreme states or limits: *full shear transfer* (also known as full composite action or monolithic limit) and *no shear transfer* (also known as no composite action or layered limit), see Figs. 9.16 and 9.17.

The most commonly used interlayer is Polyvinyl Butyral (PVB); however, other materials with different properties can be used such as ionomer or Ethylene Vinyl Acetate (EVA) (see Sect. 3.1.3). Figure 9.18 shows the temperature dependency of the complex shear modulus for various interlayers.

For example, PVB-laminated glass subjected to short-term loads (e.g. impact and wind loads) at temperatures below 50 °C responds in an almost fully composite manner. Similarly, for long-term loads (e.g. snow loads), at low temperatures (around zero degrees celsius) there is also a very high shear transfer; but only partial shear transfer is achievable at room temperature. At high temperatures, above 50 °C, the shear transfer is almost non-existent [23]. For common applications (symmetrical laminated safety glass with two plies, edge-supported, with a uniform lateral load and standard dimensions), a constant interlayer shear modulus of $G = 0.4$ MPa reduces the surface tension in the glass by up to 50 % compared to no composite action. In contrast, a shear modulus of $G \geq 10$ MPa is required to generate almost full composite action (Fig. 9.19).

Neglecting composite action generally yields a safe design for load-induced, but it is not always safe to ignore shear transfer in the engineering design of laminated glass

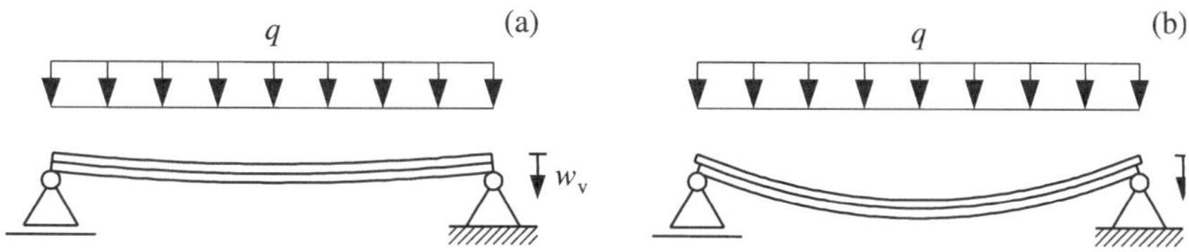

Fig. 9.16 Deformation of a laminated glass pane with **a** full shear transfer, and **b** no shear transfer

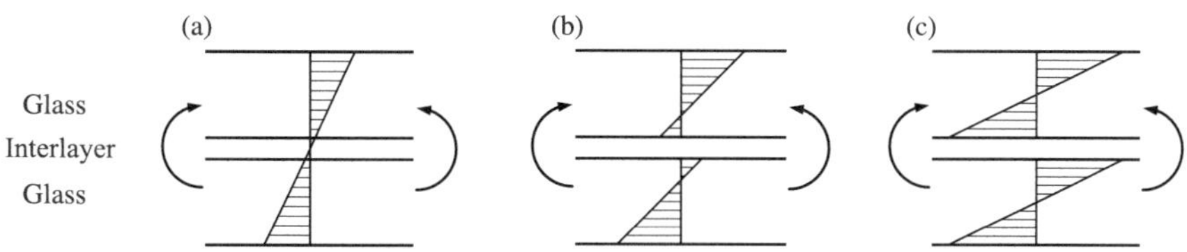

Fig. 9.17 Stress distribution across the thickness of a laminated glass pane, **a** with full shear transfer, **b** with partial shear transfer and **c** with no shear transfer

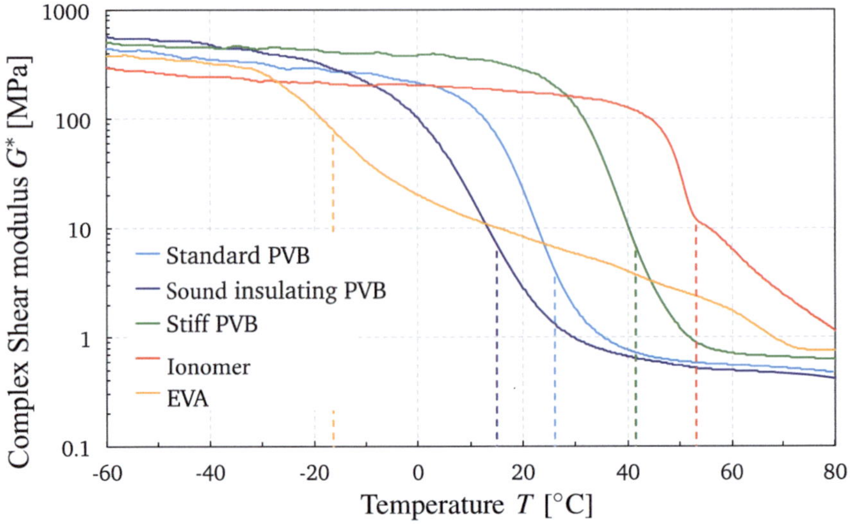

Fig. 9.18 Complex shear modulus as a function of temperature, T, for different interlayer materials determined using DMTA at a constant frequency of 1 Hz. Dotted lines indicate the glass transition temperatures [90]

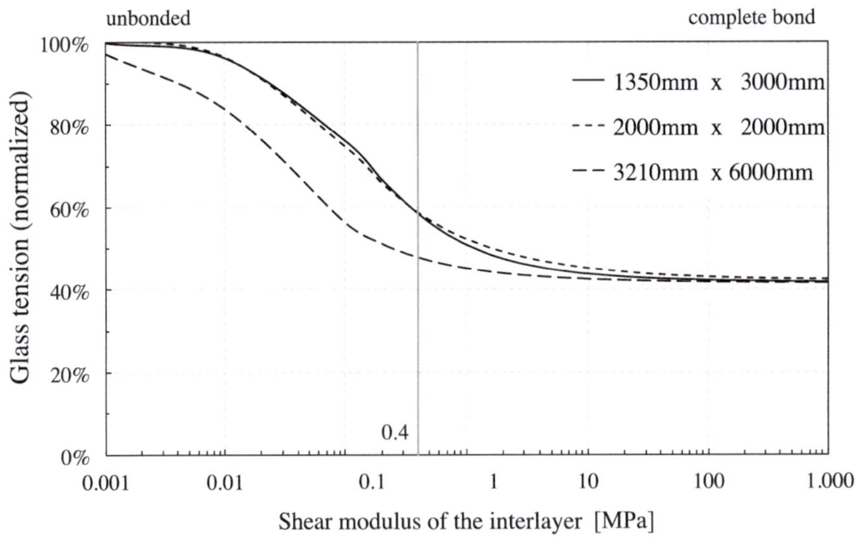

Fig. 9.19 Tensile stress in the glass in relation to the interlayer stiffness, for three different panel dimensions (four-side supported; consisting of 2 × 4 mm thermally toughened glass with a 0.76 mm interlayer; geometrically linear consideration)

components. For example, with forced displacements, a full composite response may lead to higher stresses compared to the stresses that would be generated assuming partial (or no) composite response. This scenario arises in the design of insulating

Table 9.11 Equivalent thickness t_w^* [mm] of a two-ply laminated glass pane assuming no shear transfer between the glass plies

	t_l **[mm]**			
t_u **[mm]**	4	6	8	10
4	5.04	6.54	8.32	10.2
6	6.54	7.56	9.00	10.7
8	8.32	9.00	10.2	11.5
10	10.2	10.7	11.5	12.6

glass units for climate induced loads as well as for cold bent glass. In such cases it is essential to consider the extreme condition of full composite action, in addition to the partial (or no) composite action.

In the event of full composite action, laminated glass behaves like one monolithic pane. The bending stresses and deflections of the individual plies can therefore be determined based on the full cross-sectional depth (Fig. 9.17a). In the absence of composite action, the load on the individual plies and the resulting stresses are distributed according to the individual bending stiffness of each ply (Fig. 9.17c).

In the following example, both plies of a two-ply laminate deflect by equal amounts, w. The distribution of loads on each of the plies can be determined by an equivalent thickness, according to:

$$t_w^* = \sqrt[3]{t_u^3 + t_l^3} \tag{9.20}$$

The part of the load taken by the upper glass ply is

$$\delta_u = \frac{t_u^3}{t_u^3 + t_l^3}, \tag{9.21}$$

the part of the load taken by the lower glass ply results in

$$\delta_l = \frac{t_l^3}{t_u^3 + t_l^3}. \tag{9.22}$$

For common glass thicknesses, the equivalent thicknesses are given in Table 9.11.

It is possible to consider a partial shear transfer under the condition of linear elasticity for all materials with different analytical approaches. The approach in Wölfel, allows, for example, the calculation of a simply-supported one-way spanning laminated glass panel under uniformly distributed lateral load. For this, a shear transfer coefficient [65]

$$\Gamma = \frac{1}{1 + 9.6\frac{EI_s t_v}{Gt_s^2 a^2}} \tag{9.23}$$

is determined with:

$$I_s = d_o d_{s,u}^2 + d_u d_{s,o}^2,$$
$$I_s = d_o d_{s,u}^2 + d_u d_{s,o}^2,$$
$$t_{s,u} = \frac{t_s t_u}{t_u + t_l},$$
$$t_{s,2} = \frac{t_s d_l}{t_u + t_l},$$
$$t_s = 0,5(t_u + t_l) + t_{int},$$

t_{int}	Thickness of the interlayer	[mm],
t_u	upper glass thickness	[mm],
t_u	lower glass thickness	[mm],
E	Young's modulus for glass	[MPa],
a	Span length	[mm],
G	Shear modulus	[MPa].

For a given shear transfer coefficient Γ deflections, w, and stresses, σ, can be calculated by using the following equivalent thickness and simple bending theory

$$t_w^* = \sqrt[3]{t_u^3 + t_l^3 + 12\Gamma I_s}, \tag{9.24}$$

$$t_{\sigma,u}^* = \sqrt{\frac{t_w^{*3}}{t_o + 2\Gamma t_{s,1}}}, \tag{9.25}$$

$$t_{\sigma,1}^* = \sqrt{\frac{t_w^{*3}}{t_l + 2\Gamma t_{s,u}}}. \tag{9.26}$$

The theory of Kutterer can be applied to two-way spanning simply supported laminated glass panes. Further approaches are given in [4, 5, 80] and described by Güsgen. The application of the ω procedure, for example in prEN 16612, in which a constant shear transfer factor is applied to any system, may lead to inaccurate results. The reason for this is that the shear transfer factor—as seen in Eq. 9.23—is not a material constant, but depends on the static system (laminate build-up, dimensions, support conditions, and load). Further details on the derivation of stresses and strains in multi-layer glass elements subjected to lateral loads (i.e. bending) and compressive loads (i.e buckling) are provided in [97, 107], respectively.

Numerical tools, such as the finite element method, offer the most versatile and accurate tool for designing laminated glass. Many commercial FE programs can fully capture the time and temperature-dependent shear modulus using viscoelastic material models (such as Prony series) and time-temperature displacement functions, see Sect. 4.1.5. Thus both creep and relaxation processes can be simulated. The prerequisite for this is that the material behaves in a linear viscoelastic manner and in the relevant temperature and time range. Temperature can be accounted for by the theory of thermorheologically simplicity. Two Prony series for PVB as well as the corresponding Williams-Landel-Ferry (WLF) [2] factors for the thermorheological

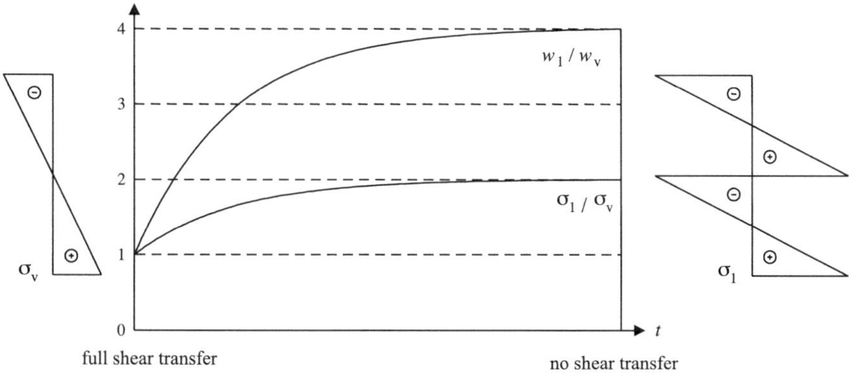

Fig. 9.20 Changes to the stress distribution (σ) and deformation (w) in a symmetrically laminated glass plate subjected to bending for varying shear transfer

behaviour are given in Table 9.12. Other Prony series can be found, for example in [25, 66, 71].

In Fig. 9.20 a qualitative behaviour of a symmetrically two-ply laminated glass pane subjected to bending is shown. In case of a total absence of shear transfer, the bending stresses increase by up to a factor of two compared to a laminated glass with a full shear transfer. The deflections increase by up to four times the initial deflection.

9.7.2 *Post-fracture Performance of Laminated Glass*

Post-fracture capacity (also known as residual load bearing capacity) is the resistance of a fractured system against a complete failure (i.e. collapse). In glass structures, this term is used in particular to describe the behaviour of laminated glass in which one or more plies are fractured. Figure 9.21 illustrates the different failure states of a two-ply laminated glass subjected to bending.

In certain situations, specific requirements can be made regarding the residual load bearing capacity. For example, in some countries, horizontal glazing above pedestrian-accessible areas needs to be retained in case of fracture. Accessible or

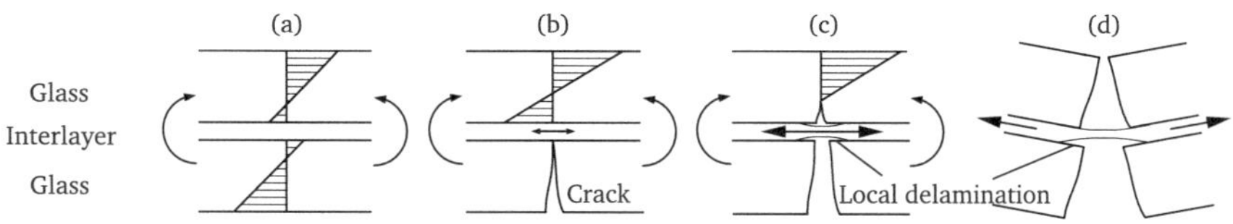

Fig. 9.21 States of a two-ply laminated glass pane: **a** both plies intact, **b** one ply fractured, **c** both plies fractured and **d** interlayer membrane forces at the fully fractured state

Table 9.12 Prony and WLF parameters for PVB, reference temperature $T_{ref} = 20\,°C$

	Kuntsche [90]		Barredo et al. [61]	
	$G_0 = 750\,MPa$		$G_0 = 428\,MPa$	
	$G_\infty = 0.181\,MPa$		$G_\infty = 0.307\,MPa$	
	$c_1 = -1.069 \times 10^6$		$c_1 = -49.81$	
	$c_2 = 6.072 \times 10^6\,°C$		$c_2 = 328.5\,°C$	
i	τ_i [s]	g_i [-]	τ_i [s]	g_i [-]
1	1×10^{-17}	1.1071×10^{-1}	3.09×10^{-7}	0.151
2	1×10^{-16}	5.9190×10^{-2}	3.08×10^{-6}	0.191
3	1×10^{-15}	1.4683×10^{-1}	3.07×10^{-5}	0.141
4	1×10^{-14}	1.2744×10^{-3}	3.066×10^{-4}	0.184
5	1×10^{-13}	1.3611×10^{-1}	3.057×10^{-3}	0.139
6	1×10^{-12}	6.0921×10^{-2}	3.049×10^{-2}	0.122
7	1×10^{-11}	6.1245×10^{-2}	0.304	0.054
8	1×10^{-10}	5.7080×10^{-2}	3.032	0.0137
9	1×10^{-9}	3.6053×10^{-2}	30.23	2.11×10^{-3}
10	1×10^{-8}	6.3861×10^{-2}	301.5	9.46×10^{-4}
11	1×10^{-7}	8.3617×10^{-3}	3007	2.65×10^{-5}
12	1×10^{-6}	6.1737×10^{-2}	3.00×10^{4}	2.75×10^{-4}
13	1×10^{-5}	3.4859×10^{-2}	1.54×10^{5}	1.54×10^{-4}
14	1×10^{-4}	4.6068×10^{-2}		
15	1×10^{-3}	4.2979×10^{-2}		
16	1×10^{-2}	3.9236×10^{-2}		
17	1×10^{-1}	2.1514×10^{-2}		
18	1.0	8.6305×10^{-3}		
19	1×10^{1}	2.0102×10^{-3}		
20	1×10^{2}	3.3657×10^{-4}		
21	1×10^{3}	2.5507×10^{-4}		
22	1×10^{4}	7.2345×10^{-5}		
23	1×10^{5}	7.7346×10^{-5}		
24	1×10^{6}	5.0524×10^{-5}		
25	1×10^{7}	7.1125×10^{-5}		
26	1×10^{8}	7.1078×10^{-5}		
27	1×10^{9}	3.8180×10^{-5}		
28	1×10^{10}	1.0876×10^{-4}		
29	1×10^{11}	7.6516×10^{-6}		
		$\sum = 0.9998$		$\sum = 0.9993$

Table 9.13 Residual load bearing capacity for laminated glass panes after fracture of all glass layers

Laminated safety glass made of annealed glass	High
Laminated safety glass made of heat-strengthened glass	Medium
Laminated glass made of a combination of thermally toughened and heat-strengthened glass	Low
Wired glass	low
Laminated safety glass made of thermally toughened glass	Very low

Table 9.14 Residual load bearing capacity for different types of supports when using laminated glass panes after fracture of all glass layers

Four-sided support	High
Point support with plate holders	Medium
Point support with recessed holders	Low
Two-sided support	Very low

walkable glazing can only be designed with certain types of glass—usually laminated safety glass.

Assessing the residual load bearing capacity must always take into account the respective component and the associated safety requirements. For example, proving the residual load capacity of walkable glass must be done under more extreme conditions than with horizontal glazing. However, in general, to assess the residual load bearing capacity one or more plies of a laminated glass are intentionally fractured and must perform satisfactorily for a specific amount of time, see Fig. 9.22 and Fig. 9.23. The residual load carrying capacity is significantly influenced by the type of glass. Laminated safety glass made of thermally toughened glass has a lower residual capacity compared to laminated float and heat-strengthened glass. This is sometimes referred to as the "wet towel" or "wet blanket" effect because the fractured laminated glass panel sags like a wet towel/blanket under its self weight. Tables 9.13 and 9.14 contain design recommendations for different support conditions.

Due to the increasing use of new interlayer materials in laminated glass,there is a growing need for experimental methods for assessing the salient mechanical properties of the interlayers that affect the the post-fracture performance. The methods are currently still in the development stage [89]. In addition to the shear stiffness of the interlayer, the adhesion to the glass has a significant influence on the residual load carrying capacity due to possible delamination, see Fig. 9.21. Possible test methods for the assessment of adhesion or delamination are described in Sect. 3.1.4. In particular, the Through Crack Tension (TCT) test provides useful results in this regard [89, 121].

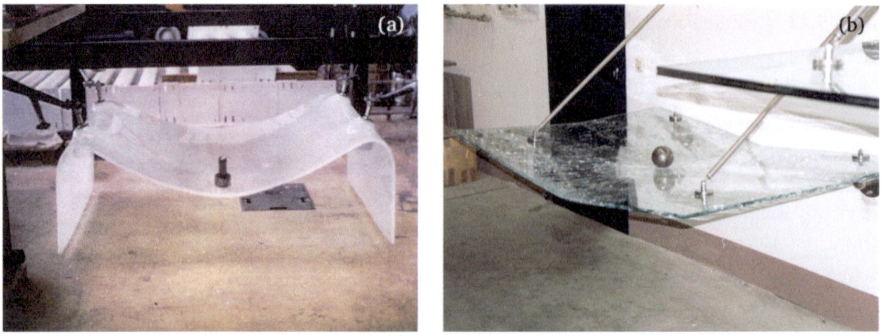

Fig. 9.22 Tests for **a** a thermally toughened point-supported laminated glass pane of the Neue Messe Leipzig *Neuen Messe Leipzig*; and **b** a heat-strengthened point supported laminated glass canopy pane

Fig. 9.23 Post-fracture load test on walkable glazing

Several research projects address the characterisation and calculation of the residual load bearing capacity of fully fractured laminated glass using engineering models [41, 44, 59, 99, 115] or finite element models [28, 73, 89]. However, a comprehensive computational model that captures post-fracture performance is not yet available.

Some residual load carrying capacity may be achieved in the absence of laminated glass, for example, insulating glass units offer some post-fracture glass retention through their bonding to the edge spacer.

9.8 Design of Insulating Glass Units

9.8.1 *Climate Loads on Insulating Glass Units*

Climatic changes lead to pressure differences between the surroundings and the hermetically sealed cavity in insulating glass units (IGU's) thereby causing stresses in the glass. In particular, the gas in the cavity changes volume when temperature and atmospheric air pressure change, leading to climatic induced stress on the individual panes. The caving or bulging of insulating glass units is also referred to as "pillowing" and is investigated in more detail by Feldmeier [9, 16, 17]. The resulting stresses depend significantly on the stiffness of the glass plates. Due to their low bending stiffness, large-format IGU's with relatively thin glass, experience little or no climatic stresses. Conversely, small panes with relatively thick glass attract significant climatic stresses. Such stresses should be considered carefully in IGU design.

The climatic influence involves various aspects, namely:

- Air pressure differences due to meteorological high and low pressure systems lead to pressure differentials Δp_{met} between the ambient air and the IGU cavity;
- Temperature changes ΔT due to fluctuations in the ambient temperature or due to incident solar radiation lead to expansion/contraction of the gas in the cavity resulting in pressure on the glass, see Fig. 9.24;
- Altitude differences, ΔH, between the production and installation locations also cause changes in the external air pressure and therefore have to be taken into account in the design.

When designing an IGU it is important to establish the duration of the respective climatic loads. The meteorological pressure difference Δp_{met} and the temperature changes ΔT represent transient actions with, typically, specified short or medium

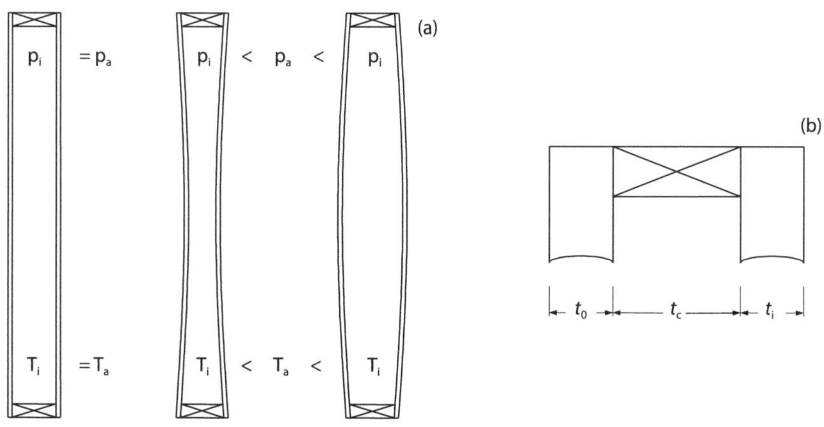

Fig. 9.24 Double glazed unit: **a** deflections due to changing the air pressure p and temperature T and, **b** edge detail schematic

term durations. However, the difference in altitude, ΔH, is a constant action leading to long term load. All of the climatic loads have an effect on glass strength, but the long term loads arsing from altitude difference can have a very significant influence on the strength, especially for annealed glass, see Sect. 9.3.

The isochoric pressure p_0 , the cavity pressure assuming that the volume of the cavity remains constant, can be determined from:

$$p_0 = c_1 \cdot \Delta T + c_2 \cdot \Delta H - \Delta p_{\mathrm{met}} \tag{9.27}$$

Where:

$$c_1 = 0.34\,\mathrm{kPa\ K^{-1}}$$
$$c_2 = 0.012\,\mathrm{kPa\ m^{-1}}.$$

With completely rigid (isochoric) systems, for example, heating in the IGU cavity around $\Delta T = 20\,\mathrm{K}$ with an altitude difference of $\Delta H = 600\,\mathrm{m}$ an $\Delta p = -2\,\mathrm{kPa}$ generates an isochoric pressure of $p_0 = 16\,\mathrm{kPa}$. However, due to the flexibility of the glass plates, the IGU cavity volume changes with pressure, therefore the actual load on the glass plates is lower, and a new equilibrium between the pressure and volume is found to satisfy the ideal gas law in an isochoric system:

$$p \cdot V/T = \text{constant}. \tag{9.28}$$

Therefore, the stiffness of the glass plates and the IGU cavity width, t_c, are important parameters. When using insulating glass units, the four-sided supported rectangular plate is the most common configuration. Due to the edge seal of insulating glass units, the support can generally be assumed as a simple support (free rotation, see Fig. 9.13). The procedure for the analytical calculation of such systems is described in Sect. 9.6. In addition, it is possible to calculate stresses and deformations using spreadsheets or the finite element method. For the calculation of climatically induced loads in rectangular double glazing units, Feldmeier defines the characteristic edge length as follows:

$$a^* = \sqrt[4]{\frac{E \cdot t_c \cdot t_o^3 \cdot t_i^3}{p_L \cdot (t_o^3 + t_i^3) \cdot B_v}} \tag{9.29}$$

where

E *Youngs* Modulus for the glass,
t_c IGU cavity width;
t_i Thickness of the inner glass ply;
t_o Thickness of the outer glass ply;
p_L Air pressure at the installation site ($p_L \approx 0.1\,\mathrm{MPa}$);
B_v Value according to Table 9.10.

The relationship between the characteristic edge length and the cavity width is shown in Fig. 9.25 for an insulating glass unit with an inner and outer pane of equal

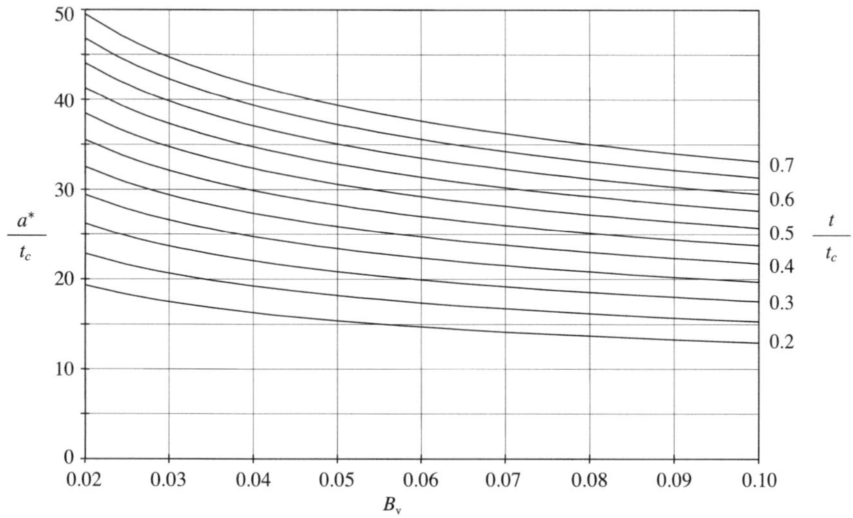

Fig. 9.25 Characteristic edge length a^* of a rectangular double glazing unit with four side support and symmetrical construction ($t_i = t_i = t$)

Table 9.15 Characteristic edge length a^* (in mm) of a rectangular two-pane insulating glass unit for a cavity $t_c = 16$ mm depending on the side ratio a/b

t_i	t_o	a/b										
[mm]	[mm]	0	0.1	0.2	0.3	0.4	0.5	0.6	0.7	0.8	0.9	1.0
4	4	250	254	262	270	280	291	304	318	334	351	369
4	6	279	284	291	301	312	324	339	355	372	391	411
4	8	289	294	302	312	323	336	351	367	386	405	426
4	10	293	298	306	316	327	341	356	373	391	411	432
6	6	339	345	354	366	379	394	412	431	453	476	500
6	8	369	376	386	398	413	429	448	470	493	518	545
6	10	384	391	401	414	429	447	466	488	513	537	566
8	8	421	428	440	454	470	489	511	535	562	590	620
8	10	451	459	472	487	504	525	548	574	602	633	665
10	10	497	506	520	536	556	578	604	633	664	698	733

thickness. Table 9.15 lists the characteristic edge lengths for insulating glass units with a 16 mm cavity width for various glass thicknesses.

From the definition of the insulating glass factor:

$$\varphi = \frac{1}{1 + (a/a^*)^4} \tag{9.30}$$

follows the actual (effective) climate pressure on the insulated glazing from the product of insulating glass factor φ and the isochoric pressure p_0:

$$q = \varphi \cdot p_0 \,. \tag{9.31}$$

According to Kirchhoff's (linear) plate theory, see Sect. 9.6, the maximum principal tensile stress at the centre of the glass ply is calculated with Eq. 9.16, see also Table 9.8, as:

$$\sigma_{\max} = \frac{a^2}{t^2} \cdot B_\sigma \cdot q \,. \tag{9.32}$$

The characteristic climatic stress is defined by:

$$\sigma^* = \frac{a^{*2}}{2t^2} \cdot B_\sigma \cdot p_0 \tag{9.33}$$

This results in the effective, climatically induced bending stress in a plate:

$$\sigma = \frac{2 \cdot (a/a^*)^2}{1 + (a/a^*)^4} \sigma^* \,. \tag{9.34}$$

Figure 9.26 represents this graphically. It is clear that the maximum load on the glass panes occurs when the smaller edge length of the glazing a equals the characteristic edge length a^*. In this case, the insulating glass factor is $\varphi = 0.5$. Climatic loads in a double glazing unit are regarded as significant if the relevant

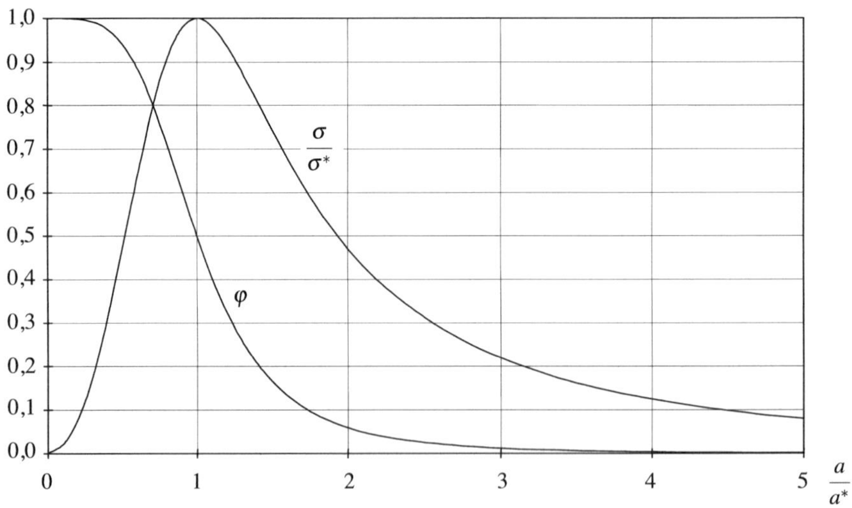

Fig. 9.26 Portion of the climatically induced bending stress and insulating glass factor φ for a rectangular double glazing unit, all-side supported (a: short edge length [17])

length is up to three times the characteristic edge length. For insulating glass units with two or more cavities, similar principles apply. The result is a linear equation system with n equations for the pressure in the n IGU cavities. This makes the problem rather complex. For further details, refer to [37, 105]. However, it should be noted that significantly higher climatic stress occurs in triple glazing units, due to the larger gas volume enclosed compared to double glazing units. Triple glazing units also have twice the amount of edge spacers compared to double glazing units. This leads to a higher failure risk when using triple glazing units compared to double glazing units and therefore requires a higher degree of quality control.

9.8.2 *Coupling Effect for Insulating Glass Units*

The hermetically sealed IGU cavity creates a coupling effect between the glass panes, also known as load-sharing, when the IGU is loaded laterally (i.e. out-of-plane). When one of the panes deflects under load, the volume of the gas filled cavity changes, thereby leading to a pressure change which in turn causes the other pane(s) to deflect, see Fig. 9.27 [9, 14, 96, 105].

The external load is thus distributed to all glass panes of the insulating glass unit. The distribution of load among the panes primarily depends on the relative stiffness of the individual panes.

This is illustrated in the simple case of a double glazed unit below (see Fig. 9.24b). Here, the lateral bending stiffness of the outer glass pane relative to the bending stiffness of the whole IGU can be expressed as:

$$\delta_\text{o} = \frac{t_\text{o}^3}{t_\text{o}^3 + t_\text{i}^3}. \tag{9.35}$$

The relative stiffness of the the inner glass pane is therefore given by:

$$\delta_\text{i} = 1 - \delta_\text{o}\,. \tag{9.36}$$

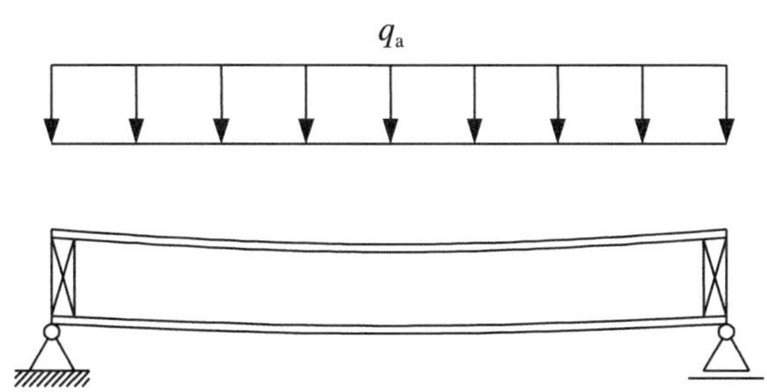

Fig. 9.27 Coupling effect for double glazing units

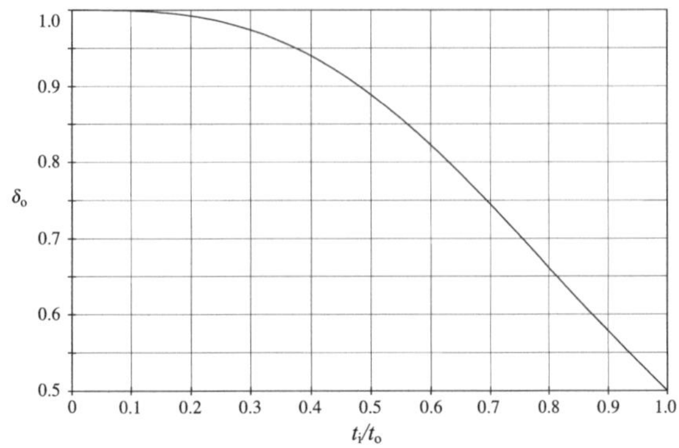

Fig. 9.28 Relative stiffness, δ_o for the outer glass plate of a double glazing unit ($t_o \geq t_i$)

Table 9.16 Relative stiffness, δ_o for the outer glass plate of a two-pane insulating glass unit

	t_i [mm]			
t_o [mm]	4	6	8	10
4	0.50	0.23	0.11	0.06
6	0.77	0.50	0.30	0.18
8	0.89	0.70	0.50	0.34
10	0.94	0.82	0.66	0.50

Table 9.17 Distribution of the effects of two pane insulating glass unit

Load applied to	**Resulting load share on**	
	The outer glass plate $\bar{q}_a$	**The Inner glass plate $\bar{q}_i$**
The outer glass plate q_a	$(\delta_a + \varphi\delta_i) \cdot q_a$	$(1 - \varphi) \cdot \delta_i \cdot q_a$
The inner glass plate q_i	$(1 - \varphi) \cdot \delta_a \cdot q_i$	$(\varphi\delta_a + \delta_i) \cdot q_i$

Figure 9.28 shows the relative stiffness of the outer glass plate, δ_o as a function of the glass thickness ratio. Table 9.16 provides the value of the relative stiffness of the outer glass plate, δ_o, for common glass thicknesses.

The distribution of the load on the outer and inner glass plates is given in Table 9.17. It should be noted that some pressures (e.g. suction from wind load) do not actually consist of suction on one of the two panes, but are rather an indoor-outdoor pressure difference across the IGU.

Similar principles apply for insulating glass units with two or more IGU cavities, [37]. In addition to the analytical approach described above, it is also possible to analyse insulating glass units by using numerical methods (e.g. finite element analysis). The added benefit of this is that it is possible to assess the coupling effects for large deflections, by accounting for the geometrically nonlinear response as described in Section 9.6.

9.9 Edge Protection

Freely accessible, edges (end faces) of glass panes may need protection from hard impacts by design measures. The edge protection can be in the form shielding provdied by other building components such as mullions, transoms or adjacent glass panes or by installation into adjacent components such as walls or ceilings. For example, the German standard, [76], states that edges of glass used for fall-through protection are adequately protected if the distance to e.g. a wall or another glass is no more than 30 mm. If this distance cannot be provided, the glass edge must be equipped with a robust edge protector. The effectiveness of the edge protection shall be demonstrated experimentally by component tests if the geometries shown in Fig. 9.29 cannot be achieved. At parapet corners or final edges which are not effectively protected by posts, solid components or by permanently attached edge protection, the total failure of the parapet, must be considered.

Figure 9.30a shows the edge protection device schematically shown in Fig. 9.29 on a free glass edge. In the area of all-glass corners of the construction, one of the laminated safety glass elements was designed as stepped glass, see Fig. 9.30b. To protect the edges of the adjacent glass edge, this stepped glass completely overhangs the corner area.

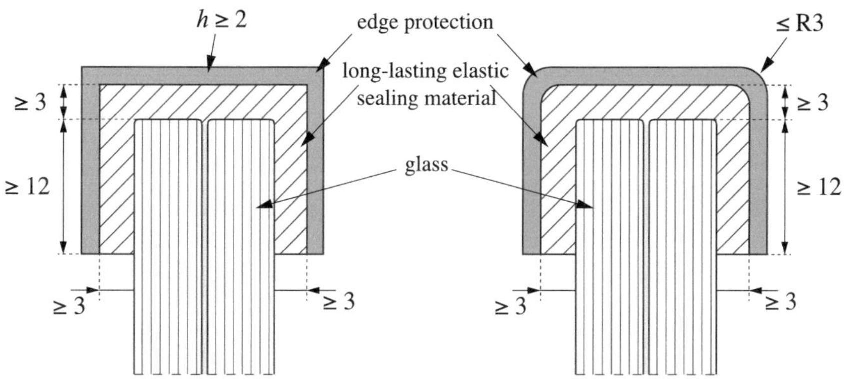

Fig. 9.29 According to DIN 18008-4 tested and effective edge protection (data in mm)

Fig. 9.30 Examples of edge protection designs: **a** adhesively bonded U-shaped stainless steel profiles, **b** use of stepped glass to protect the glass edges arranged behind them

9.10 Designing Environmentally Sustainable Glass

9.10.1 Assessing Environmental Performance

Life-cycle assessment (LCA) methods are used to quantitatively evaluate the environmental performance of products and services. This approach is particularly useful for designing and specifying environmentally sustainable glass solutions, but its use in practical glass design is still in its infancy. The first step is to define a functional unit as a reference, for example, 1 m^2 of flat glass. A system boundary that includes predefined LCA stages, as shown on Fig. 9.31, can then be selected, and subsequently evaluated with reference to the functional unit. The output from the LCA can be in the form of environmental impact indicators, such as global warming potential (GWP) in terms of greenhouse gas emissions (GHG), ozone depletion, water use, etc. [38].

The energy inputs and emissions associated with stages A1-A5 are often referred to as embodied. In the context of glass, life-cycle stages A1-A3 include all the energy utilised for raw material extraction/sourcing/processing, transportation and primary- and secondary-manufacture into the final glass product, for example, 1 m^2 of thermally toughened glass.

Stages B1-B7 relate to the energy and emissions associated with the use phase of the glass product including maintenance, repair, replacement, refurbishment, water use and operational energy. Therefore, the energy and emissions associated with

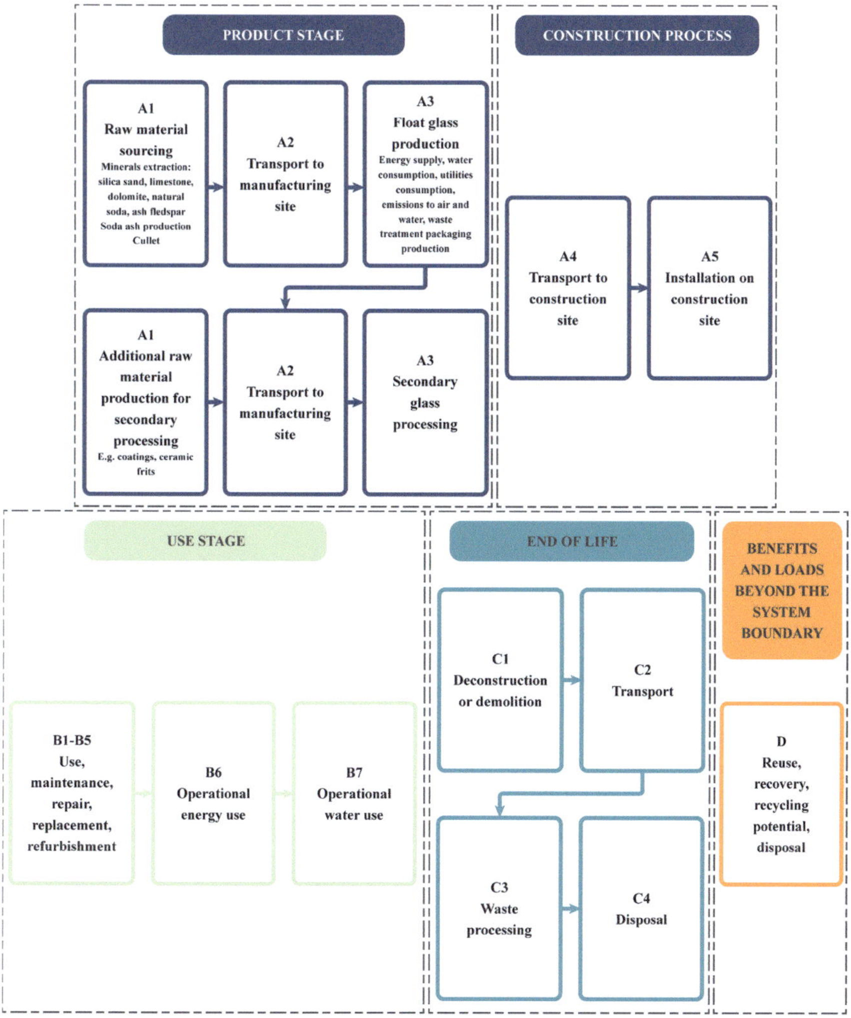

Fig. 9.31 Life cycle assessment stages for a flat glass product based on ISO14040 [38]

heating, cooling, lighting and ventilating of the indoor environment enclosed by the glazing would for example, fall under life-cycle stage B6. In the context of glass, typically B6 would evaluate the performance of the glass component within its final product, e.g. as part of a double- or triple-glazing unit.

Life-cycle stages C1-C4 refer to the energy and emissions associated with deconstruction or demolition, transportation of materials from site, waste processing and disposal. The corresponding energy and emissions can vary significantly depending on the recovery route, for example, recycling to flat-glass quality cullet in domestic locations compared with recycling to an aggregate-grade application in international locations.

Stage D is associated with the energy and/or emission savings that may be achieved by some form of recovery process when the unit is removed at the end of its initial service life. It sits outside the system boundary of the life-cycle assessment, to avoid double-counting energy and/or emission savings. However, stage D deserves equal levels of attention at the design stage (e.g. design for disassembly/re-use/recycling strategies), to ensure that the highest value recovery method can be achieved at the end-of-life stage. This, in turn, opens up available resource for new glass products through reuse and/or recycling.

Comparative LCAs are performed to assess alternative products/solutions with reference to a functional unit. Such evaluation methods should avoid focusing on one life-cycle stage in isolation, to ensure that the impact of the product/solution choices made are holistic and effective. For example, a higher embodied energy glass product (e.g. high-performance glazing) may sometimes be justified, if the upfront energy required in production is outweighed by proven reductions in operational energy over its lifetime in use.

There is currently a lack of established tools and methods for evaluating reuse and recycling (stage D) options. This would ensure that design decisions carefully balance optimisations in life-cycle performance (stages A-C) and benefits beyond the life-cycle (stage D). These methods and tools are subject of ongoing research [94, 106, 119].

9.10.2 Embodied Energy and Carbon

9.10.2.1 Energy Inputs and Corresponding CO_2 Emissions

The production of glass products consumes energy through the energy-intensive process of melting constituent raw materials and cullet, and subsequent primary and secondary forming/processing methods. The secondary processing methods that are required for different glass products vary, see Fig. 2.1 and the corresponding energy inputs and associated emissions used in an LCA will also differ accordingly, see Fig. 9.32.

Embodied carbon refers to the (CO_2-equivalent) greenhouse gas (GHG) emissions associated with the sourcing and processing of raw materials (24 %) and subsequent glass manufacturing processing steps. Emissions of GHGs from glass manufacturing arise due to the combustion of fuel used to power the glass furnace (49 %), often referred to as combustion emissions, and the decomposition of calcinated raw materials (15 %), commonly referred to as process emissions [63, 112, 117]. Glass furnaces require significant amounts of energy to heat and melt the raw materials to form molten glass. Float glass furnaces are typically powered through the combustion of natural gas, although other fuels and/or electricity may be used as a back-up to natural gas [49]. The actual mix of fuels will be site-specific and some alternative

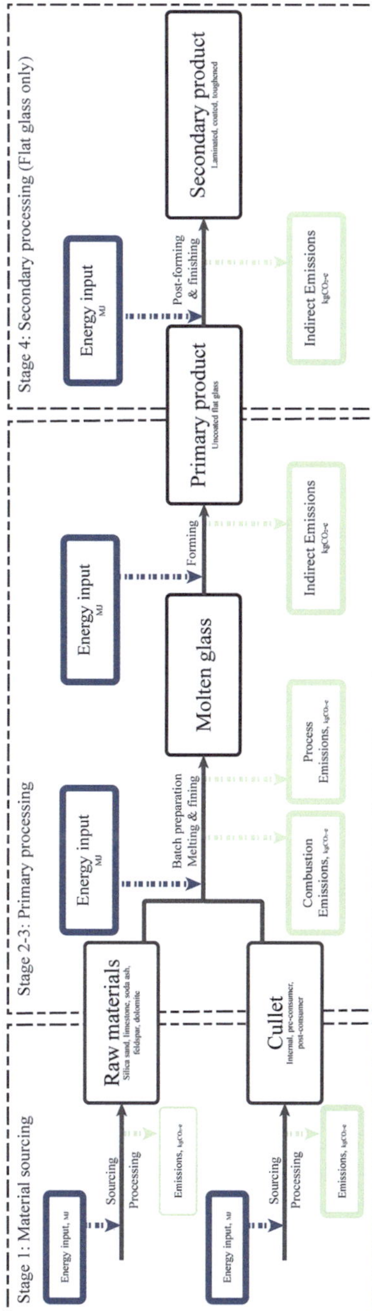

Fig. 9.32 Flat glass processing: relevant energy inputs and associated emissions [117]

Table 9.18 Typical values of embodied energy and carbon for stages life-cycle stages A1-A3, raw materials to final glass product

Product	Embodied energy	Embodied carbon
[%RC = % Recycled Content]	[MJ/kg]	[$kgCO_2 - eq/kg$]
Annealed glass (23 % RC) [7]	14.32	1.09
Low-E coated glass (23 % RC) [7]	15.99	1.35
PVB Laminated glass (23 % RC) [6]	22.61	1.39
Thermally pre-stressed glass (23 % RC) [9]	23.77	1.35
Concrete 32/40 MPa [10]	0.88	0.13
Extruded Aluminium (0 % RC) [10]	214	12.5
Extruded Aluminium (100 % RC) [10]	34	2.12

low-carbon fuel sources are now being considered [101]. Process emissions occur as a consequence of the carbonate raw materials used in glass production such as limestone ($CaCO_3$), dolomite ($CaMg(CO_3)_2$), and soda ash (Na_2CO_3). When heated in glass melting furnaces, these materials release CO2, which is subsequently emitted from the furnace [49]. Downstream emissions (12 %)[63, 112, 117] - attributed to electricity usage in powering equipment to process molten glass into its final product - constitute the remaining total embodied carbon emissions, see Table 9.18.

9.10.2.2 Use of Cullet in New Production

Waste glass, known as cullet, is often remelted in new glass production. The benefits of cullet usage are three-fold. Firstly, there is a saving in the energy and emissions associated with sourcing and processing of the equivalent quantity of raw materials multiplied by a factor of approximately 1.2. The 20 % input mass difference is a consequence of the decomposition losses incurred during the melting process of carbonate raw materials. Secondly, it allows for the furnace to operate at lower temperatures which consequentially reduces the energy consumption of the primary processing stage by 2.5–3 % for every 10 % increase in cullet [62]. This, in turn, reduces the corresponding combustion emissions. Finally, process emissions are reduced, due to the substitution of raw materials that need to be calcined, with cullet that has already undergone thermal decomposition. Thus, a 10 % increase in cullet, equates to a 3.9 % emissions in the primary processing stage alone, see Fig. 9.32 [117]. Glass manufacturers typically use 10–25 % of cullet in new glass production. The majority of this cullet is internal and/or pre-consumer cullet. A clear distinction must be made between internal/pre-consumer and post-consumer cullet.

Internal and pre-consumer cullet is generated internally during glass production and/or through the manufacturing of products that contain glass, becoming waste before ever reaching the consumer market [64]. An example of pre-consumer cullet is the off-cuts of jumbo sheet glass, which can then be re-melted for the production of new glass without even leaving the factory. Post-consumer cullet is waste glass arising when a glass product has reached its end-of-life after a period of time in use. At present, it is used in marginal quantities, typically <1 %.

9.10.3 Glass Recovery Hierarchy

In the European Union a Waste Framework Directive was developed to prevent or reduce the negative impacts caused by the generation and management of waste and to improve resource efficiency [46]. Located at the top of this hierarchy is the prevention of waste and minimisation of energy use by design. In order to achieve this, it is necessary to optimise the use of material, minimise internal waste and maximise the durability of the final glass products at the design stage. Beyond the first use of a glass product, there are several recovery methods to consider as illustrated in Fig. 9.33 in order of priority, with the prevailing aim of minimising resource use and environmental impact.

Reuse avoids the use of raw materials and energy-intensive processes and is generally the most environmentally favourable recovery method for flat glass. Depending on environmental exposure during use, product reconditioning may be required to

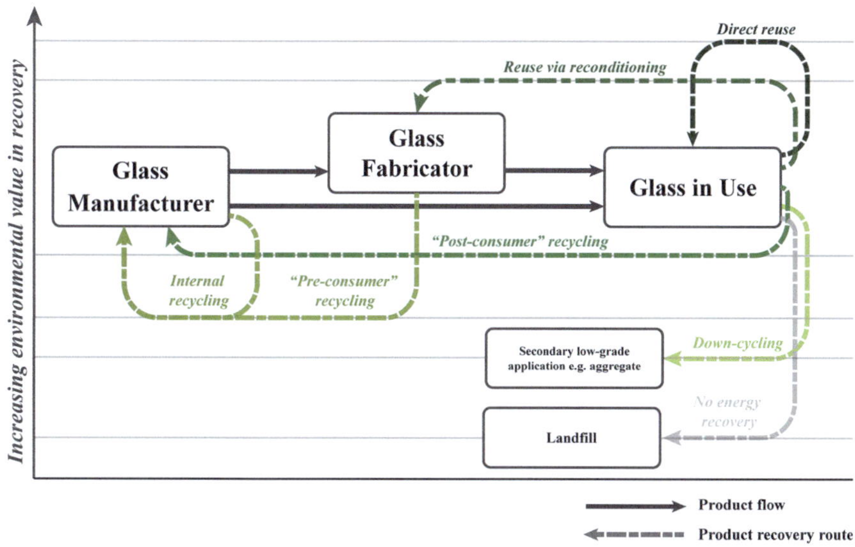

Fig. 9.33 Recovery hierarchy of flat glass products from manufacture to end-of-life application

provide performance assurance for the glass product beyond its first use. Despite the recognised environmental benefits of reuse [100], few secondary markets for reused flat glass products exist. In addition to technological challenges for assessing the residual performance of glass products, there is also a lack of business models and systems for reverse logistics. The latter are essential to avoid the high costs associated with transportation and storage and to ensure that reuse is both a financially- and environmentally-viable recovery option [113]. Various options for recycling exist. The preferred option is closed-loop recycling, where cullet is effectively separated and processed to produce "furnace-ready" cullet be used in place of primary raw materials. The environmental benefits of using cullet in new glass production are well-established, see Section 9.10.2. The second option for recycling is via the open-loop process, also known as down-cycling, where glass waste is considered as an additive into other manufactured goods, and the material undergoes a change to its characteristics [39]. Open-loop recycling processes are likely to yield less environmental savings than closed-loop recycling [33, 100]. The key challenges to increasing rates of closed-loop recycling lie in recovering flat glass cullet that is free from contaminants, with an acceptable quality, at a reasonable cost. Flat glass as a primary product, prior to any secondary processing methods, can be recycled in a closed-loop manner with no loss of quality. However flat glass products often incorporate other materials in the form of coatings, frits and interfaces with other materials such as adhesive polymers and metal parts. The acceptability criteria for the use of cullet in new flat glass production are very strict in order to avoid optical blemishes and maintain strict compliance with relevant performance standards [64]. The incorporation of unknown and uncontrolled compositions of cullet can result in high production losses and disrupt glass production for several days. A contaminant of concern is nickel-sulphide which can be formed by the presence of stainless steel flake contaminants in the glass melt, see Sect. 6.2.6.

Decisions taken during the early design stages can significantly influence the recycling options available and thus it is beneficial to consider end-of-life options—module D in Fig. 9.31—at the design stage. For example, cleanly collected thermally pre-stressed glass free from interface materials can be readily recycled in the float glass furnace. Conversely, ceramic-fritted glass is difficult to recycle due to the high melting temperature of ceramics leading to ceramic inclusions in the recycled glass product, which may in turn lead to glass fracture. The interlayers, such as PVB, used in laminated glass are also a barrier to direct recycling because the organic polyvinyl butyral interlayer, which can account for 10 % by weight of the laminated glass sheet, can alter the finely controlled redox state of the glass [22, 62]. Thus, laminated glass requires more specialised reprocessing techniques to separate the glass from the polymer interlayer in order to produce cullet suitable for new glass production. This usually involves breaking up the glass using a roller or similar mechanical action, followed by removal of the glass [60, 84]. The technical issues associated with the use of contaminated cullet in glass production are well detailed in [22, 62, 72, 81].

Some glass production processes and glass products are more tolerant to contaminants than the float process and are therefore more accepting of post-consumer glass cullet. For example, there are several sites for reprocessing end-of-life flat glass into "furnace-ready" cullet for use in container glass (bottles and jars) production, glass-fibre production for insulation products or foam glass or glass abrasives [32, 40, 64, 102]. Glass casting is also very well placed for recycling the difficult-to-recycle glass into high-value glass products, but techniques for doing so are not yet well established in industry and are the subject of ongoing research, see Sect. 2.2.6. Despite its recyclability, end-of-life (post-consumer) building glass is practically never recycled into new flat glass products. As an alternative it is either down-cycled by crushing it together with other building materials to be used in aggregate production or hard core for use on-site, or sent to landfill disposal. These recovery methods are deemed permissible in most regulatory frameworks due to the inert characteristics of glass. In addition, the costs for disposing of building glass is often less than the cost of collecting and treating it to make it suitable for recycling, particularly in countries with low landfill taxes/fees.

9.10.3.1 External Influence on Flat Glass Recycling

The principal barriers to the recycling of flat glass in the construction and demolition industries are the time, cost, safety risks and cross-contamination problems associated with recovery and collection [34]. Waste glass has a relatively low market value, due to the low cost of raw materials used in glass production and a lack of organized collection systems to efficiently generate clean "furnace-ready" cullet. Increased recycling rates will require better collection practices, easier separation of glass composite materials, and products that are designed for re-use/disassembly. To achieve this, changes are required in practices in the demolition and the replacement glazing industries [34]. Organisations such as VlakGlas in the Netherlands are attempting to create a financially-sustainable recovery supply-chain by setting up an extended producer responsibility scheme where fabricators pay a fee on the new production of flat glass products [114]. The revenue from this scheme goes towards the establishment of a network of collection, storage and processing facilities for post-consumer flat glass waste to be efficiently recycled into new glass products.

References

1. R. Courant, K. Friedrichs, *Supersonic Flow and Shock Waves* (Springer, New York, 1948)
2. M.L. Williams, R.F. Landel, J.D. Ferry, The temperature dependence of relaxation mechanisms in amorphous polymers and other glass-forming liquids. J. Am. Chem. Soc. **77.14**, 3701–3707 (1955)
3. S.M. Wiederhorn, Influence of water vapor on crack propagation in soda-lime glass. J. Am. Ceramic Soc. **50.8**, 407–414 (1967)
4. J.A. Hooper, On the bending of architectural laminated glass. Int. J. Mech. Sci. **15.4**, 309–323 (1973)
5. K. Stamm, H. Witte, *Sandwichkonstruktionen: Berechnung, Fertigung, Ausfü hrung* (Springer, Berlin, 1974)

6. B. Aalami, D.G. Williams, *Thin Plate Design for Transverse Loading* (Halsted Press Book, New York, 1975)
7. G. König, D. Hosser, W. Schobbe, Sicherheitsanforderungen für die Bemessung von baulichen Anlagen nach den Empfehlungen des NABau – eine Erläuterung, in *Bauingenieur*, vol. 57.2, pp. 69–78 (1982)
8. W.L. Beason, J.R. Morgan, Glass failure prediction model. J. Struct. Eng. **110.2**, 197–212 (1984)
9. F. Feldmeier, Belastung von Isoliergläsern durch Klimaschwankungen. Fenster und Fassade **2**, 41–52 (1984)
10. DIN 52338, *Pr ü fverfahren f ü r Flachglas im Bauwesen – Kugelfallversuch f ü r Verbundglas* (1985)
11. ETB. *Einheitliche Technische Baubestimmungen: Bauteile, die gegen Absturz sichern* (1985)
12. E. Wölfel, Nachgiebiger Verbund: Eine Näherungslösung und deren Anwendungsmöglichkeiten, in*Stahlbau*, vol. 56.6, pp. 173–180 (1987)
13. G. Spaethe, *Die Sicherheit tragender Baukonstruktionen* (Springer, Wien, 1992)
14. J.-D. Wörner, X. Shen, B. Sagmeister, Determination of load sharing in insulating glass units, in *Engineering Mechanics*, vol. 19.2, pp. 386–392 (1993)
15. J. Schneider, H.P. Schlatter, *Sicherheit und Zuverl ä ssigkeit im Bauwesen: Grundwissen f ü r Ingenieure* (VDF Hochschulverlag AG, Zürich, 1994)
16. F. Feldmeier, Zur Berücksichtigung der Klimabelastung bei der Bemessung von Isolierglas bei Überkopfverglasungen, in *Stahlbau*, vol. 65.8, pp. 285–290 (1996)
17. F. Feldmeier, Die Statik von Isolierglas, in *Glas im Bauwesen*. ed. by O. Technologie-Kolleg (Eigenverlag, 1997), pp.1–15
18. X. Shen, *Entwicklung eines Bemessungs- und Sicherheitskonzeptes f ü r den Glasbau*, vol. 138 (VDI, Düsseldorf, 1997)
19. J. Güsgen, Bemessung tragender Bauteile aus Glas. Dissertation. RWTH Aachen (1998)
20. H.W. Krewinkel, *Glasarchitektur* (Birkhäuser Verlag, Basel, 1998)
21. NF DTU 39, *Travaux de bâtiment – travaux de miroiterie-vitrerie – partie 1: cahier des clauses techniques* (1998)
22. R. Beerkens, Redox and sulphur reactions in glass melting processes. eng, in *Ceramics - Silikaty*, vol. 43.3, pp. 123–131 (1999)
23. A. Schutte, W. Hanenkamp, Zum Tragverhalten von Verbund- und Verbundsicherheitsglas bei erhöhten Temperaturen unter Einwirkungen von statischen und dynamischen Lasten. Bautechnik **76**, 49–63 (1999)
24. G. Sedlacek, K. Blank, W. Laufs, J. Güsgen, *Glas im konstruktiven Ingenieurbau* (Ernst and Sohn, Berlin, 1999)
25. A. Van Duser, A. Jagota, S.J. Bennison, Analysis of glass/polyvinyl butyral laminates subjected to uniform pressure. J. Eng. Mech. **125.4**, 435–442 (1999)
26. J. Schneider, Festigkeit und Bemessung punktgelagerter Gläser und stoßbeanspruchter Gläser. Dissertation. Technische Universität Darmstadt (2001)
27. EN356:2002, *Glass in building—Security glazing—Testing and classification of resistance against manual attack* (2002)
28. M. Seshadri, S.J. Bennison, A. Jagota, S. Saigal, Mechanical response of cracked laminated plates. Acta Mater. **50**, 4477–4490 (2002)
29. EN 12600, *Glass in building—Pendulum tests—Impact test method and classification for flat glass* (2003)
30. *Glass in building*—Pendulum test—*Impact test method for float glass*, Feb. 2003
31. TRAV, *Technische Regeln f ü r die Verwendung von absturzsichernden Verglasungen, Deutsches Institut f ü r Bautechnik (DIBt), Fassung Januar 2003* (2003)
32. G. Brusatin, E. Bernardo, G. Scarinci, Production of foam glass from glass waste. eng, in *Proceedings of the International Conference on Sustainable Waste Management and Recycling: Glass Waste*, pp. 67–82 (2004)
33. J. Flanagan, M. Davies, *Glass Recycling—Life Cycle Carbon Dioxide Emissions* (Technical report, Enviros Consulting Ltd, 2004)

34. P.J. Mahon, A.J. Dawe, The impact of government policy on glass recycling rates. eng, in *Proceedings of the International Conference on Sustainable Waste Management and Recycling: Glass Waste*, pp. 17–29 (2004)
35. EN 14449, *Glass in building—Laminated glass and laminated safety glass—Evaluation of conformity/Product standard* (2005)
36. M. Kutterer, Verbundglasplatten – Schubverbund und Membrantragwirkung, in *Stahlbau*, vol. 74.1–2, pp. 39–46 (2005)
37. F. Feldmeier, Klimabelastung und Lastverteilung bei Mehrscheiben-Isolierglas, in *Stahlbau*, vol. 75.6, pp. 467–478 (2006)
38. *ISO 14040:2006 Environmental management—Life cycle assessment—Principles and framework* (2006)
39. *ISO/IEC 14044:2006 Environmental management—Life cycle assessment—Requirements and Guidelines* (2006)
40. A. Kasper, Recycling of cullet into flat glass melting furnaces. eng, in *Ceramic Engineering and Science Proceedings*, vol. 27.1, pp. 169–179 (2006)
41. A. Kott, Zum Trag- und Resttragverhalten von Verbundsicherheitsglas. Dissertation. ETH Zürich (2006)
42. TRLV, *Technische Regeln f ü r die Verwendung von linienf ö rmig gelagerten Verglasungen, Deutsches Institut f ü r Bautechnik (DIBt), Schlussfassung Ausgust 2006* (2006)
43. TRPV, *Technische Regeln f ü r die Bemessung und Ausf ü hrung punktf ö rmig gelagerter Verglasungen, Deutsches Institut f ü r Bautechnik (DIBt), Schlussfassung Ausgust 2006* (2006)
44. M. Fahlbusch, Zur Ermittlung der Resttragfähigkeit von Verbundsicherheitsglas am Beispiel eines Glasbogens mit Zugstab. Dissertation. Technische Universität Darmstadt (2007)
45. M. Haldimann, A. Luible, M. Overend, *Structural use of glass internat* (Association for Bridge and Structural Engineering, Zürich, 2008)
46. E. Parliament, O.T. Council, *Directive 2008/98/ec On Waste And Repealing Certain Directives*, Nov. 2008
47. B. Weller, K. Härth, S. Tasche, S. Unnewehr, *Konstruktiver Glasbau* (Institut für internationale Architektur-Dokumentation, München, 2008)
48. prEN 13474-3, *Glass in building—Determination of the strength of glass panes—Part 3: General method of calculation and determination of strength of glass by testing* (2009)
49. *Technical Support Document for the Glass Manufacturing Sector: Proposed Rule for Mandatory Reporting of Greenhouse Gases*. Technical report Office of Air and Radiation, U.S. Environmental Protection Agency (2009)
50. CWCT, *Technical Note 66: Safety and Fragility of Overhead Glazing: Guidance on Specification*. Technical report. Centre for Windows and Cladding Technilogy (2010)
51. CWCT, *Technical Note 67: Safety and Fragility of Glazed Roofing: Testing and Assessment*. Technical report. Centre for Windows and Cladding Technology (2010)
52. DIN 18008-1, *Glas im Bauwesen – Bemessungs- und Konstruktionsregeln – Teil 1: Begriffe und allgemeine Grundlagen* (2010)
53. EN 1990, *Eurocode: basis of structural design* (2010)
54. EN 1990/NA, *National Annex—Nationally determined parameters—Eurocode: Basis of structural design* (2010)
55. EN 1991-1-1, *Eurocode 1: Actions on structures—Part 1-1: General actions—Densities, self-weight, imposed loads for buildings* (2010)
56. EN 1991-1-3, *National Annex—Nationally determined parameters—Eurocode 1: Actions on structures—Part 1-3: General actions—Snow loads* (2010)
57. EN 1991-1-4, *National Annex—Nationally determined parameters—Eurocode 1: Actions on structures—Part 1-4: General actions—Wind actions* (2010)
58. EN 1991-1-5, *National Annex—Nationally determined parameters—Eurocode 1: Actions on structures—Part 1-5: General actions—Thermal actions* (2010)
59. S. Feirabend, Steigerung der Resttragfähigkeit von Verbundsicherheitsglas mittels Bewehrung in der Zwischenschicht. Dissertation. Universität Stuttgart (2010)

60. C.F. Acevedo, L.M.D.M. Osaba, E.O. Arteta, G.M. Herrera, A.F. María, Polyvinyl Butyral Recycling Method. PCT/ES2008/000175 (2011)
61. J. Barredo, M. Soriano, M.S. Gómez, L. Hermanns, Viscoelastic vibration damping identification methods. Application to laminated glass, in *Procedia Engineering*, vol. 10, pp. 3208–3213 (2011)
62. R. Beerkens, G. Kers, E. Van Santen, Recycling of post-consumer glass: Energy savings, CO2 emission reduction, effects on glass quality and glass melting. eng, in *Ceramic Engineering and Science Proceedings*, vol. 32.1, pp. 167–194 (2011)
63. A. Schmitz, J. Kamiński, B. Maria Scalet, A. Soria, Energy consumption and CO2 emissions of the European glass industry. eng, in *Energy Policy*, vol. 39.1, pp. 142–155 (2011)
64. E.R. Vieitez, P. Eder, A.V. Krzyzaniak, H. Saveyn, *End-of-Waste Criteria for Glass Cullet: Technical Proposals* (Publications Office of the European Union, Technical report, 2011)
65. ASTM E1300, *Standard practice for determining load resistance of glass in buildings* (2012)
66. D. Callewaert, J. Belis, D. Delincé, R. Van Impe, Experimental stiffness characterisation of glass/ionomer laminates for structural applications. eng, in *Construction and Building Materials*, vol. 37, pp. 685–692 (2012)
67. EN 1863-1, *Glass in building—Heat strengthened soda lime silicate glass—Part 1: Definition and description* (2012)
68. EN 572-1, *Glass in building—Basic soda-lime silicate glass products—Part 1: Definitions and general physical and mechanical properties*, Beuth (2012)
69. ETAG 002, *Guideline for european technical approval for structural sealant glazing kits (SSGK)— Part 1: Supported and unsupported systems, EOTA* (2012)
70. W. Graf, S. Freitag, J.-U. Sickert, M. Kaliske, Structural analysis with fuzzy data and neural network based material description, in *Computer-Aided Civil and Infrastructure Engineering*, vol. 27.9, pp. 640–654 (2012)
71. P. Hooper, R.A.M. Sukhram, B.R.K. Blackman, J.P. Dear, On the blast resistance of laminated glass. Int. J. Solids Struct. **49**, 899–918 (2012)
72. IMPEL, *Summary of Energy-Related Information for the Manufacture of Glass and Proposal for the Sector Specific Supplement to the Draft Application Form for Energy Efficiency*, July 2012
73. S. Kolling, J. Schneider, N. Gebbeken, M. Larcher, C. Alter, J. Kuntsche, Deformations- und Bruchverhalten von Verbundsicherheitsglas unter dynamischer Beanspruchung, in *Stahlbau*, vol. 81.3, pp. 219–225 (2012)
74. G. Siebert, I. Maniatis, *Tragende Bauteile aus Glas. Grundlagen, Konstruktion, Bemessung, Beispiele* (Ernst and Sohn, Berlin, 2012)
75. DIN 18008-3, *Glas im Bauwesen—Bemessungs- und Konstruktionsregeln—Teil 3: Punktf ö rmig gelagerte Verglasungen*. Beuth (2013)
76. DIN 18008-4, *Glas im Bauwesen—Bemessungs- und Konstruktionsregeln—Teil 4: Zusatzanforderungen an absturzsichernde Verglasungen* (2013)
77. DIN 18008-5, *Glas im Bauwesen—Bemessungs- und Konstruktionsregeln—Teil 5: Zusatzanforderungen an begehbare Verglasungen* (2013)
78. J. Kuntsche, J. Schneider, Explosion resistant glazing—experimental tests and numerical simulation, in *Glass Performance Days 2013*. Tampere, pp. 85–89 (2013)
79. prEN 16612, *Glas im Bauwesen – Bestimmung des Belastungswiderstandes von Glasscheiben durch Berechnung und Pr ü fung* (2013)
80. prEN 16613, *Glas im Bauwesen – Verbundglas und Verbundsicherheitsglas – Bestimmung der mechanischen Eigenschaften von Zwischenschichten* (2013)
81. A. Sissa, L. Delgado Sancho, S. Roudier, B. Scalet, M. Garcia Muñoz, *Best available techniques (BAT) reference document for the manufacture of glass : industrial emissions Directive 2010/75/EU: integrated pollution prevention and control*. Technical report. Joint Research Centre and Institute for Prospective Technological Studies (2013)
82. M. Feldmann, R. Kasper, *Guidance for European Structural Design of Glass Components—Support to the implementation, harmonization and further development of the Eurocodes (SaT-report)* (2014)

83. C. O'Regan, *Structural Use of Glass in Buildings*. eng. ed. by I. of Structural Engineers. 2nd ed. (Institution of Structural Engineers, 2014)
84. M. Tupy, P. Mokrejs, D. Merinska, P. Svoboda, J. Zvonicek, Windshield recycling focused on effective separation of PVB sheet". eng. J. Appl. Polym. Sci. **131.4**, 39879 (2014)
85. M. Vandebroek, Thermal Fracture of Glass. Ph.D. thesis. Ghent University (2014)
86. M. Vandebroek, C. Louter, R. Caspeele, F. Ensslen, J. Belis, Size effect model for the edge strength of glass with cut and ground edge finishing. Eng. Struct. **79**, 96–105 (2014). (Nov.)
87. DIBt, *Muster-Liste der Technischen Baubestimmungen—Fassung Juni 2015* (2015)
88. EN 12150-1, *Glass in building—Thermally toughened soda lime silicate safety glass—Part 1: Definition and description* (2015)
89. J. Franz, Untersuchungen zur Resttragfähigkeit von gebrochenen Verglasungen. Dissertation. Technische Universität Darmstadt (2015)
90. J. Kuntsche, Mechanisches Verhalten von Verbundglas unter zeitabhängiger Belastung und Explosionsbeanspruchung. Dissertation. Technische Universität Darmstadt (2015)
91. J. Pelfrene, J. Kuntsche, S. Van Dam, W. Van Paepegem, J. Schneider, Critical assessment of the post-breakage performance of blast-loaded laminated glazing: experimental and numerical simulation. Int. J. Impact Eng. **88**, 61–71 (2015)
92. S. Schula, Charakterisierung der Kratzanfälligkeit von Gläsern im Bauwesen. Dissertation. Technische Universität Darmstadt (2015)
93. ASTM E1300, *Standard practice for determining load resistance of glass in buildings* (2016)
94. P. Beurskens, M. Bakx, M. Ritzen, E. Durmisevic, J. Lichtenberg, A morphological design and evaluation model for the development of circular facades. English, in *Sustainable built environment: Transition zero*. Sustainable Built Environment: Transition Zero 2016 : The Utrecht SBE16 Conference ; Conference date: 07 Apr. 2016 Through 08 Apr. 2016. Apr. 2016, pp. 252–268
95. R. Green, *The challenges of writing a structural standard for glass*. eng. (2016)
96. S.M. Morse, H.S. Norville, Comparison of methods to determine load sharing of insulating glass units for environmental loads, in *Glass Structures and Engineering*, vol. 1.1, pp. 315–329, June 2016
97. C. Pascual, J. Montali, M. Overend, Adhesively-bonded GFRP-glass sandwich components for structurally efficient glazing applications. eng, in *Composite Structures*, vol, 160, pp. 560–573 (2016)
98. DIN 18008-6, *Glas im Bauwesen—Bemessungs- und Konstruktionsregeln—Teil 6: Zusatzanforderungen an zu Instandhaltungsmaßnahmen betretbare Verglasungen und und an durchsturzsichere Verglasungen* (2018)
99. S.C. Angelides, J.P. Talbot, M. Overend, The effects of high strain-rate and in-plane restraint on quasi-statically loaded laminated glass: a theoretical study with applications to blast enhancement. eng, in *Glass Structures and Engineering*, vol. 4.3, pp. 403–420 (2019)
100. R. Hartwell, M. Overend, Unlocking the re-use potential of glass Façade systems, in *Glass Performance Days* (2019)
101. R. Ireson, A. Fuller, J. Woods, R. Simon, G. Andrews, S. H. P. Bingham, M. Davies, *Alternative Fuel Switching Technologiesfor the Glass Sector*. Technical report. BEIS Industrial Fuel Switching Phase 2, 2019
102. R. Lebullenger, F.O. Mear, Glass recycling. eng, in *Springer Handbook of Glass* (Springer International Publishing, 2019), pp. 1355–1377
103. DIN 18008-1, *Glas im Bauwesen—Bemessungs- und Konstruktionsregeln—Teil 1: Begriffe und allgemeine Grundlagen* (2020)
104. DIN 18008-2, *Glas im Bauwesen – Bemessungs- und Konstruktionsregeln – Teil 2: Linienfoermig gelargerte Verglasungen* (2020)
105. L. Galuppi, G. Royer-Carfagni, Betti's Analytical Method for the load sharing in double glazed units, in *Composite Structures*, vol. 235, p. 111765, Mar. 2020
106. R. Hartwell, M. Overend, End-of-Life challenges in facade design: a disassembly framework for assessing the environmental reclamation potential of facade systems, in *Facade Tectonics World Congress* (2020)

107. C. Pascual, M. Overend, Buckling of sandwich struts with a particular application in composite multi-layer glazing. eng., in *Engineering Structures*, vol. 230, p. 111545 (2020)
108. S.C. Angelides, J.P. Talbot, Blast response of laminated glass panels: a critical review of analysis and design methods. eng., in *Proceedings of the Institution of Civil Engineers: Structures and Buildings*, pp. 1–13 (2021)
109. *CEN/TS-19100-1:2021 Design of glass structures—Part 1: Basis of design and materials*. eng. Technical report 2021
110. *CEN/TS-19100-2:2021 Design of glass structures—Part 2: Design of out-of-plane loaded glass components*. eng. Technical report (2021)
111. *CEN/TS-19100-3:2021 Design of glass structures—Part 3: Design of in-plane loaded glass components and their mechanical joints*. eng. Technical report (2021)
112. G. Europe, *EPD Flat glass: Uncoated flat glass, laminated safety glass and coated flat glass* (2021)
113. R. Hartwell, S. Macmillan, M. Overend, Circular economy of façades: real-world challenges and opportunities. eng., in *Resources, Conservation and Recycling*, vol. 175, p. 105827 (2021)
114. Vlakglas, *Vlakglasrecycling Nederland* (2021). Accessed 25 Oct. 2021. `https://www.vlakglasrecycling.nl/`
115. S.C. Angelides, J.P. Talbot, M. Overend, The influence of fracture pattern on the residual resistance of laminated glass at high strain-rates: an experimental investigation of the post-fracture bending moment capacity based on time-temperature mapping of interlayer yield stress. eng., in *Glass Structures and Engineering*, vol. 7.4, pp. 549–568 (2022)
116. G. Coult, M. Overend, Developments in structural glass. eng., in *Structural Engineer*, vol. 100.2, pp. 22–24 (2022)
117. R. Hartwell, G. Coult, M. Overend, Mapping the flat glass value-chain: a material flow analysis and energy balance of UK production. eng., in *Glass Structures and Engineering*, pp. 1–26 (2022)
118. M. Feldmann et al., The new CEN/TS 19100: design of glass structures, in *Glass Structures and Engineering* (2023)
119. R. Hartwell, M. Overend, Reclamation potential in the built environment: a method and metric for assessing environmental benefits beyond first use, in *Building and Environment*, vol. 263, 111866 (2024)
120. M. Overend, Structural design of buildings: elemental design, in *Glass in facades and other applications* (ICE Publishing, Leeds, July 2024)
121. D. Delincé, Experimental approaches for assessing time and temperature dependent performances of fractured laminated safety glass. eng., Ph.D. thesis. Ghent University, XXVIII, 279 (2014)

Chapter 10
Maintenance and Damage Analysis

This Chapter discusses maintenance and damage analysis of glass. After a general introduction in Sect. 10.1, *the subsequent Sect.* 10.2 *deals with cleaning, maintenance and repair. Additionally, Sect.* 10.3 *presents the basics of fractography and glass failure assessment.*

10.1 General

Glass components require particular and careful handling throughout the stages of the design, execution, repair and dismantling. Additionally, appropriate cleaning and maintenance procedures must be considered during service life to ensure that the serviceability and load-bearing capacity of glass structures are not impaired.

When damage occurs, the question often arises as to the cause. In case of glass fracture, fractography can be used to deduce the root cause e.g. the fracture origin and the related stress level at fracture. Furthermore, the fracture patterns can in themselves provide clues on the probable causes of faiure. Aside from failure by fracture, glass damage may also manifest itself in the form of optical defects. Such optical or aesthetic defects may occur instantly or after a certain period of time after installation, which may require replacement of the glass.

10.2 Cleaning, Maintenance and Repair

10.2.1 Cleaning of Glass Constructions

During the construction of buildings, window and façade elements are often installed to make the building water and air tight before the interior finishes are installed.

J. Belis et al., *Engineering Glass Structures*,
https://doi.org/10.1007/978-3-032-03233-1_10

Fig. 10.1 Example of carelessness with glass surfaces on a building site

The large variations in temperature (e.g. before the heating and cooling systems are operational), and humidity (e.g. from screeding or plastering), may expose the glass to worse conditions during the temporary construction phase rather than during the operational service life. This may warrant the use of preventive protective measures to avoid damage to the glass surface (e.g. application of adhesive films, see Fig. 10.1).

Particularly when construction is still ongoing, the finishing works constantly may cause sticky contamination of the glass with mineral building materials (e.g. cement), lime, silicones and adhesive residues; in addition, through ongoing construction, dust-like, granular contamination with mineral substances (e.g. sand) may occur [27]. This can become critical during the final cleaning after completion of the construction phase, as impurities, such as concrete residues, are generally completely hardened at this time and, in practice, are often removed with the help of a glass scraper and razor-like blade, see Fig. 10.2. The glass scraper consists of a handle and a broad, sharp blade, for removing the deposited particles from the surface (with a forward movement). In the scraping process, some impurities can get caught under the blade, causing scratches on the glass [19]. The risk of scratching with the glass scraper is widely known and documented in numerous reports [11, 12, 14]. Although its use is prohibited in some guidelines for glass cleaning [27, 30], as well as in glass cleaning instructions by the glass producers, it is still regularly applied. The impact of inappropriate cleaning operations on the glass strength (along with other forms of damage) is shown in Fig. 10.3. This demonstrates that annealed float glass in particular is more sensitive to surface damage than thermally pre-stressed glass [35].

Routine window cleaning during the operational stages of a building normally follows a standardised scheme [16, 27]. First of all, the surface to be cleaned is brushed, blow-cleaned or vacuum cleaned. Persistent impurities are usually pre-wetted. When cleaning framed windows, the frame is cleaned first. Then the glass surfaces are moistened with water and detergent using a thick velour cloth or sponge. In the next step the glass surface is wiped with a window cloth or squeegee. The remaining water residues are removed with a chamois (or similar) cloth. During the washing process, hard elements end up on the surface of the glass in the cloth or sponge and so also in the washing water. Scratches can therefore be induced in the surface

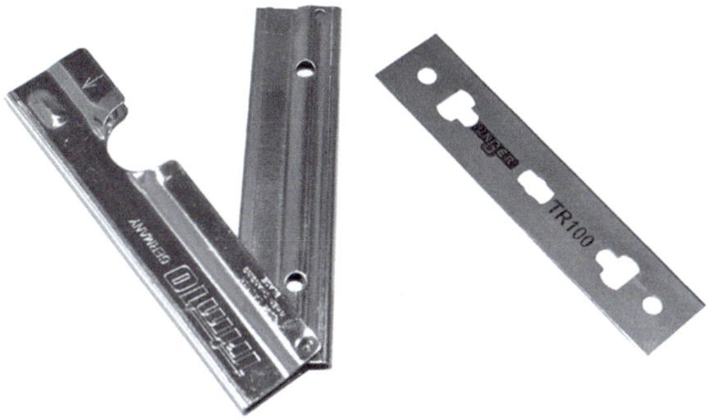

Fig. 10.2 Commercial glass scraper and blade

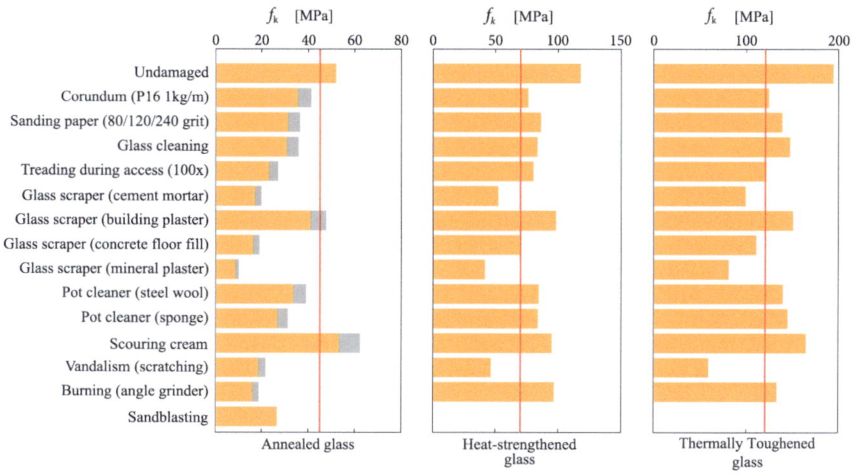

Fig. 10.3 Impact of real damage patterns on the characteristic strength of glass in the building industry (red line: declared norm, [35])

unless the washing water is changed regularly and the velour or sponge is rinsed often. Furthermore, When wiping away the water, dirt particles can end up under the rubber lip of the squeegee, thus creating clearly visible scratch marks. Further risks of scratching arise if the cleaning tools are improperly stored on contaminated surfaces (e.g. windowsills).

The cause of the scratch marks can be identified by their shape: while damage caused by a glass scraper is often characterised by many parallel scratch marks up to 10 cm long, scratch marks caused by a washer are up to 20 cm long and occur when residual dirt is pulled over the glass panel. Often these scratches are bow-shaped, with the radius of the length of a forearm over the glass surface [19]. In contrast, scratches caused by a squeegee wiper are often longer (up to 30 cm) and, indicating the direction of the work, run parallel to the glass edges. If they stretch to the vicinity of the glass borders, they often exhibit sharp direction changes of up to 180°.

Fig. 10.4 Intensive organic contamination of a mullion-and-transom roof construction in the area of the intersection points

10.2.2 Repair and Maintenance of Glass

Glass used in constructions typically requires significant cleaning, repair[1] and maintenance costs. This applies in particular to barrier, accessible and walkable glazing, as well as for structural glass components such as glass fins and beams. Here, proper inspection and maintenance is required to ensure the functionality throughout the service life. Due to the diverse uses of glass in construction, there is no single set of universal measures. However, key components (e.g. fittings, bolts, brackets, laminates, etc.) should be inspected and if necessary repaired or replaced.

Glazing in façade areas normally have a reduced lifespan, that falls short of the 50 years service life typically used for other construction elements in the construction industry. For vertical façades, the expected lifespans [15] are typically:

- 25 years for insulating glass units;
- 20 years for glass joint seals consisting of dry seals (e.g. gaskets);
- 12 years for glass joint seals consisting of wet seals (e.g. sealants).

It is worth noting that it is often the other materials used in the facade system rather than the glass itself that cause the reduced lifespan.

For glass roofs, the lifespans are even shorter, since the stresses, strains and dirt and damage accumulation are significantly higher in comparison with vertical façades. In particular, insulating glass units in roofs experience higher loads, so that replacements every 15 years should be planned. Regular cleaning has a positive effect on functional lifetime, due to the removal of organic impurities preventing clogging of the drainage system. Reference values for maintenance intervals are defined in material standards [28] and depend on the damage class and the robustness class of the building or the construction. For the mullion-and-transom roof construction, of which the intersection point details are shown in Fig. 10.4, proper cleaning has not

[1] The term maintenance refers to a measure to preserve the specified condition of buildings and structures.

Fig. 10.5 Results of poor maintenance: **a** weathered and brittle sealing joint; **b** damaged secondary seal of the double glazing unit; and, **c** in the space between the panes of the glazing, large amounts of condensed water accumulated in the cavity of an IGU

been carried out. Therefore the intended drainage of the roof construction (in this case, by means of narrow channels between the cover strips of the mullions and transoms) is not working properly. As a result, limited durability and impermeability are to be expected. The relevant standards for the repair of roofs with seals [20, 26] can be transferred to glass roof structures:

- Roof constructions should be inspected and repaired regularly (at least once or twice a year);
- The condition of roof seals, joints and borders, as well as the water tightness should be assessed and documented during the inspection;
- Unwanted deposits and biological growth should be removed as part of repair and maintenance.

Depending on the elements' location and orientation, joints in glass construction (see Sect. 8.3), should be classified as *maintenance joints*, since these may be exposed to chemical, biological, physical and mechanical influences, so that a limited durability must be assumed. This is particularly true for surfaces that are exposed

Fig. 10.6 Temperature differences in annealed float glass can trigger so-called thermal fracture

to direct weathering and atmospheric influences (UV radiation, humidity and temperature effects). These include, in particular, glass roof structures. Therefore, the functionality of this type of joint must be checked at regular intervals. If there are signs of malfunction, the joints should be replaced because otherwise damage in the form of leakage is likely to occur. The sealing joint of the horizontal glazed mullion-and-transom beam construction shown in Fig. 10.5a-b was about 15 years old at the time of the damage assessment. Regular maintenance had not taken place, so that defects in tightness in the roof construction only became noticeable when water penetrated to the interior. As the joint became brittle, see Fig. 10.5a the rebate was permanently exposed to high vapour pressure and moisture, the sealing of the insulating glass unit was under significant strain and experienced long-term damage, see Fig. 10.5b, so that large amounts of water penetrated into the cavity between the glass panes, see Fig. 10.5c. To repair the damage all roof glazing had to be replaced.

Glazing with coatings on the outside surface, (including self-cleaning glass), should receive regular maintenance, including regular cleaning of the glass surface; with self-cleaning glass the intervals between cleaning are longer than with ordinary glazing. Coatings made of titanium oxide must not come into contact with silicone-containing materials during their expected lifespan. This applies, for example, to silicone sealants for wet sealing or silicone sealing lips of squeegees, which are used for glass cleaning.

Care must also be taken when applying adhesive films, paints or prints on glazing or when partial shading occurs, especially if the glazing is highly absorbent, as this can lead to different/local heating of the glazing in direct sunlight, which in turn can result in critical stress conditions in the glass, see e.g. Section 10.3.2.

This is especially important for annealed glass, which has a significantly lower thermal shock resistance than thermally pre-stressed glass, and can result in glass

Table 10.1 Reference values for the thermal shock resistance of different types of glass*

Glass type	Thermal shock resistance (K)
Soda-lime silica float glass	40
Borosilicate float glass	100
Soda-lime silica thermally toughened glass	200
Soda-lime silica heat-strengthened glass	100
Chemically pre-stressed glass	100–200

*Values may be influenced by the glass edge finishing used

fracture (also known as *thermal fracture*), see Table 10.1 and Fig. 10.6. This also applies to the cleaning of glass, so that glazing should not be cleaned with large amounts of cold water while it is heated from direct exposure to the sun.

10.2.3 Repair of Glass Surfaces

Scratches on glass surfaces can be repaired only to a limited extent, depending on the extent of the damage. On-site repair is generally preferable because the size and installation circumstances of glass make it difficult and costly to temporarily remove the glass from its location. Often aesthetic reasons are considered, so that *polishing* or *grinding*, as discussed in [21], prevail. While the regulations in [30], in its references to the cleaning of glass, prevents such surface treatment, the regulations in [27] in this context refer explicitly to a repair method [18], with which scratches are removed using various grinding discs with different abrasiveness. According to the manufacturer's specifications, scratches that can be felt with the fingernails and surface damage occurring from large groups of scratches cannot be removed without subsequent detriment to the appearance. All abrasive techniques are similar in that increased material removal with a diagonal approach leads to clear visibility of the treated area through the disappearance of plane-parallel surfaces (referred to as the *lens effect*), which is why the the grinding depth should be no larger than 30 μm. This means that only the actual visual part of the scratch is removed; below this, deep cracks, see Section 5.2 and internal cracks remain in the glass, so that polishing or grinding cannot serve to restore the load-bearing capacity of a glass construction component, and instead it is far more likely that the strength will be further reduced by the additional damage [35].

As of yet, there is no suitable method to restore the load-bearing capacity of glass with surface defects when already installed. Some procedures have been tested on a laboratory scale, e.g. ion exchange, thermal treatment (fire polishing), etching, thermal and chemical pre-stressing, as well as application of coatings [7, 10, 22, 34, 36, 38]. Their application is limited to glass that is already in-service. Chemical etching with hydrofluoric acid could be suitable for local surface treatment. In the field of professional glass cleaning, [14] recommends removing more serious scratch damage by wetting the glass surface with water and subsequently applying three percent hydrofluoric acid. Adjacent and acid-sensitive components should be adequately protected here. The potential for strength improvement through etching has been experimentally proven in [35]. However, relevant safety regulations (the regulations on hazardous substances, [29]) for working with hydrofluoric acid have to be taken into account, so that the use of hazardous substances in the façade area will only be possible on a limited basis and when using elaborate protection measures for health and safety.

10.3 Damage Analysis

10.3.1 Fractography Basics

10.3.1.1 General Information

Fractography is the macroscopic and microscopic examination of the fracture structure and the crack surface morphology [4, 24]. On the basis of distinct physcial characteristics, it is often possible to identify the cause of the fracture of the component. In addition to a basic understanding of the stress distribution acting in the component, depending on the degree of detail of the investigation, appropriate experimental facilities are required. The minimum facilities include magnifiers with a range of magnifications, cleaning utensils, adequate lighting and camera equipment for taking close-ups of the object. Ordinary digital cameras are often sufficient.

Microscopes are often used for more detailed investigations. The basic equipment includes optical microscopes with magnifications of up to 1600X. Satisfactory three-dimensional measurements with standard optical microscopes are not possible due to the transparency of the glass; therefore the use of stereo microscopes is recommended. These have two eyepieces, through which the specimen can be viewed at different angles, thus achieving a greater depth effect. The magnification required is generally below 400X because the depth of focus is rapidly reduced with increasing magnification and so a spatial representation is no longer possible. In addition to the optical features of the microscope, the choice of lighting is also important. Conventional microscopes require an incident light source or, for transparent specimens, a transmitted light source is possible. Another option is the use of a ring light, which illuminates the sample surface evenly. For an assessment of the structure of a fractured surface all these variants are often unsuitable because the irregularities will be too evenly lit. A helpful solution here can be the use of external cold light sources with fibre-optic light cable, which can make the lighting targeted-oriented. In particular, a flat angle to the sample surface is useful for displaying the surface structure. Another possibility is the use of a dark-field illumination, in which light illuminates the sample circumferentially similar to the normal ring light, but from the side of the illumination device so that the surface structures are accentuated.

A polarisation microscope is a special form of optical microscope which functions according to the principle of photoelasticity. According to this, two polarisation filters are integrated above the transmitted light source in the beam path. The first polarisation filter is designated as the polariser and is directly in the path of the light source (in front of the specimen). The second is located behind the test object and is referred to as the analyser. The principle of photoelasticity (see e.g. Sect. 6.4) uses the effect of *birefringence* in the specimen. Through the effect of stress in a transparent, optically isotropic glass body (e.g. glass), the body obtains optically anisotropic properties that can be observed at specific angles. This will partially polarise the transmitted light. The birefringence is thus a direct measure of the type and magnitude of the stress, so the main changes in stress can be made visible. In fractographic analysis, this procedure is very well suited for observation of the stress state in singular geometries (e.g. crack tips). The basics of photoelasticity are explained in Sect. 6.4.

In optical microscopy, high magnification is achieved at the expense of smaller depth of field, see Fig. 10.8. Therefore , it is not possible to make satisfactorily detailed observations (e.g. of crack tips). Scanning Electron Microscope (SEM), offers a much higher magnification (up to 20 000X) and at the same time a greater depth of field (> 10 mm), see Fig. 10.7a. Particularly in the case of fine structures, such as a deep crack, the SEM produces exceptionally good images, see Fig. 5.1. It is, however, recommended to first examine the samples with an optical microscope and follow with further investigations with SEM, since valuable information, such as colour, reflectivity and internal defects (e.g. air bubbles, inclusions) cannot be captured by a scanning electron microscope. In addition, the sample handling is more time-consuming, due to required placement and removal processes and any previously necessary *sputtering* of the sample and the the sample dimensions are restricted. However, the samples can be rotated or tilted during the viewing. The images in the SEM are produced in a sample chamber under high-vacuum conditions. On the one hand this is necessary because the electron source (usually of tungsten) can be operated only in an inert environment and on the other hand an interaction between the atoms of the sample and the molecules of the air is avoided. The energy-intensive, focused electron beam (primary electrons) is moved over the sample surface in a grid-like pattern to generate the image. The primary electrons on the sample surface interact with the atoms of the object under investigation and are emitted from the surface in the form of secondary electrons and recorded by a detector. Since the generated secondary electrons have a low energy level, they originate from the top nanometers of the sample surface and thus depict the topography of the object very precisely. The characteristic X-ray radiation of the smallest sample areas in a SEM can characterise the elements contained in the sample by means of Energy Dispersive X-ray (EDX) analysis. Further developments of the standard scanning electron microscopes are the so-called field emission scanning electron microscopes, which enable optical magnifications of up to 600 000X; the Environmental Scanning Electron Microscope (ESEM), in which the electron beam generation takes place in a high vacuum while the sample is maintained in a low vacuum; and the Environmental Scanning Electron Microscope (TEM), which can X-ray very thin samples (<500 nm) with its electron beams.

Detailed information on fractographic investigations of glass and ceramics can be found in [24].

10.3.1.2 Fracture Mirror

A characteristic feature of the origin of fracture in ceramic materials (including glass) is the formation of a so-called *fracture mirror*, see Fig. 10.7. This is a specific area around the original flaw which develops an almost smooth surface. The fracture mirror indicates the fracture origin and is often also macroscopically visible.

Due to the presence of the acting influence of a crack in the area of a critical imperfection, a crack will grow with increasing stress intensity at the crack tip, see Section 5.3. The smooth surface of the mirror, where also *Wallner lines* (see Sect. 10.3.1.3) could be present, marks the area in which the velocity of crack growth accelerates to the full velocity in glass of about 1500 m/s, see e.g. [25]. This process takes place in an interval of milliseconds. The theoretical maximum attainable crack velocity of glass corresponds to *Rayleigh wave velocity*, of which in glass, however, only up to 40 % to 50 % is achieved (which corresponds to a maximum velocity of 1500 m/s [17, 25]), since, instead of further crack acceleration, energy is dissipated by roughening of the crack surfaces, micro-crack branching and finally crack branching. In the fracture mirror area, these zones can be observed very well with a microscope, because the smooth surface is surrounded by a fine *mist*, which in turn is finally surrounded by a zone of coarser roughness called Hackle that is caused by in- and out of plane micro branching, see Figs. 10.7 and 10.8. The fracture mirror radius, r_{m}, or, r_{h}, is inversely proportional to the square of the fracture stress σ_{f}, which can be calculated using the empirical equations:

$$\sigma_{\mathrm{f}} = \frac{A_{\mathrm{m}}}{\sqrt{r_{\mathrm{m}}}} \tag{10.1}$$

and

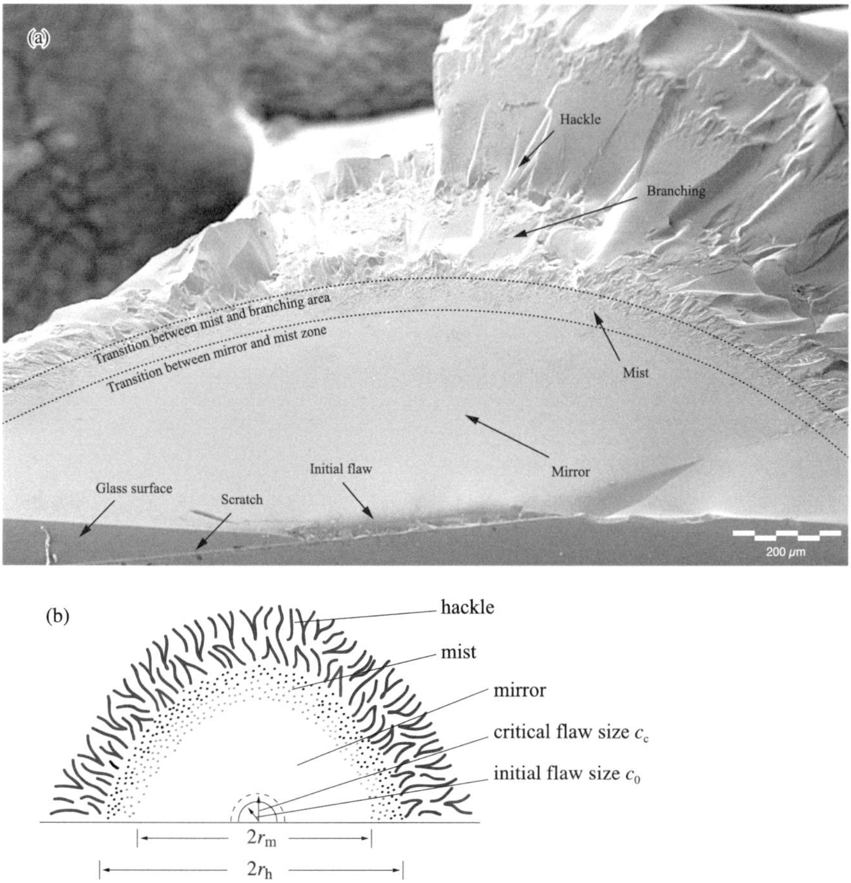

Fig. 10.7 **a** Scanning electron microscope image of a fracture mirror with identification of the different areas [35], as well as **b** a schematic representation of the zones developing around the initial crack [33]

$$\sigma_f = \frac{A_h}{\sqrt{r_h}} . \tag{10.2}$$

The parameters A_m and A_h are so-called fracture mirror constants and for soda-lime silica glass are $A_m = 1.8\ \text{MPa}\,\text{m}^{1/2}$ to $1.9\ \text{MPa}\,\text{m}^{1/2}$ and $A_h = 2.0\ \text{MPa}\,\text{m}^{1/2}$ to $2.1\ \text{MPa}\,\text{m}^{1/2}$ [6, 13].

The Eqs. 10.1 and 10.2 neither take the stress gradient, the crack geometry, nor the location of the crack in the structural component into account. A far more fundamental description of the relationship between tensile stress and fracture mirror radius is possible with the introduction of the stress intensity factor:

$$K_{Im} = \sigma_f\, Y \sqrt{\pi r_m} . \tag{10.3}$$

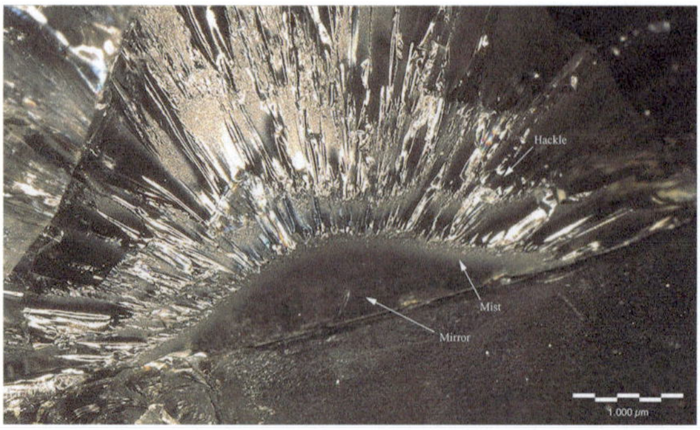

Fig. 10.8 Image of an elliptical fracture mirror in thermally toughened glass with different areas identified

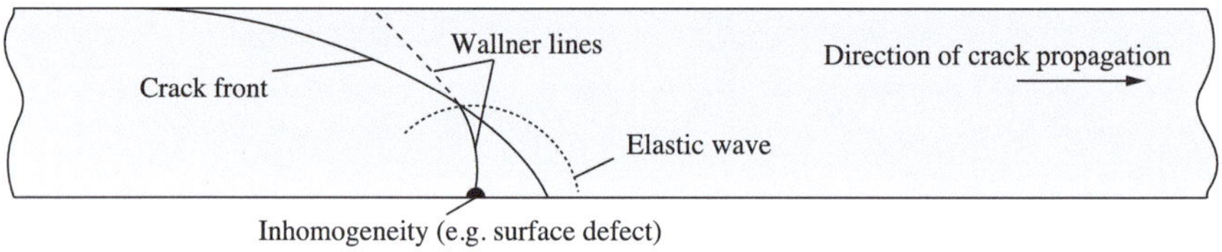

Fig. 10.9 Schematic diagram of the Wallner lines forming on the fracture surface [33]

When this stress intensity is exceeded, the mist zone starts to form, see Sect. 4.2.2. In [5, 8, 9] it is shown that this method often delivers more precise results, although a modified stress intensity factor for soda-lime silica glass at the mist zone $K_{\mathrm{Im}} = 2.33\mathrm{MPa\,m}^{1/2}$ is to be applied in the equation. The shape of the fracture mirror also depends on the stress gradient and the geometry of the crack, as well as the location in the component [24]. For larger surface cracks under bending stress, the roughness appears at the edges only, while it disappears in the through-thickness direction. Also due to residual stresses, the fracture mirror becomes flatter and/or more elliptical under the influence of compressive stress near the surface, as shown in Fig. 10.8 for a fracture mirror in thermally toughened glass.

10.3.1.3 Wallner Lines

In addition to the formations on the smooth surface of the fracture mirror, fine lines can also be seen along the direction of movement of a crack. These features denoted as *Wallner lines* arise when the crack front encounters an inhomogeneity on the sample surface (e.g. scratches) or within the sample (e.g. bubbles or structural imperfections). This contact induces an elastic wave, which spreads faster than the crack front in a circular direction around the point of contact. The meeting of the crack front and the elastic wave causes the former to rise a little out of plane, whereby the features named Wallner lines are formed, see Fig. 10.9.

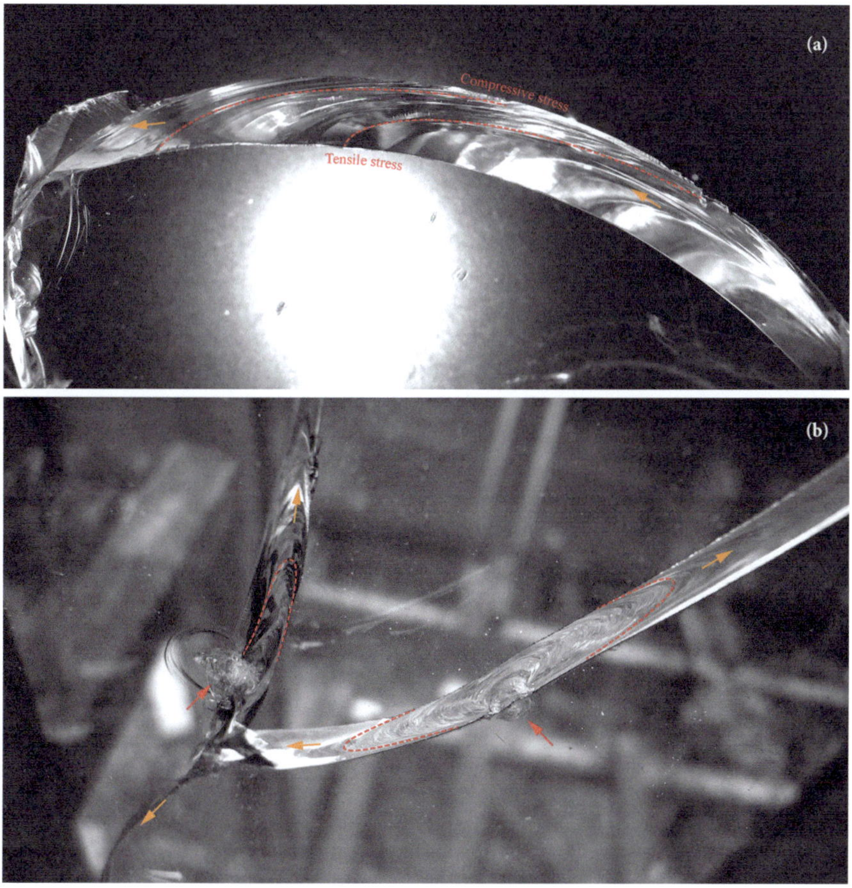

Fig. 10.10 Wallner lines on the fracture surface of a crack in **a** annealed glass with identification of the existing stress conditions on the surface at the time of the fracture and **b** Wallner lines along fracture patterns in heat-strengthened glass (fracture originated from hard impacts, indicated by red arrows); the crack propagation directions are indicated by arrows in each case

Wallner lines exhibit a parabolic curvature, where the vertex points in the direction of crack propagation. Wallner lines characteristically appear on annealed float glass perpendicular to the surface that is undergoing tensile stress, see Fig. 10.10a. Along the compressed glass surface, in contrast, an asymptotic convergence of the lines can be observed, which is also seen in heat-strengthened glass, see Fig. 10.10b. Therefore, with the aid of Wallner lines, the direction of crack growth and the prevailing stress states on the glass surfaces during the fracture event can be clearly determined with fractography. This can be used in the event of damage for identification of the origin of the fracture.

10.3.1.4 Fracture Characteristics of Thermally Pre-Stressed Glass

In accordance with EN 12150-1 and EN 1863-1 the fracture pattern must be evaluated, see Sects. 6.2.8 and 2.3.4, to comply with the classification of thermally toughened or heat-strengthened glass. The quantitative evaluation is used to check whether thermally toughened glass demonstrates a sufficiently fine and heat-strengthened glass a sufficiently coarse fracture pattern. The number of fragments per unit area can also be used as an indication for the residual stress due to thermal pre-stressing. By means of a spring loaded punch and counting of the fragments within a defined area, the thermally-induced residual stress of thermally pre-stressed glass can be estimated. However, this is a rather course measure and local variations may not be found. Akeyoshi et al. [2] counted the fragments in an area of 5 cm × 5 cm after impacting test specimens ($a \times b = 10\,\text{cm} \times 10\,\text{cm}$) of various thicknesses with a sharp object. As shown in Fig. 6.26, the number of fragments per unit area depends on the elastic strain energy and thus also on the thermal residual stress as well as the glass thickness. This study was recently redone and further developed by Pourmoghaddam et al. [37].

Barsom [3] investigated the relationship between the thermal residual stress and the average weight of a fragment. From this he could derive the empirical approach

$$\sigma_{\text{e,center}} \left(\frac{m}{d}\right)^{1/4} = \text{const.} \tag{10.4}$$

whereby $\sigma_{\text{e,center}}$ is the thermal residual stress in the centre of the glass pane, m the average mass of a fragment, and d the glass thickness.

For thermally toughened glass, the stress at the fracture origin can be determined on the basis of the fragment geometry. For this purpose, the crack branching length at the fracture origin should be determined, see Fig. 10.11. According to [6] and [13], the following empirical relationship exists between the fracture stress σ_f and the crack branching length, r_b:

$$\sigma_f = \frac{A_b}{\sqrt{r_b}}\,. \tag{10.5}$$

Here, A_b is the branching constant, which can depend on the material properties, the stress state and the geometric boundary conditions. Marshall, Lawn, and Mecholsky [6] and Mecholsky [13] quantify the constant for soda-lime silica glass to be $A_b = 2.3\,\text{MPa}\,\text{m}^{1/2}$. Zaccaria and Overend [39] extended Eq. 10.5, so that for known fracture stress σ_f, the internal stress due to thermal pre-stressing σ_s can be determined by the relationship

$$\sigma_f = \frac{A_b}{\sqrt{r_b}} + \sigma_s \tag{10.6}$$

One of the dangers of annealed glass is its ability to produce large dagger-shaped fragments upon failure. The size of the individual fragment is dependent on the

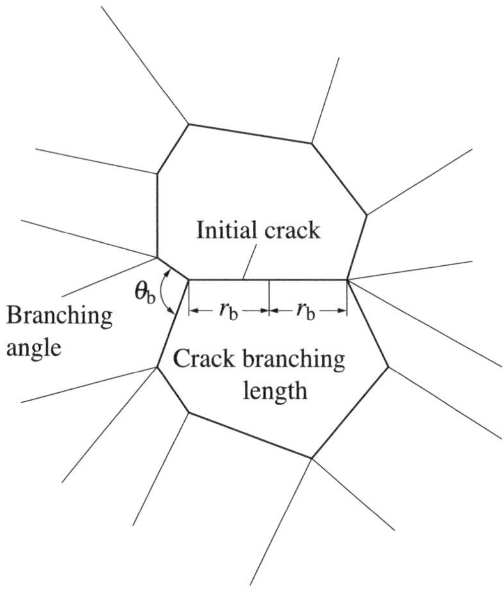

Fig. 10.11 Crack branching length r_b and branching angle θ_b: schematic representation of the fracture centre of thermally toughened glass (after [24])

strength of the glass itself, however, the angle of the area where these fragments are formed is primarily dependent on the stress state [1]. In annealed glass sheets such as windows it is reasonable to describe any stress state by the two orthogonal principal stresses, σ_1 and σ_2. In order to drive the crack, at least one of these should be in tension and it is reasonable to select the largest of the two as σ_1. For annealed glass the graph depicted in Fig. 10.12 shows the so-called "forking angle" as a function of the plane stress state. As an example, a glass plate in pure bending (where $\sigma_2 = 0$) will have a forking angle of approximately 45° as shown in Fig. 10.12, while a biaxial stress state (for example, in the coaxial double-ring bending test) results in an angle of approximately $\theta_{\text{fork}} = 180°$.

10.3.2 Glass Failure Assessment

In the event of damage to glass components, the engineer is often asked to determine the root cause of the damage. Was the damage caused by specific load, such as the effect of wind, snow and dead load, or by avoidable local constraints, vandalism, etc.? Frequent causes of damage are hard contact and impacts on the glass surface through hail, stones, balls, bird flight and human impact.

The fracture pattern of the glass pane provides valuable information on the cause of the damage [23]. The fracture origin can usually be easily determined visually

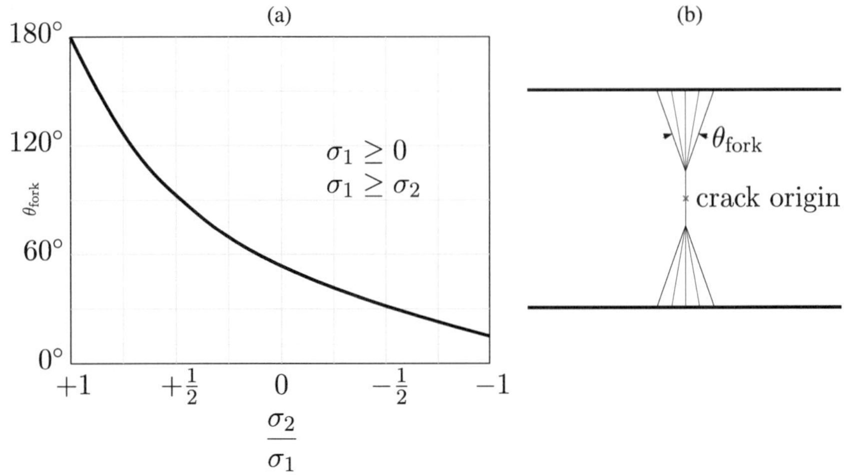

Fig. 10.12 **a** The "forking" angle in annealed glass as a function of stress state. **b** sketch (top view) of bending tests in annealed glass where $\sigma_{fork} \approx 45°$ (after [1])

from the nature of the crack branching and the structure of the fracture surface and morphology. Moreover, it is possible to determine the fracture stress in certain circumstances, see Sect. 10.3.1. Some typical fracture patterns of annealed glass panes are shown in Fig. 10.13.

Thermal effects often lead to unexpected fracture in glass. If a building component cannot freely expand with increasing temperature, stresses are generated in the glass pane depending on the mechanical boundary conditions, see Sect. 9.4. Uneven heating of the glass (e.g. from partial shading) can cause fracture-inducing stresses even when external constraints are not present. Partial shading occurs when part of the glass pane is exposed to direct sunlight or local heating from lamps and heating devices. Equally, adhesive films, prints on glass or glazing bites can also lead to shading of the glass and thus become the trigger for thermal fracture, see Fig. 10.6.

As shown in Fig. 10.14, the cracks in thermally-induced edge fractures usually run from the glass edge and approximately perpendicular to it. This is caused by the prevailing stress state, i.e. the principal tensile stress, which acts parallel to the edge. If the inner part of the glass pane is heated up and the perimeter is kept cold, the principal stress, that drives the crack, will be causing a crack perpendicular to the edge.

Unsuitable or incorrect setting blocks, bending of panels, excessive clamping or deformations of the supporting substructure may also lead to large stresses in the glass and thus to glass fracture, see Fig. 10.15. Similarly, local contacts, such as those from steel-to-glass or glass-to-glass contact, can be the cause of fracture.

In the case of wired glass, the different coefficients of thermal expansion of metal and glass can lead to stress in the glass, which can induce glass fracture. When heated, the wire inlay expands more than the glass and may fracture the glass from

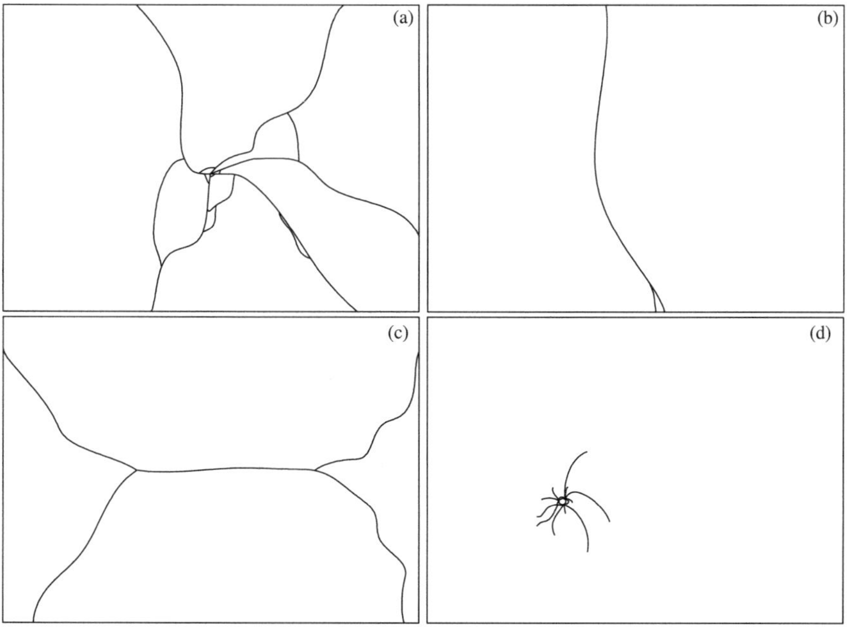

Fig. 10.13 Fracture patterns in annealed float glass: **a** fracture due to local impact in the centre of the panel, **b** bending fracture of a two-side supported panel, **c** bending fracture in a four-side supported panel with fracture originating in the centre and **d** bullet hole

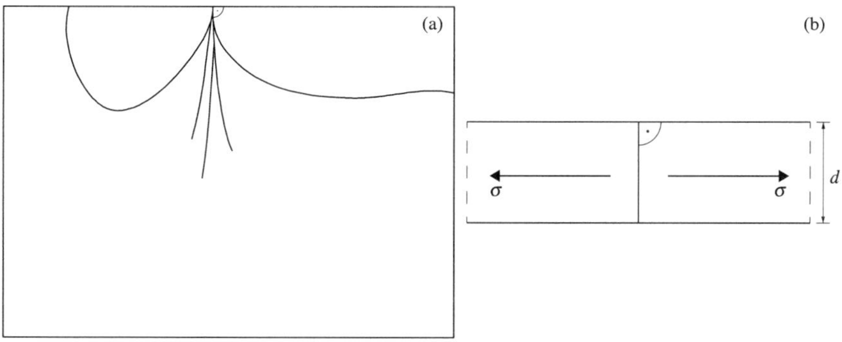

Fig. 10.14 Thermal crack in annealed glass: **a** top view and **b** side view of the glass edge

the inside. Particularly in the case of overhead glazing with partial shading, this effect is superimposed by the stresses from temperature differences and is a frequent cause of damage.

Another type of danger in wired glass is caused by the corrosion of the wire mesh at the glass edge. The local corrosion of the metal will cause a considerable increase in volume, which may also lead to glass fracture.

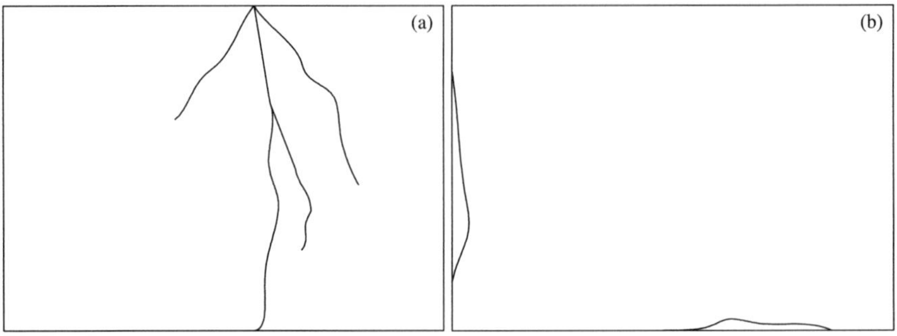

Fig. 10.15 Fracture patterns in annealed float glass: **a** fracture at the edge from local stress concentration points, e.g. due to unsuitable or incorrect blocking; **b** edge fractures from constraints in the frame construction

Pre-stressed glass is significantly less sensitive to thermal fracture or local contact stresses. As long as the surface tensile stress does not exceed the magnitude of the compressive surface pre-stress, no tensile stresses occur at the glass surface. In the event of failure, thermally toughened glass fractures into many small fragments, see Sects. 6.2.8 and 10.3.1.4. The cracks radiate from the origin of the fracture and are usually finest branched in the region of the fracture origin where external energy is added, see Fig. 6.28. The minimum number of fragments in thermally toughened glass are set out in Table 6.3.

Thermally toughened glass is also associated with "spontaneous fracture". A characteristic of spontaneous fractures in thermally toughened glass, see Sect. 6.2.6 are two larger butterfly-shaped fragments at the origin of the fracture between which is the inclusion, see Fig. 6.22. The fracture origin, which generally is related to the pane thickness, lies in the middle of the pane i.e. in the tension zone. The inclusions can be detected by microscopic examinations and, in some cases, are also visible with the naked eye.

Material impurities other than nickel sulphide inclusions can also lead to spontaneous fracture. The butterfly-shaped fragments are not always related to inclusions of nickel sulphide, but are typical for high local stresses. Only a chemical analysis of the inclusion provides certainty of its presence.

The fracture pattern of heat-strengthened glass is similar to the fracture pattern of annealed float glass, although the latter typically is more dense (i.e. larger number of fragments). Cracks appear that radiate from the fracture origin. The crack progression is, however, often more wave-like than with annealed glass. Also heat-strengthened glass is more prone to so-called island formations of smaller fragments compared to annealed glass, see Sect. 6.2.7.

As shown in Fig. 10.16a, the cracks at the edges of heat-strengthened glass run parallel to the edge more frequently. Due to the heat treatment process, the membrane stresses at the edge area are often higher than the surface compressive stress in the glass panel centre and thus influence the crack progression.

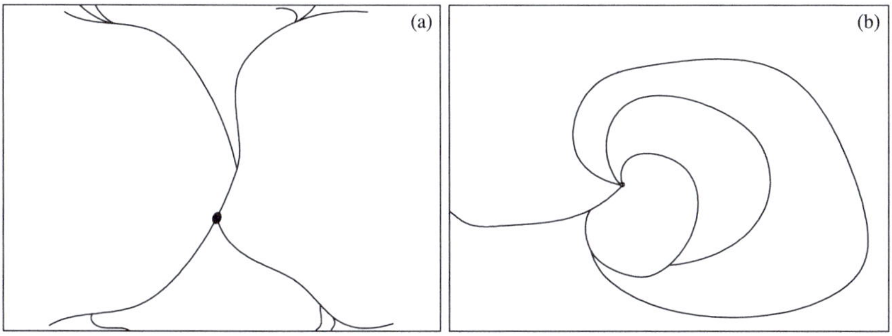

Fig. 10.16 Fracture as a result of impact: **a** heat-strengthened monolithic glass and **b** laminated safety glass of heat-strengthened glass

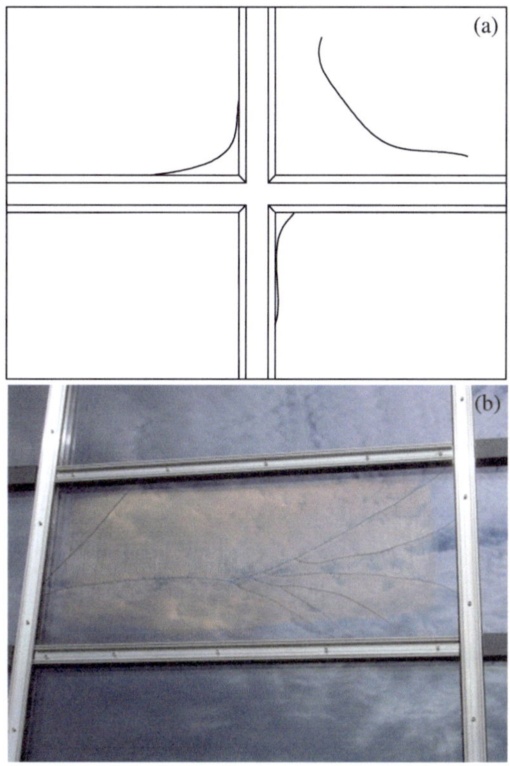

Fig. 10.17 **a** Schematic fracture pattern of an insulating glass unit with glued-in crossbars and **b** damage as a result of cavity over-pressure in an insulating glass unit

Particularly with laminated glass made of heat-strengthened glass, the fracture pattern often shows completely different characteristics, see Fig. 10.16b. Islands appear, which cannot be compared with the stipulated standard fracture pattern.

The residual load-bearing capacity of fractured glass is governed by the size of the fragments. Larger fragments enable a sufficient load transfer between the different

Fig. 10.18 Delamination as a damage type in laminated glass

Fig. 10.19 Imperfections in the base glass in the application of wavy thermally toughened glass

glass plies and provide a local bending strength by interaction of the fragments and the polymer interlayer.

In laminated glass, the glass can detach from the polymer interlayer. This damage type, known as delamination, may occur several years after installation, see Fig. 10.18. Possible causes for this are moisture in the interlayer, imperfections in the base glass material, material incompatibilities or errors in the lamination process, see Fig. 10.19.

The problem of optical anisotropy in thermally toughened glass has already been discussed in Sect. 6.2.4.

References

1. F. W. Preston. "The angle of forking of glass cracks as an indicator of the stress system". In: *Journal of the American Ceramic Society* 18.1-12 (1935), pp. 175–176
2. K. Akeyoshi, E. Kanai, K. Yamamoto, and S. Shima. "Report of the research lab". In: *Asahi Garasu Kenkyu Hokoku* 17.1 (1967), pp. 23–26
3. J. M. Barsom. "Fracture of tempered glass". In: *Journal of the American Ceramic Society* 51.2 (1968), pp. 75–78

4. F. Kerkhof. *Bruchvorg ä nge in Gl ä sern*. Deutsche Glastechnische Gesellschaft, 1970
5. H. P. Kirchner and J. W. Kirchner. "Fracture mechanics of fracture mirrors". In: *Journal of the American Ceramic Society* 62.3-4 (1979), pp. 198–202
6. D. B. Marshall, B. R. Lawn, and J. J. Mecholsky. "Effect of residual contact stresses on mirror/flaw-size relations". In: *Journal of the American Ceramic Society* 63.5-6 (1980), pp. 358–360
7. H. Franz. "Glass surface improvements by chemical reactions and thin film deposition". In: *Glastechnische Berichte* 60.5 (1987), pp. 182–186
8. H. P. Kirchner and J. C. Conway. "Criteria for crack branching in cylindrical rods – Part I: Tension". In: *Journal of the American Ceramic Society* 70.6 (1987), pp. 413–418
9. H. P. Kirchner and J. C. Conway. "Criteria for crack branching in cylindrical rods – Part II: Flexure". In: *Journal of the American Ceramic Society* 70.6 (1987), pp. 419–425
10. I. Donald. "Methods for improving the mechanical properties of oxide glasses". In: *Journal of Materials Science* 24.12 (1989), pp. 4177–4208
11. W. Grönegräs. "Der Glashobel, ein Gerät des Gebäudereinigers – Der tägliche Schadensfall: 11". In: *Glaswelt* 42.11 (1989), p. 32
12. W. Grönegräs. "Kratzer und kein Ende – Der tägliche Schadensfall: 27". In: *Glaswelt* 45.1 (1992), p. 58
13. J. J. Mecholsky. "Quantitative fractographic analysis of fracture origins in glass". In: *Fractography of Glass*. Springer US, 1994, pp. 37–73
14. W. Lutz. "Reibspuren und Verätzungen – Der tägliche Schadensfall: 56". In: *Glaswelt* 48.7 (1995), pp. 52–55
15. BBR. *Bundesamt f ü r Bauwesen und Raumordnung: Leitfaden Nachhaltiges Bauen – Anlage 6: Bewertung der Nachhaltigkeit von Geb ä uden und Liegenschaften*. 2001
16. W. Lutz. *Fachbuch Geb ä udereinigung*. 36. Dettingen/Ems: Lutz-Fachbücher, 2001
17. J. Schneider. "Festigkeit und Bemessung punktgelagerter Gläser und stoßbeanspruchter Gläser". Dissertation. Technische Universität Darmstadt, 2001
18. 3M. *3M Glas-Reparatursystem Trizcat Glas-Schleifwerkzeug*. 2002
19. H. Pfeifer, R. Krugmann, and M. Neumann. *ESG und seine Kratzanf ä lligkeit*. 2002
20. DIN 1986-3. *Entw ä sserungsanlagen f ü r Geb ä ude und Grundst ü cke – Teil 3: Regeln f ü r Betrieb und Wartung*. 2004
21. B. Abegg, B.H. Dür, *Argumentation f ü r Kratzersanierung auf thermisch vorgespanntem Glas* (Tech. rep. Vetrox AG, Schindellegi/CH, 2007)
22. G. Carturan, N. Khandelwal, L. Tognana, and V. Sglavo. "Strengthening of soda-lime-silica glass by surface treatment with sol-gel silica". In: *Journal of Non-Crystalline Solids* 353.16–17 (2007), pp. 1540–1545
23. M. Overend, S. De Gaetano, and M. Haldimann. "Diagnostic Interpretation of Glass Failure". eng. In: *Structural Engineering International* 17.2 (2007), pp. 151–158
24. G. D. Quinn. *Fractography of ceramics and glasses*. 960-16. Washington, DC: U.S. Government Printing Office, 2007
25. J. H. Nielsen, J. F. Olesen, and H. Stang. "The Fracture Process of Tempered Soda-Lime-Silica Glass". In: *Experimental Mechanics* 49.6 (2009), pp. 855–870
26. DIN 18531-4. *Dachabdichtungen – Abdichtungen f ü r nicht genutzte D ä cher – Teil 4: Instandhaltung*. 2010
27. RAL. *G ü tegemeinschaft Geb ä udereinigung e.V.: Reinigung von vorgespannten ESG- und beschichteten Gl ä sern im Architekturbereich*. 2010
28. VDI 6200. *VDI-Richtlinie – Standsicherheit von Bauwerken – Regelm ä ßige Ü berpr ü fung, Verein Deutscher Ingenieure*. 2010
29. BGI 576. *BG RCI – Berufsgenossenschaft Rohstoffe und chemische Industrie: Merkblatt M 005: Fluorwasserstoff, Flusss ä ure und anorganische Fluoride*. 2012
30. Bundesverband Flachglas. *Reinigung von Glas*. 2012
31. EN 1863-1. *Glass in building - Heat strengthened soda lime silicate glass - Part 1: Definition and description*. 2012

32. EN 12150-1. *Glass in building - Thermally toughened soda lime silicate safety glass - Part 1: Definition and description*. 2015
33. J. Hilcken. "Zyklische Ermüdung von thermisch entspanntem und thermisch vorgespannten Kalk-Natron-Silikatglas". Dissertation. Technische Universität Darmstadt, 2015
34. M. Overend, C. Louter, The effectiveness of resin-based repairs on the inert strength recovery of glass. Constr. Build. Mater. **85**, 165–174 (2015)
35. S. Schula. "Charakterisierung der Kratzanfälligkeit von Gläsern im Bauwesen". Dissertation. Technische Universität Darmstadt, 2015
36. K. C. Datsiou, D. Hall, and M. Overend. "Repair of soda–lime–silica glass". In: *Glass Structures & Engineering* 3.2 (2018), pp. 125–137
37. N. Pourmoghaddam, M. A. Kraus, J. Schneider, and G. Siebert. "Relationship between strain energy and fracture pattern morphology of thermally tempered glass for the prediction of the 2D macro-scale fragmentation of glass". In: *Glass Structures & Engineering* 4.2 (2019), pp. 257–275
38. G. Mariggiò, S. Dalle Vacche, R. Bongiovanni, C. Louter, and M. Corrado. "Enhancing the design bending strength of new and aged glass with a functional coating". In: *Glass Structures and Engineering* 5.2 (2020), pp. 135–146
39. M. Zaccaria and M. Overend. "Nondestructive Safety Evaluation of Thermally Tempered Glass". In: *Journal of Materials in Civil Engineering* 32.4 (2020), p. 04020043

Appendix A
Cable Design

For the design of the cables the following relationships can be used for a cable under uniform transverse load, see Fig. A.1. It applies in general [1]:

$$H^3 + H^2 EA \left(1 - \frac{l}{L}\right) = \frac{EA \cdot q^2 l^3}{24\,L} \tag{A.1}$$

where

H	Horizontal component of the cable strength;
EA	tensile rigidity of the cable;
l	span;
L	true cable length;
q	shear (here: wind force).

The pre-tensioned straight cable is $l = L$ and thus the horizontal force can be determined directly from:

$$H^3 = \frac{EA \cdot q^2 l^2}{24}. \tag{A.2}$$

With the addition of the transverse load, the horizontal load with pre-tensioning is composed of:

$$H = H_V + H_q \tag{A.3}$$

with

H_V	horizontal component from pre-tensioning,
H_q	Horizontal component from transverse load.

If a pre-tension of H_V is given and the pre-stretched cable is installed in this manner, this results in elastic length deviation from elastic cable elongation ΔL_e due to external load q from:

$$\Delta L_e \approx \frac{H_q \cdot l}{EA} = \frac{(H - H_V) \cdot l}{EA}. \tag{A.4}$$

J. Belis et al., *Engineering Glass Structures*,
https://doi.org/10.1007/978-3-032-03233-1

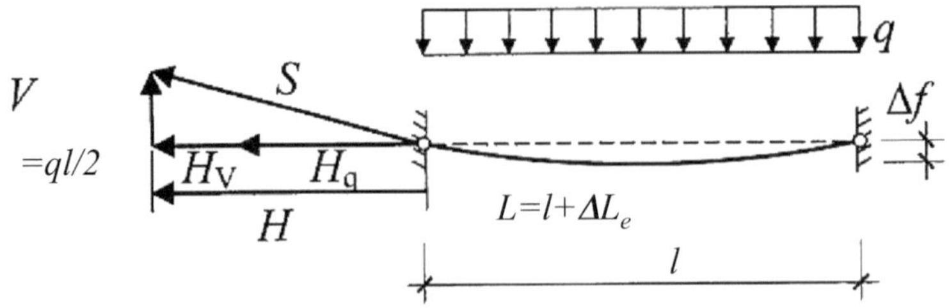

Fig. A.1 Pre-tensioned cable affected by transverse load

Under effect of the transverse load, the sag, f, also changes for the pre-stretched state in the equation:

$$\Delta f = \frac{ql^2}{8H}. \tag{A.5}$$

The cable length can be closely estimated by a ratio $f/l = 0,2$:

$$L = l + \frac{8}{3}\frac{f^2}{l} = l\left(1 + \frac{8}{3}\frac{f^2}{l^2}\right) = l\left(1 + \frac{q^2l^2}{24H^2}\right). \tag{A.6}$$

This results in:

$$l\left(1 + \frac{q^2l^2}{24\,H^2}\right) = l + \Delta L_e = l + \frac{H_q \cdot l}{EA} = l + \frac{H - H_V \cdot l}{EA}. \tag{A.7}$$

The horizontal force in the cable from pre-tensioning and transverse load can now be determined using:

$$H^2(H - H_V) = \frac{EA \cdot q^2l^2}{24}. \tag{A.8}$$

Reference

1. S. Palkowski, *Statik der Seilkonstruktionen* (Springer, Berlin, 1989)

Index

J. Belis et al., *Engineering Glass Structures*,
https://doi.org/10.1007/978-3-032-03233-1

MIX
Papier aus verantwortungsvollen Quellen
Paper from responsible sources
FSC® C105338

If you have any concerns about our products,
you can contact us on
ProductSafety@springernature.com

In case Publisher is established outside the EU,
the EU authorized representative is:
Springer Nature Customer Service Center GmbH
Europaplatz 3, 69115 Heidelberg, Germany

Printed by Libri Plureos GmbH
in Hamburg, Germany